Stochastic Systems: Estimation, Identification, and Adaptive Control

PRENTICE-HALL INFORMATION AND SYSTEM SCIENCES SERIES

Thomas Kailath, Editor

Stochastic Systems: Estimation, Identification, and Adaptive Control

P.R. Kumar

University of Illinois

Pravin Varaiya

University of California

Prentice Hall, Inc., Englewood Cliffs, New Jersey, 07632

Library of Congress Catalog Card Number 86-60085

Editorial/production supervision: Gloria Jordan
Interior design: Pravin Varaiya
Cover design: Robert Yagura and Ben Santora
Manufacturing buyer: Gordon Osbourne

Printed in the United States of America.

10   9   8   7   6   5   4   3   2   1

ISBN 0-13-846684-X   025

Prentice-Hall International (UK) Limited, *London*
Prentice-Hall of Australia Pty. Limited, *Sydney*
Prentice-Hall Canada Inc., *Toronto*
Prentice-Hall Hispanoamericana, S.A., *Mexico*
Prentice-Hall of India Private Limited, *New Delhi*
Prentice-Hall of Japan, Inc., *Tokyo*
Prentice-Hall of Southeast Asia Pte. Ltd., *Singapore*
Editora Prentice-Hall do Brasil, Ltda., *Rio de Janeiro*
Whitehall Books Limited, *Wellington, New Zealand*

for my parents and Jaya
—PRK

for Papa, Leona, Oscar
and in Mummy's memory
—PV

CONTENTS

PREFACE

This book is concerned with the questions of modeling, estimation, optimal control, identification, and the adaptive control of stochastic systems. The treatment of these questions is unified by adopting the viewpoint of one who must make decisions under uncertainty.

Since its origins in the 1940s, the subject of decision making under uncertainty has grown into a ramified area with applications in several branches of engineering and in those areas of the social sciences concerned with policy analysis and prescription. The theoretical approaches to the subject derive from the invention in the 1950s of dynamic programming and the rich body of research in statistical time series analysis.

Although theoretically appealing, these approaches found little practical use since they demanded a computing capacity that was too expensive at that time. The tremendous advances in information technology (the ability to collect and process huge quantities of data) provided the motivation for subsequent impulses that advanced the subject. Recent years have witnessed an explosion of work aimed at applications in signal processing, control, operations research, and economics. As a result there has been a proliferation of specialized journals, books, conferences, and inevitably—vocabularies—that makes it difficult for the newcomer to grasp the unity of the subject.

This book is an attempt to provide such a newcomer with a unified treatment of the most important aspects of the subject to the level from where more specialized contemporary research may then become accessible. Within the scope of this book, we have adopted several restrictions. First, we have aimed our discussion at those who have already had an introduction to probability theory and stochastic processes, including multivariate random variables, conditional expectation, and Gaussian and Markov processes. Second, we have decided to restrict our attention to stochastic systems that evolve in discrete rather than continuous time in order to keep the level of mathematical sophistication consistent with our intended audience.

Third, we have made liberal use of "Exercises" in the text. Many assertions, examples, and counterexamples are relegated as exercises, and we have provided hints or references that should help in solving them. These exercises form an integral part of the text and provide the reader with a standard for judging whether he or she understands the preceding material; subsequent material as well builds on these exercises. Finally, each chapter concludes with "Notes" which provide a guide to the relevant literature.

Used as a textbook, the material here can be covered either in a leisurely two-semester or in an intensive one-semester course. For a two-semester course we would suggest covering Chapters 1-7 in the first semester, and Chapters 8-13 in the second. In a single semester, we suggest skipping Sections 4.6-4.8, 7.6, 7.9, Chapter 8, and parts of Chapters 10, 11, and 12.

The following scheme is used in referencing sections, equations, exercises, and chapters. In each chapter, sections are numbered 1, 2, and so on. In each section, equations and exercises are numbered sequentially with two digits, the first denoting the section, so that (6.2) refers to equation (or exercise) 2 in section 6. When reference is made to a section or equation in another chapter, then an additional digit denoting the chapter is appended. Thus, Section 6.4 denotes section 4 in Chapter 6, and equation (6.4.2) denotes equation 2 in Section 6.4.

We are grateful to many colleagues and students for discussions which have increased the clarity of our presentation.

P.R. Kumar
University of Illinois (Urbana)

Pravin Varaiya
University of California (Berkeley)

Stochastic Systems: Estimation, Identification, and Adaptive Control

CHAPTER 1
INTRODUCTION

In this chapter, we briefly summarize the aim of this book and present an example of the kind of problems that will concern us. We also indicate decision-making situations that cannot be satisfactorily addressed within our framework.

1. Making decisions under uncertainty

We will be concerned with controlling a dynamic system whose behavior is not completely forseeable. That is, we will deal with a system whose future evolution is not completely determined by its present state and future control actions as is the case with deterministic systems. The uncertainty associated with the future behavior can be quantified in several ways. Our approach assumes that this quantification is probabilistic or stochastic, and we say that we are dealing with stochastic dynamic systems. Two different types of models of such systems are useful for most purposes. One type of model consists of a first-order difference equation involving the state, input, and output variables. This gives rise to the description of the system as a controlled Markov chain. The second type of model is a high-order difference equation involving only input and output variables.

In practice the analyst faces two different issues. The first issue concerns the use of the information obtained from the observed outputs to reduce the uncertainty about the system behavior. Second, the analyst has to assess the effect on the system behavior of alternative control policies and to select the most desirable policy. The first concern leads to the problem of *state estimation* and to the problem of *system identification*. The second is formulated as a problem of *stochastic control*. This book provides an account of three aspects of decision-making: estimation, identification, and control. An example will help indicate the kind of problems that can be usefully analyzed with the methods introduced here.

2. Control of water in a reservoir

Water is stored in a reservoir and replenished by rain. The stored water is used for irrigation and other purposes. Neither the quantity of rainfall nor the demand for water in the future is known in advance. Because of

1

this uncertainty, it is natural to model rainfall and demand both as random variables. In each period the water authority must decide how much of the current demand should be met from the available water.

In period k denote by x_k the water in the reservoir, by r_k the amount of rainfall, by d_k the demand for water, and by u_k the amount of water made available by the authority. Then we get the relation

$$x_{k+1} = x_k + r_k - u_k,$$

where the decision variable u_k is constrained to

$$0 \leq u_k \leq x_k.$$

Since the authority wants to meet the demand, it will try to select u_k to be close to d_k. However, its current decision will depend also on its assessment of the levels of future rainfall and demand. This assessment is a problem of state estimation, and the the decision to be made by the authority can be formulated as a stochastic control problem. Observe that the control problem requires a solution of the estimation problem. This is to be expected since the current decision must take into account the future evolution of the system, and since the future is not known in advance, an estimate of it must be obtained.

Let us ignore the decision problem and concentrate on estimation of the future rainfall. To simplify matters further, suppose that we want to predict the amount of rainfall in the next period only. Since the rainfall is modeled as a random variable, or rather as a sequence of random variables, one for each period's rainfall, this prediction will involve the probability distribution of these random variables. For instance, the prediction may be simply $E\ r_{k+1}$, the expected value of next period's rainfall. Or it may be $E\ \{r_{k+1} \mid r_k\}$ the conditional expectation of next period's rain given the present period's rain. The latter prediction is superior if the rainfall in successive periods is correlated. Evidently both predictions require the knowledge of the probability distribution of the rainfall.

How do we know this probability distribution? It may have been given by a hydrologist. If it is not known, then we have to estimate it on the basis of the recorded pattern of past rainfalls. This estimation of the relevant probability distribution is called the problem of identification or system identification. We see then how the problem of estimation often involves that of identification. In fact, in most situations of decision-making in a stochastic environment, the problems of identification, estimation, and control are all tied together.

As we shall see, computational considerations force us to treat the three aspects of identification, estimation, and control separately. In the final section of the book, however, we introduce the method of *adaptive control*, which is a way of considering all three aspects simultaneously.

3. Other situations of making decisions

Consider a situation where the uncertainty is associated with the outcome of a coin toss. To suppose that the outcome of the toss will be a head with probability p has an objective meaning as the frequency with which heads will appear in a series of repeated trials will be p.

Now consider a situation where a frequency interpretation is inappropriate. Suppose you are betting in a horse race. In placing your bet you will follow your belief as to which horse will place first, which one will place second, and so on. It is not farfetched to suppose that you can quantify your beliefs. Consistency demands that this quantitative specification have all the properties of a probability distribution. However, it is unreasonable to interpret these probabilities as the frequencies with which horses will win repeated races. Rather the probabilities are simply the representation of your subjective beliefs. (The subjective character is evident from the fact that different persons will have different probability distributions; if all agreed there would be no horse races.)

For purposes of this book it does not matter whether the probabilities that characterize uncertainty are objective or subjective.

However, there are classes of situations where a probabilistic characterization appears artificial. One such class arises in dealing with rare events. For instance, what is the meaning of this scientific-sounding statement: The probability of a catastrophic meltdown in a nuclear reactor is 10^{-12}? The small number suggests that the accident is very unlikely. But can we go beyond that and interpret the actual number? A frequency interpretation is clearly not sensible. A subjective interpretation would imply that if the number had been 10^{-11} instead, then the speaker believes the chance of the accident to be 10 times as great. But surely in dealing with such small numbers one cannot place much reliance on the subjective ability to make quantitative comparisons, although rank ordering may make sense. In this case it is meaningful to say that under certain conditions an accident is more likely. Whereas it sounds precise to say that it is 10 times as likely, this precision is illusory.

Another class of situations where quantification is illusory concerns an uncertain event which is so complex that even a verbal description is difficult. An economist predicts that the government's policy is likely to lead to a recession but unlikely to end up in a depression. This is a meaningful statement. But if the adjectives likely and unlikely are replaced by probability estimates, no real precision is gained because recession and depression denote complex realities and convey different meaning to different people.

Situations like those described above where probabilistic specification of uncertainty is not possible cannot be usefully addressed in the framework of this book.

4. Notes

1. The origins of the subject matter treated in this book lie in statistical decision theory and the theory of games, and in the theory of filtering noisy data. Classic expositions of these subjects are von Neumann and Morgenstern (1947), Wald (1950), Blackwell and Girshick (1954), Luce and Raiffa (1957), and Wiener (1949).

2. Savage (1954) gives consistency conditions under which subjective beliefs can be interpreted as probabilities.

3. For a recent attempt at formalizing imprecise, verbal descriptions of complex situations and using this formalism for decision analysis, see Gupta and Sanchez (1982a,b). Simon has argued forcefully that real world decision problems are too complex to permit mathematical models of the kind proposed in this book. In addition, he argues that people do not attempt to find the best or optimal decision, they merely try to "satisfice." See Simon (1975).

4. Harvey (1985) surveys the literature that quantifies the impact of very unlikely catastrophic events like a nuclear meltdown.

5. References

[1]   D. Blackwell and M.A. Girshick (1954), *Theory of games and statistical decisions,* John Wiley, New York.

[2]   M.M. Gupta and E. Sanchez (eds.) (1985a,b), *Approximate reasoning in decision analysis,* and *Fuzzy information and decision processes,* North Holland, Amsterdam.

[3]   C.M. Harvey (1985), "Preference functions for catastrophe and risk inequity", *Large Scale Systems,* Vol 8(2), April, 131-146.

[4]   R.D. Luce and H. Raiffa (1957), *Games and decisions,* John Wiley, New York.

[5]   L.J. Savage (1954), *The foundations of statistics,* John Wiley, New York.

[6]   H. Simon (1975), "Theories of decision-making in economic and behavioral science," reprinted in E. Mansfield (ed), *Microeconomics: Selected Readings,* 2nd Ed., Norton, New York.

[7]   J. von Neumann and O. Morgenstern (1947), *Theory of games and economic behavior,* 2nd Ed., Princeton University Press, Princeton, NJ.

[8]   A. Wald (1950), *Statistical decision functions,* John Wiley, New York.

[9]   N. Wiener (1949), *Extrapolation, interpolation and smoothing of stationary time series,* MIT Press, Cambridge, MA.

CHAPTER 2
STATE SPACE MODELS

The chapter begins with a description of a deterministic system model. By making an analogy with this case, and with the aid of several examples, a state space model of stochastic systems is proposed. The model permits the definition of the state, output, and control processes. The crucial distinction between open loop and closed loop policies is introduced. It is shown that when the input noise is independent, the model is a controlled Markov process.

1. Deterministic system model

We shall consider discrete time systems exclusively. The dynamic behavior of a deterministic system is usually modeled by an equation of the form

$$x_{k+1} = f_k(x_k, u_k), \quad k = 0, 1, ..$$
(1.1)

where $x_k \in R^n$ is the state and $u_k \in R^m$ is the input at time k. Usually there is also an output $y_k \in R^p$ modeled by the equation

$$y_k = h_k(x_k, u_k), \quad k = 0, 1, ...$$
(1.2)

For the moment we ignore the output equation and we focus on the state equation (1.1).

An obvious but important property of (1.1) is that the current state x_k and the input sequence $u_k, u_{k+1}, ..., u_{k+m}$ determine the state x_{k+m+1} independently of the past values of state and input $x_{k-1}, x_{k-2}, ..., u_{k-1}, u_{k-2}, ...$ That is, there is a function $f_{k+m+1,k}$ such that

$$x_{k+m+1} = f_{k+m+1,k}(x_k, u_k, ..., u_{k+m}).$$
(1.3)

These (state) **transition functions** can be obtained by recursive substitution:

$$f_{k+1,k}(x_k, u_k) := f_k(x_k, u_k),$$
(1.4)

$$f_{k+m+1,k}(x_k, u_k, ..., u_{k+m})$$

$$:= f_{k+m}(f_{k+m,k}(x_k, u_k, ..., u_{k+m-1}), u_{k+m}), \quad m > 1.$$
(1.5)

In the case that (1.1) is time-invariant, that is, $f_k(x, u) \equiv f(x, u)$, then the transition functions are also time-invariant in the sense that $f_{k+m,k}$ depends only on $(k+m) - k = m$.

2. Stochastic system model

We want to write a model for stochastic systems similar to that of (1.1). It is instructive to consider first some examples of more concrete situations.

A gasoline dealer has x_k gallons on hand at the beginning of day k. Suppose the dealer has ordered an additional u_k gallons to be delivered at the beginning of the same day. Then the stock of gasoline at the beginning of day $k+1$ is

$$x_{k+1} = x_k + u_k - w_k =: f(x_k, u_k, w_k), \tag{2.1}$$

where w_k is the amount of gasoline sold on day k. Equation (2.1) is similar to (1.1) if we regard as input not u_k alone but the pair (u_k, w_k). The u_k in (2.1) is an input in the sense that it is chosen by the dealer. Hence its value is known to the dealer on day k. On the other hand w_k is not known and its value cannot be controlled by the dealer. In this example w_k will be known on day $k+1$.

Here is another similar example. Let x_k be the amount of corn that a farmer grows and plants on a particular plot in the kth planting season. Let u_k be the additional inputs used, such as fertilizer and labor. Define x_{k+1} to be the amount of corn grown. Then x_{k+1} depends upon x_k, u_k, as well as the weather (amount of rainfall, hours of sunshine, etc.) that prevails during the growing season. This relationship can plausibly be expressed as

$$x_{k+1} = (1+w_k) g(x_k, u_k) =: f(x_k, u_k, w_k), \tag{2.2}$$

where w_k represents the effect of the weather: During a normal year $w_k = 0$ and so $x_{k+1} = g(x_k, u_k)$; during a good year $w_k > 0$, and in a poor year $w_k < 0$. Again we may interpret w_k as an input similar to u_k; and again it is distinguished from the latter by the fact that at time k it is not known, whereas u_k is known and, in fact, controlled.

The next example is of a different kind. Consider a system whose behavior is known to be approximately of the form

$$x_{k+1} \approx g(x_k, u_k).$$

A common occurrence of this kind is when a nonlinear system is approximated by a linear model. Define the model error as

$$w_k := x_{k+1} - g(x_k, u_k),$$

so that the exact behavior is given by

$$x_{k+1} = g(x_k, u_k) + w_k =: f(x_k, u_k, w_k), \tag{2.3}$$

which formally is similar to (2.1) and (2.2). Once again we may treat w_k as an input which is not known at time k.

In summary, these three examples lead to a state equation of the form

$$x_{k+1} = f_k(x_k, u_k, w_k), \quad k = 0, 1, \ldots,$$

in which x_k is the state, u_k is the known input or control, and w_k is the unknown input that we will also call input disturbance or noise or error. What does it mean to say that w_k is unknown? Let us reexamine the examples. The gasoline dealer would know from experience that if k happens to be a weekday, then the demand w_k is likely to be higher than if it falls on the weekend. The dealer may be able to specify the character of the unknown demand even further: that the chance of w_k exceeding 1000 gallons or being less than 100 gallons are each 1 in 10 and the chance that $400 < w_k < 600$ is 4 out of 10, etc. In other words, the dealer may be able to structure the uncertainty about the demand probabilistically, i.e., w_k is modeled as a random variable. More precisely, the sequence $\{w_k, k \geq 0\}$ is modeled as a stochastic process.

Similarly, in the second example, the farmer or an agronomist may study the soil conditions and the historical pattern of rainfall and sunshine in the area and specify as a stochastic process the sequence $\{w_k, k \geq 0\}$ representing the effect of the unknown weather. In the last example the analyst may specify the structure of errors $\{w_k, k \geq 0\}$ as a stochastic process also. This time the specification of the probability law which defines the process may be based on simulation experiments or on educated guesswork.

We thus arrive at the following formulation. A stochastic system model is an equation of the form

$$x_{k+1} = f_k(x_k, u_k, w_k), \quad k = 0, 1, \ldots, \tag{2.4}$$

where x_k is the state, u_k is the known input or control, and w_k is the unknown disturbance or noise at time k. The sequence $\{w_k, k \geq 0\}$ is a stochastic process with known probability law; that is, the joint probability distribution of the random variables $w_0, w_1, \ldots, w_k$ is known for each k. The functions $f_0, f_1, \ldots$ are also assumed known. What about the state x_k at time k? Can we assume that it is known? The answer depends upon the situation. Suppose in the first example the stock of gasoline is measured by an instrument that has some random errors, so that the information about the state x_k that is actually available is

$$y_k = x_k + v_k, \quad k = 0, 1, \ldots,$$

where y_k is the actual instrument reading and v_k is the error. Similarly, the amount of corn actually grown can only be indirectly known by measuring the amount y_k of corn that is actually harvested. The two quantities are related by

$$y_k = (1+v_k)\, x_k,$$

where v_k gives the unknown variation in the ratio of harvest to growth. Finally, in the last example, suppose that only p of the components of the n-dimensional state x_k are directly measurable. Call $y_k \in R^p$ the vector of these components. Then one gets

$$y_k = H\, x_k,$$

where H is an appropriate $p \times n$ matrix.

In summary, when the state is not directly observed it is necessary to augment the state equation (2.4) with the measurement equation

$$y_k = h_k(x_k, v_k), \quad k = 0, 1, \ldots, \tag{2.5}$$

where $y_k \in R^p$ is the observation or output and v_k is the unknown measurement error or noise. As in the case of the input noise sequence, it will be supposed that $\{v_k, k \geq 0\}$ is a stochastic process with known probability law. If in (2.5) one has $h_k(x, v) \equiv x$, then $y_k \equiv x_k$ which means that the state is observed directly. We shall refer to this as the case of **complete observation.** When $y_k \not\equiv x_k$, it is the case of **partial observation.** In the latter case, it is unreasonable to suppose that the initial state is known exactly, and so x_0 will also be considered a random variable.

The preceding discussion leads us to propose the following definition. The definition is still preliminary since some additional conditions will be imposed later.

Definition (2.6)
A **stochastic system** or **model** is given by specifying
(1) the functions f_k, $k \geq 0$, and h_k, $k \geq 0$, which define the **state equation** (2.4) and the **observation equation** (2.5), respectively, and
(2) the joint probability distribution or pd of the random variables

$$x_0, w_0, w_1, \ldots, v_0, v_1, \ldots, \tag{2.7}$$

where x_0 is the initial state, $w_0, w_1, \ldots$ is the sequence of **input disturbances,** and $v_0, v_1, \ldots$ is the sequence of measurement errors, or **measurement noise.**

Remark (2.8)
Sometimes it is necessary to describe more explicitly the probability structure underlying the random variables in (2.7). Recall that a specification of these random variables presupposes a probability space (Ω, F, P) where Ω is the sample space, F is the collection of events— i.e.,— a σ-algebra of subsets of Ω—and P is a probability measure on F.

The random variables x_0, w_k, and v_k are then functions defined on Ω, and $x_0(\omega)$, $w_k(\omega)$, $v_k(\omega)$ denote the particular values taken by these random variables when the sample point ω is realized. The pd of these random variables is induced by P,

$$Prob \{x_0 \in X_0, w_k \in W_k, v_k \in V_k; k \geq 0\} =$$

$$P \{\omega \mid x_0(\omega) \in X_0, w_k(\omega) \in W_k, v_k(\omega) \in V_k; k \geq 0\}. \tag{2.9}$$

Here X_0, W_k, V_k are arbitrary (Borel) subsets .

3. State, observation, and control processes

Consider the stochastic system (2.4) and (2.5) and the random variables (2.7). We call these **basic** random variables, since all other random variables will be constructed from them.

Suppose $u_0, u_1, ..$ is a specified deterministic (that is, nonrandom) sequence of inputs. For ease of reference denote this sequence by u. Given u, we can use (2.4) to define the sequence $x_1, x_2, ..$ recursively, as follows:

$$x_1 := f_1(x_0, u_0, w_0), \tag{3.1}$$

$$x_2 := f_2(f_1(x_0, u_0, w_0), u_1, w_1),$$

and so on. Notice that x_k depends upon the input sequence u as well as the basic random variables $w_0, w_1, ...$ Sometimes we need to emphasize this dependence on the input and so we write $x_1^u, x_2^u, ...$ This will be necessary only when we are considering alternative inputs $u, u', ...$ But if the context makes it clear, the superscript u will be omitted as in (3.1). From (3.1) we can see that $x_1^u, x_2^u, ..$ are random variables. Indeed, we can use the representation introduced in Remark (2.8) to express x_k^u as a function on Ω by

$$x_1^u(\omega) := f_1(x_0(\omega), u_0, w_0(\omega)), \tag{3.2}$$

$$x_2^u(\omega) := f_2(x_1^u(\omega), u_1, w_1(\omega)),$$

and so on. The stochastic process $\{x_0^u \equiv x_0, x_1^u, x_2^u, ..\}$ is called the **state process** corresponding to the input sequence u. Note that the pd of the state process can be obtained directly from the pd of the basic random variables, without reference to Ω. For example,

$$Prob \{x_1^u \in X_1, x_2^u \in X_2\}$$

$$= Prob \{f_1(x_0, u_0, w_0) \in X_1, f_2(f_1(x_0, u_0, w_0), u_1, w_1) \in X_2\},$$

and the right hand side can be evaluated from the pd of x_0, w_0, w_1.

Similarly, the **observation process** corresponding to u is the stochastic process $\{y_0^u, y_1^u, ..\}$ defined by

$$y_0^u := h_0(x_0, v_0), \quad y_1^u := h_1(x_1^u, v_1), \ldots \tag{3.3}$$

We now introduce a concept of fundamental importance in control.

The elements of the control sequence $u = \{u_0, u_1, ..\}$ considered earlier were fixed in advance. This is called **open loop** control, as distinct from **closed loop** or **feedback** control, in which the actual value of control input at time k is a function of the observation at time k.

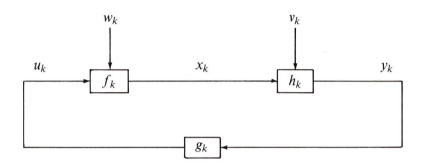

We define closed loop controls more precisely. Let $g = \{g_0, g_1, ..\}$ be a sequence of functions from R^p, the space of observation values, into R^m, the space of control values. We call g_k the feedback function or law at time k and the entire sequence g the **feedback law** or **policy**. The actual value of control u_k used at time k is given by

$$u_k = g_k(y_k), \quad k = 0, 1, \ldots \tag{3.4}$$

Thus a feedback policy prescribes for each time k the control action to be taken as a function of the observation made at k.

The equations describing the system behavior for a given feedback policy g are

$$x_{k+1} = f_k(x_k, u_k, w_k), \quad k \geq 0,$$

$$y_k = h_k(x_k, v_k), \quad k \geq 0,$$

$$u_k = g_k(y_k), \quad k \geq 0.$$

Recall that the state x_k and output y_k are random variables, and so u_k defined by (3.4) is also random. All these random variables will depend on the feedback law g. To emphasize this dependence we sometimes write x_k^g, y_k^g, u_k^g. The values of these random variables for a realization $\omega \in \Omega$ are recursively determined by

$$x_{k+1}^g(\omega) = f_k(x_k^g(\omega), u_k^g(\omega), w_k(\omega)), \quad k \geq 0, \tag{3.5a}$$

$$y_k^g(\omega) = h_k(x_k^g(\omega), v_k(\omega)), \quad k \geq 0, \tag{3.5b}$$

$$u_k^g(\omega) = g_k(y_k^g(\omega)), \quad k \geq 0. \tag{3.5c}$$

Following our notational convention, we call $x^g :=$ $\{x_0^g \equiv x_0, x_1^g, ..\}$ and $y^g := \{y_0^g \equiv y_0, y_1^g, ..\}$ the **state** and **observation** processes, respectively, and $u^g := \{u_0^g, u_1^g, ..\}$ the **control process** corresponding to the feedback law g.

The probability distribution of these processes clearly depends on g, and so we will denote it by P^g. The distribution of the basic random variables does not depend on g and we continue to denote it by P. Using (3.5) one can obtain P^g from P and g. For example,

$$P^g \{x_1^g \in X_1\} = P \{f_0(x_0, g_0(h_0(x_0, v_0)), w_0) \in X_1\}.$$

Above, the choice of control u_k at time k depends only on the observation y_k through $u_k = g_k(y_k)$. Now at time k the controller also knows the previous observations $y_0, ..., y_{k-1}$; moreover, and especially in the case of partial observations as we shall see later, a better performance can be achieved by utilizing all the available information. Therefore we want to permit g_k to be a function of $y_0, ..., y_k$. To simplify notation let

$$y^k := (y_0, ..., y_k) \in R^{(k+1)p}.$$

By a feedback law we will henceforth mean a sequence of functions $g := \{g_0, g_1, ..\}$, where $g_k : R^{(k+1)p} \rightarrow R^m$. Given such a feedback law the processes $\{x_k^g\}, \{y_k^g\}, \{u_k^g\}$ are defined exactly as in (3.5) except that (3.5c) should now read

$$u_k^g(\omega) = g_k(y^{g,k}(\omega)) = g_k(y_0^g(\omega), ..., y_k^g(\omega)), \quad k \geq 0. \tag{3.5c}$$

4. Open loop vs. closed loop

The distinction between open loop and feedback control is fundamental to much of what follows, so it is worth further elaborating upon it.

In the first place, an open loop control sequence $\{u_0, u_1, ..\}$ is deterministic, whereas a feedback law g determines a random control process $\{u_0^g, u_1^g, ..\}$.

Second, an open loop control is a trivial special case of closed loop control. To see this, take a particular open loop sequence $\{u_0, u_1, ..\}$ and now consider the feedback law $g = \{g_0, g_1, ..\}$ where the feedback functions are constant: $g_k(y^k) := u_k$ for all y^k. The random variable u_k^g corresponding to g is now degenerate, and $u_k^g(\omega) = u_k$ with probability 1.

Thus an open loop control is formally a special case of a closed control. For a deterministic system the reverse also holds. The important point is that this is *not* true for stochastic systems. To appreciate this, consider first the case of the deterministic system (1.1) and (1.2). (Equivalently, suppose in (3.5) that the basic random variables are constant with probability 1.) Let $g = \{g_0, g_1, ..\}$ be a feedback law. The corresponding sequence of states x_k^g, $k \geq 0$, is determined by $x_0^g = x_0$ and $x_{k+1}^g = f_k(x_k^g, g_k(y_k^g))$.

Therefore the sequence of control values $\{u_0^g, u_1^g, ..\}$ corresponding to g is given by $u_k^g = g_k(y^{g,k})$. Hence if the open loop control sequence $\{u_0 = u_0^g, u_1 = u_1^g, ..\}$ had been used instead of the closed loop control g, the resulting state sequence would have been the same. Thus for a deterministic system, to every closed loop control there corresponds an open loop control sequence that produces the same state sequence. (Of course, it may be preferable to implement the control in closed loop.)

Now consider the one-dimensional stochastic system

$$x_{k+1} = x_k + u_k + w_k,$$

and suppose that the state is observed, i.e., $y_k \equiv x_k$. Suppose that the basic random variables $x_0, w_0, w_1 ..$ are all independent with zero mean and the same variance $\sigma^2 > 0$. Consider the feedback law $g = \{g_0, g_1, ..\}$, where $g_k(y^k) := -x_k$. The resulting state process $\{x_k^g\}$ is then given by

$$x_0^g = x_0, \tag{4.1a}$$

$$x_{k+1}^g = x_k^g + g_k(y^{g,k}) + w_k = w_k. \tag{4.1b}$$

The resulting control process is $u_0^g = -x_0$, $u_k^g = -w_{k-1}$, $k > 0$, which is random. Consider next any open loop control sequence $u = \{u_0, u_1, ..\}$. The corresponding state process is

$$x_0^u = x_0, \tag{4.2a}$$

$$x_{k+1}^u = x_k^u + u_k + w_k = x_0 + \sum_{m=0}^{k} u_m + \sum_{m=0}^{k} w_m. \tag{4.2b}$$

From (4.1) we see that $\{x_k^g\}$ is a sequence of independent random variables with zero mean and variance σ^2. Whereas from (4.2), $\{x_k^u\}$ is a sequence of random variables with mean $\sum_{m=0}^{k-1} u_m$ and variance $E(x_0^2) + \sum_{m=0}^{k-1} E(w_m^2) = (k+1)\sigma^2$. Thus there is *no* open loop control sequence u that can produce the state process $\{x_k^g\}$.

Let us develop this example further so as to draw another important conclusion. Suppose that the aim of control is to maintain the state process as close to zero as possible during the time interval $k = 0, 1, .., K$. This objective makes sense if, for example, the state x represents the deviation of the system from a desired operating condition, and the numerical measure of this deviation is taken to be

$$E \sum_{k=0}^{K} x_k^2.$$

In such a case, if a feedback law g is used, we can evaluate its performance by the function

$$J(g) := E \sum_{k=0}^{K} (x_k^g)^2.$$

For the feedback law g and the open loop control u introduced earlier, we get

$$J(g) = E \{x_0^2 + \sum_{k=0}^{K-1} w_k^2\} = (K+1)\sigma^2,$$

$$J(u) = \sum_{k=0}^{K} E (x_0 + \sum_{m=0}^{k-1} u_m + \sum_{m=0}^{k-1} w_m)^2$$

$$= \sum_{k=0}^{K} \{\sigma^2 + (\sum_{m=0}^{k-1} u_m)^2 + k\sigma^2\}$$

$$= \tfrac{1}{2} (K+1)(K+2) \sigma^2 + \sum_{k=0}^{K} (\sum_{m=0}^{k-1} u_m)^2 \geq \tfrac{1}{2} (K+1)(K+2) \sigma^2,$$

which is clearly greater than $J(g)$. Thus the performance of the feedback law g is superior to any open loop control. It is this decisive advantage that makes feedback so valuable.

5. Notation

It is convenient at this stage to fix some notational conventions.

For $i = 1, 2$ let z_i be a n_i-dimensional random variable associated with a stochastic system. The **probability distribution** or **pd**, of (z_1, z_2) is denoted by $P_{z_1,z_2}(\ . \ )$. This is a probability measure on $R^{n_1} \times R^{n_2}$ such that for subsets Z_i of R^{n_i}

$$P_{z_1,z_2}(Z_1 \times Z_2) = Prob \{z_1 \in Z_1, z_2 \in Z_2\}.$$

We can use this to calculate the **mean**, or **expected value** of any function q of (z_1, z_2) by

$$E \ q(z_1, z_2) := \int_{R^{n_1}} \int_{R^{n_2}} q(z_1, z_2) \ P_{z_1,z_2}(dz_1 \times dz_2).$$

In particular, the mean of z_i, denoted by $\bar{z}_i$, is given by $\bar{z}_i = E \ z_i$, and its **variance** is $\sigma^2(z_i) := E \ | z_i - \bar{z}_i |^2$, where $| x | := (x_1^2 + .. + x_n^2)^{1/2}$

denotes the Euclidean norm of a vector $x \in R^n$. The **covariance** of z_i is the $n_i \times n_i$ matrix

$$cov(z_i) := E \ (z_i - \bar{z}_i)(z_i - \bar{z}_i)^T,$$

where T denotes transpose. Finally, the covariance of z_1 and z_2 is the $n_1 \times n_2$ matrix

$$cov(z_1, z_2) := E \ (z_1 - \bar{z}_1)(z_2 - \bar{z}_2)^T.$$

Note that $cov(z_i)$ is a symmetric, positive semidefinite matrix, while $cov(z_1, z_2) = [cov(z_2, z_1)]^T$.

Often (z_1, z_2) has a probability density, denoted by $p_{z_1,z_2}(\ . \ )$. It is related to P_{z_1,z_2} by

$$P_{z_1,z_2}(Z_1 \times Z_2) = \int_{Z_1} \int_{Z_2} p_{z_1,z_2}(z_1, z_2) \ dz_1 dz_2 ,$$

and then

$$E \ q(z_1, z_2) = \int_{R^{n_1}} \int_{R^{n_2}} q(z_1, z_2) \ p_{z_1,z_2}(z_1, z_2) \ dz_1 dz_2.$$

Similarly, the marginal probability densities are given by

$$p_{z_1}(z_1) = \int_{R^{n_2}} p_{z_1,z_2}(z_1, z_2) \ dz_2, \quad p_{z_2}(z_2) = \int_{R^{n_1}} p_{z_1,z_2}(z_1, z_2) \ dz_1.$$

One can also get the marginal distribution P_{z_i} from P_{z_1,z_2} by

$$P_{z_1}(Z_1) = P_{z_1,z_2}(Z_1 \times R^{n_2}), \quad P_{z_2}(Z_2) = P_{z_1,z_2}(Z_2 \times R^{n_1}).$$

From P_{z_1,z_2} one obtains the **conditional distribution** of z_1 given z_2, denoted by $P_{z_1|z_2}(\ . \ | \ . \ )$. For each value of the random variable z_2, $P_{z_1|z_2}(\ . \ | \ z_2)$ is a probability measure on R^{n_1} such that for subsets Z_1 and Z_2

$$P_{z_1,z_2}(Z_1 \times Z_2) = \int_{Z_2} P_{z_1|z_2}(Z_1 \ | \ z_2) P_{z_2} \ (dz_2);$$

and in case the relevant densities exist, then

$$p_{z_1|z_2}(z_1 \ | \ z_2) = \frac{p_{z_1,z_2}(z_1, z_2)}{p_{z_2}(z_2)}$$

is the conditional density of z_1 given z_2. The last two relations are versions of Bayes' rule. From them one can obtain the conditional expectation of $q(z_1, z_2)$ given z_2. This is denoted by $E \ \{q \ | \ z_2\}$, and it is defined by

$$E \ \{q \ | \ z_2\} := \int q(z_1, z_2) \ P_{z_1|z_2}(dz_1 \ | \ z_2)$$

$$= \int q(z_1, z_2) \, p_{z_1 | z_2}(z_1 \mid z_2) \, dz_1.$$

Notice that the variable z_1 has been integrated out. Hence $E \{q \mid z_2\}$ is a function of z_2, and it is also a random variable.

Usually the probability distribution of (z_1, z_2) depends on the feedback law g being used. To emphasize this we sometimes write P_{z_1,z_2}^g, E^g, etc. This notation is obviously cumbersome and so when the context makes it clear, the subscript and superscript will be omitted.

6. Transition probability and the Markov property

Recall that for the deterministic system (1.1), the state x_{k+1} at time $k+1$ is completely determined once we know x_k and u_k. More generally, x_{k+m+1} is determined given $x_k, u_k, \ldots, u_{k+m}$ independently of the previous values $x_{k-1}, x_{k-2}, \ldots, u_{k-1}, u_{k-2}, \ldots$ We want to see the extent to which this property holds for the stochastic system

$$x_{k+1} = f_k(x_k, u_k, w_k), \quad k \ge 0,$$

$$y_k = h_k(x_k, v_k), \quad k \ge 0.$$

The basic random variables are $\{x_0, w_0, w_1, \ldots, v_0, v_1, \ldots\}$. Suppose the feedback law $g = \{g_0, g_1, \ldots\}$ is used. Corresponding to g are the stochastic processes $\{x_k^g\}$, $\{y_k^g\}$ and $\{u_k^g\}$, defined by

$$x_{k+1}^g = f_k(x_k^g, u_k^g, w_k), \quad x_0^g \equiv x_0 \tag{6.1a}$$

$$y_k^g = h_k(x_k^g, v_k), \tag{6.1b}$$

$$u_k^g = g_k(y^{g,k}). \tag{6.1c}$$

We drop the superscript g from these variables but use P^g to remind ourselves that their probability distributions do depend on g. Suppose we know the values realized by the random variables x_k, u_k. These values by themselves are insufficient to determine the value of x_{k+1} since w_k is not known. The value of x_{k+1} is statistically determined by the conditional distribution of x_{k+1} given x_k, u_k, namely, by

$$P_{x_{k+1} | x_k, u_k}^g (\cdot \mid x_k, u_k).$$

From (6.1) we get for any subset X_{k+1} of R^n

$$P_{x_{k+1} | x_k, u_k}^g (X_{k+1} \mid x_k, u_k) = P_{w_k | x_k, u_k}^g (W_k \mid x_k, u_k), \tag{6.2}$$

where

$$W_k := \{w \mid f_k(x_k, u_k, w) \in X_{k+1}\}. \tag{6.3}$$

We will drop the subscripts from P when the arguments provide the necessary information.

Suppose we know the previous values of the random variables x_m, u_m, $m \leq k-1$. The conditional distribution of x_{k+1} given these values is similar to (6.2),

$$P^g(X_{k+1} \mid x_k, \ldots x_0, u_k, \ldots, u_0) =$$
$$P^g(W_k \mid x_k, \ldots x_0, u_k, \ldots, u_0). \tag{6.4}$$

We would like to retain a property analogous to the one enjoyed by the deterministic system model (1.1), namely, we want the conditional distribution of x_{k+1} given x_k, u_k to be independent of previous values of the state and control. From (6.4) we see that one way to secure this is to guarantee that for every law g, w_k is independent of the random variables x_m, u_m, $m \leq k-1$.

It is desirable to obtain such a guarantee by imposing a condition directly on the basic random variables themselves. Here is the most natural condition that does the job.

Assumption (6.5)
The basic random variables

$$x_0, w_0, w_1, \ldots, v_0, v_1, \ldots$$

are all independent.

Lemma (6.6)
Suppose (6.5) holds, then, for any feedback law g,

$$P^g(X_{k+1} \mid x_k, \ldots, x_0, u_k, \ldots, u_0) = P^g(X_{k+1} \mid x_k, u_k).$$

Moreover these conditional probabilities do not depend on g.

Proof
From (6.1) we see that the random variables $x_0, \ldots, x_k$, and $u_0, \ldots, u_k$, are functions of the basic random variables $x_0, w_0, \ldots, w_{k-1}, v_0, \ldots, v_{k-1}$ which by (6.5) are independent of w_k. Hence w_k is also independent of $x_k, \ldots, x_0, u_k, \ldots, u_0$. Therefore the right hand sides of both (6.2) and (6.4) are equal to $P_{w_k}(W_k)$, where P_{w_k} is the pd of w_k. It follows that the left-hand sides of (6.2) and (6.4) are equal and do not depend on g, and so we may drop the superscript g. Moreover we have

$$P(X_{k+1} \mid x_k, \ldots, x_0, u_k, \ldots, u_0) = P(X_{k+1} \mid x_k, u_k)$$
$$= P_{w_k}(w_k \mid f_k(x_k, u_k, w_k) \in X_{k+1}), \tag{6.7}$$

and the proof is complete.   □

Equation (6.7) is the counterpart of the one-step transition function (1.4) for deterministic systems. It is tempting to conjecture that under assumption (6.5) we can extend this to several steps analogous to (1.5). For instance, for the two-step case, we could try to prove that

$$P^g(X_{k+2} \mid x_k, \ldots, x_0, u_{k+1}, \ldots, u_0) =$$

$$P^g(X_{k+2} \mid x_k, u_k, u_{k+1}), \tag{6.8}$$

and that the two sides do not depend on g. One can attempt to prove this as follows. From (6.1a) we see that there is a function $f_{k+2,k}$, not depending on g, such that

$$x_{k+2} = f_{k+2,k}(x_k, u_k, u_{k+1}, w_k, w_{k+1}),$$

and so the left-hand side of (6.8) equals

$$P^g(W_{k,k+1} \mid x_k, \ldots, x_0, u_{k+1}, \ldots, u_0) \tag{6.9}$$

where

$$W_{k,k+1} = \{(w_k, w_{k+1}) \mid f_{k+2,k}(x_k, u_k, u_{k+1}, w_k, w_{k+1}) \in X_{k+2}\}.$$

We want to see whether under (6.5) the probability distribution (6.9) depends only on x_k, u_k, u_{k+1}, and not on $g, x_{k-1}, \ldots, x_0, u_{k-1}, \ldots, u_0$. This will be true provided that w_k and w_{k+1} are independent of $x_{k-1}, \ldots, x_0, u_{k-1}, \ldots, u_0$ given x_k, u_k, u_{k+1}. We saw before that w_k, w_{k+1} are indeed independent of $x_k, \ldots, x_0, u_k, \ldots, u_0$. Hence the conjecture will be true if w_k, w_{k+1} are independent of u_{k+1} as well. We can also see from Exercise (6.13) that if this independence is not valid, then the conjecture will not generally be true either. Now

$$u_{k+1} = g_{k+1}(y^{k+1})$$

$$= g_{k+1}(y^k, h_{k+1}(x_{k+1}, v_{k+1}))$$

$$= g_{k+1}(y^k, h_{k+1}((f_k(x_k, u_k, w_k)), v_{k+1}))$$

and so we see that u_{k+1} and w_k will usually be dependent. The conjecture is therefore not generally true. One case when it will be true is if g_{k+1} is a constant function so that u_{k+1} is a degenerate random variable, hence independent of all other random variables including, in particular, w_k. Note that choosing g_{k+1} to be a constant function is the same as requiring that the control at time $k+1$ is chosen open loop. In this case we get the next result, whose proof is similar to Lemma (6.6).

Lemma (6.10)

Suppose (6.5) holds. Then, for any feedback law $g = \{g_0, g_1, \ldots\}$ for which $g_{k+1} \equiv \bar{u}_{k+1}, \ldots, g_{k+m} \equiv \bar{u}_{k+m}$ are constant functions,

$$P^g(x_{k+m+1} \in X_{k+m+1} \mid x_k, \ldots, x_0, \bar{u}_{k+m}, \ldots, \bar{u}_{k+1}, u_k, \ldots, u_0)$$

$$= P^g(X_{k+m+1} \mid x_k, u_k, \bar{u}_{k+1}, \ldots, \bar{u}_{k+m}) \tag{6.11}$$

$$= P_{w_k, \ldots, w_{k+m}}(W_{k, \ldots, k+m}), \tag{6.12}$$

where

$$W_{k,\ldots,k+m} := \{(w_k, \ldots, w_{k+m}) \mid$$

$$f_{k+m+1,k}(x_k, u_k, w_k, \bar{u}_{k+1}, w_{k+1}, \ldots, \bar{u}_{k+m}, w_{k+m}) \in X_{k+m+1}\}$$

and $f_{k+m+1,k}$ is the transition function defined in (1.5).

Exercise   (6.13)
Here is an example which shows that (6.8) is not generally valid if u_{k+1} is not deterministic. Consider the system

$$x_1 = w_0, \quad x_2 = w_1, \quad x_3 = x_2 + u_2,$$

with $y_k \equiv x_k$. Take the feedback law,

$$u_0 = u_1 \equiv 0, \quad \text{and} \quad u_2 = g_2(x_0, x_1, x_2) = x_0 + x_2.$$

Show that

$$P^g(X_3 \mid x_1, x_0, u_2, u_1, u_0) = \begin{cases} 1 & \text{if } 2u_2 - x_0 \in X_3 \\ 0 & \text{if } 2u_2 - x_0 \notin X_3 \end{cases}$$

whereas

$$P^g(X_3 \mid x_1, u_1, u_2) = P(2w_1 + x_0 \in X_3 \mid x_0 + w_1).$$

The two expressions are not equal.

Observe that since (6.12) does not depend on g, the conditional probability in (6.11) does not depend on g either. These lemmas suggest the following definition.

Definition   (6.14)
Suppose (6.5) holds. The conditional probabilities $P(X_{k+1} \mid x_k, u_k)$ and $P(X_{k+m} \mid x_k, u_k, \ldots, u_{k+m-1})$ of (6.7) and (6.11) are called the one-step and the m-step **transition probability**, respectively.

Lemma   (6.15)
The m-step transition probability can be obtained recursively by

$$P(X_{k+m+1} \mid x_k, u_k, u_{k+1}, \ldots, u_{k+m}) =$$

$$\int P(X_{k+m+1} \mid x_{k+1}, u_{k+1}, \ldots, u_{k+m}) \, P(dx_{k+1} \mid x_k, u_k). \quad (6.16)$$

Proof
From (6.12) the left hand side equals $P(W_{k,\ldots,k+m})$, and since w_k and $w_{k+1}, \ldots, w_{k+m}$ are independent, this probability can be expressed as

$$\int P(W_{k+1,\ldots,k+m}(w_k)) \, P(dw_k), \quad (6.17)$$

where

$$W_{k+1,\ldots,k+m}(w_k) :=$$

$$\{(w_{k+1}, \ldots, w_{k+m}) \mid (w_k, \ldots, w_{k+m}) \in W_{k,\ldots,k+m}\}.$$

From (6.7)

$$P(dw_k) = P(dx_{k+1} \mid x_k, u_k),$$

and from (6.12)

$$P(W_{k+1,...,k+m}(w_k)) = P(X_{k+m+1} \mid x_{k+1}, u_{k+1}, \ldots, u_{k+m}).$$

Substituting the last two relations in (6.17) shows that it equals the right hand side of (6.16), and the lemma is proved.   □

It turns out that in the study of estimation and control problems, the transition probability functions are indispensable. Therefore, in our definition of a stochastic system model, we have to impose the condition that the basic random variables are independent. This leads to the following definition.

Definition (6.18)
A **stochastic system** is given by specifying
(i) the state equation f_k, $k \geq 0$, the observation equation h_k, $k \geq 0$, and
(ii) the joint probability distribution of the independent basic random variables

$$x_0, w_0, w_1, \ldots, v_0, v_1, \ldots$$

Recall that a stochastic process $\{z_0, z_1, ..\}$ is said to be Markovian or a Markov chain if for all subsets Z_{n+k}, and $k > 0$,

$$P(Z_{n+k} \mid z_n, \ldots, z_0) \equiv P(Z_{n+k} \mid z_n).$$

For this reason we refer to (6.11) as the **Markov property.**

Observe that the transition probability, which is defined by the pd of the random variable w_k and the function f_k, does not explicitly depend on them. Suppose that instead of being given the functions f_k and the pd of the process $\{w_k\}$ explicitly, we are given the transition probability $P_{x_{k+1} \mid x_k, u_k}$, the observation functions h_k, and the distribution of the independent basic random variables $x_0, v_0, v_1, \ldots$. Let $g = \{g_k\}$ be any feedback policy and $\{x_k\}$, $\{y_k\}$, $\{u_k\}$ the processes corresponding to g. Then we can obtain the pd of these processes as follows

$$P(X_{k+1} \mid x_k, \ldots, x_0, y_k, \ldots, y_0, u_k, \ldots, u_0) =$$
$$P(X_{k+1} \mid x_k, u_k) \tag{6.19a}$$

$$P(Y_k \mid x_k, \ldots, x_0, y_{k-1}, \ldots, y_0, u_k, \ldots, u_0)$$
$$= P_{y_k \mid x_k}(Y_k \mid x_k) = P_{v_k}\{v_k \mid h_k(x_k, v_k) \in Y_k\} \tag{6.19b}$$

$$u_k = g_k(y^k) = g_k(y_0, \ldots, y_k) \text{ with } P^g\text{-probability 1.} \tag{6.19c}$$

Thus we have an alternative description of a stochastic system.

Definition (6.20)

A **controlled Markov chain** description of a stochastic system is given by specifying

(1) the (state) transition probability $P_{x_{k+1} | x_k, u_k}(\, . \, | \, . \,)$, the functions h_k, $k \geq 0$ which define the observation equations, and

(2) the pd of the independent basic random variables $x_0, v_0, v_1, \ldots$.

Given such a description, we can obtain the pd of the state, output, and control processes corresponding to any feedback law g using (6.19). Moreover, it suffices, from (6.19b), to specify the **output** or **observation probability** $P_{y_k | x_k}(\, . \, | \, . \,)$ instead of the observation functions $\{h_k\}$.

As we will see in Chapter 4, this alternative model description is often more natural when considering systems whose state takes on only a finite number of values.

7. Notes

1. The control of discrete time deterministic systems is discussed in Bellman and Dreyfus (1962) and in Bryson and Ho (1969).

2. An elementary treatment of the probability structure introduced in Remark (2.8) and the properties discussed in Section 5 can be found in Wong (1983). For good accounts of more advanced probability theory, see Chung (1974) and Loeve (1960). Doob (1953) is a good reference for the theory of stochastic processes.

8. References

[1] R.E. Bellman and S.E. Dreyfus (1962), *Applied dynamic programming,* Princeton University Press, Princeton, NJ.

[2] A.E. Bryson, Jr. and Y.-C. Ho (1969), *Applied optimal control,* Ginn, Waltham, MA.

[3] K. L. Chung (1974), *A course in probability theory,* Academic Press, New York.

[4] J. L. Doob (1953), *Stochastic Processes,* Wiley, New York.

[5] M. Loeve (1960), *Probability theory,* 2nd Edition, Van Nostrand, Princeton, NJ.

[6] E. Wong (1983), *Introduction to random processes,* Springer, New York.

CHAPTER 3
PROPERTIES OF LINEAR STOCHASTIC SYSTEMS

This chapter collects basic results for linear stochastic system models with input disturbance and measurement errors. Special attention is given to the case when the disturbances are Gaussian. Recursive equations are derived for the mean and covariance of the state and observation processes. Their asymptotic properties are related to the stability of the state equations. Finally, the state processes of linear Gaussian systems are related to Gauss-Markov processes.

1. Linear Gaussian systems

For a n-dimensional random variable z, we write

$$z \sim N(\bar{z}, \Sigma) \text{ or } p(z) \sim N(\bar{z}, \Sigma) \tag{1.1}$$

to indicate that z is Gaussian or normal with mean $\bar{z}$ and covariance Σ. Recall that the pd of z is given by

$$p(z) = [(2\pi)^n \mid \Sigma \mid]^{-\frac{1}{2}} \exp[-\frac{1}{2} (z-\bar{z})^T \Sigma^{-1} (z-\bar{z})] \tag{1.2}$$

where $\mid \Sigma \mid := \det \Sigma$.

Consider the **linear stochastic system**

$$x_{k+1} = A \, x_k + B \, u_k + G \, w_k , \tag{1.3a}$$

$$y_k = C \, x_k + H \, v_k , \tag{1.3b}$$

where $x_k \in R^n$, $u_k \in R^m$, $y_k \in R^p$, $w_k \in R^g$, $v_k \in R^h$ and A, B, G, C, H are fixed matrices of appropriate dimension. The independent basic random variables $x_0, w_0, w_1, .., v_0, v_1, ..$ are all assumed to be Gaussian:

$$x_0 \sim N(\bar{x}_0, \Sigma_0), \quad w_k \sim N(0, Q), \quad v_k \sim N(0, R). \tag{1.4}$$

We calculate the (one-step) transition probability for this system. To do this it is possible to apply directly the formula (2.6.7), but it is more instructive to proceed from first principles. Since w_k is Gaussian, $G \, w_k$ is also Gaussian. It has mean $E \, G \, w_k = G \, E \, w_k = 0$ and covariance $E \, (G \, w_k)(G \, w_k)^T = G \, E \, w_k w_k^T \, G^T = GQG^T$. Therefore $G \, w_k \sim N(0, GQG^T)$.

By assumption, the basic random variables are independent. Therefore, for any feedback law g, w_k is independent of x_k and u_k. In particular, we see from (1.3a) that the conditional distribution of x_{k+1} given x_k and u_k is the same as the pd of $G w_k$ with mean shifted by $A x_k + B u_k$. So the state transition probability is

$$p(x_{k+1} \mid x_k, u_k) \sim N(A x_k + B u_k, GQG^T). \tag{1.5}$$

Substitution in (1.2) gives the explicit expression

$$p(x_{k+1} \mid x_k, u_k) = K \exp[-\tfrac{1}{2}(x_{k+1} - A x_k - B u_k)^T$$
$$\times (GQG^T)^{-1}(x_{k+1} - A x_k - B u_k)],$$

where K is the normalizing constant,

$$K = [(2\pi)^n \mid GQG^T \mid]^{-\frac{1}{2}}.$$

Similar reasoning shows that the output or observation probability is

$$p(y_k \mid x_k) \sim N(C x_k, HRH^T). \tag{1.6}$$

Next we calculate the m-step transition probability. Let g be any feedback law for which $g_{k+1} \equiv u_{k+1}, \ldots, g_{k+m-1} \equiv u_{k+m-1}$, are constant functions. Let $\{x_k\}$, $\{y_k\}$ denote the corresponding state and output processes. By Definition (2.6.14), the m-step transition probability is the conditional distribution of x_{k+m} given $x_k, u_k, \ldots, u_{k+m-1}$, and it does not depend on g. Instead of using (2.6.16) it is easier to calculate this conditional distribution directly. From (1.3a)

$$x_{k+m} = A^m x_k + \sum_{j=0}^{m-1} A^{m-1-j} B u_{k+j} + \sum_{j=0}^{m-1} A^{m-1-j} G w_{k+j}. \tag{1.7}$$

The random variables $\{w_{k+j}, j \geq 0\}$ are Gaussian and independent of x_k and u_k, as well as of $u_{k+j}, j \geq 1$, since the latter are constant. Therefore, from (1.7),

$$p(x_{k+m} \mid x_k, u_k, \ldots, u_{k+m-1}) \sim$$
$$N(A^m x_k + \sum_{j=0}^{m-1} A^{m-1-j} B u_{k+j}, \Sigma_{k+m \mid k}), \tag{1.8}$$

where $\Sigma_{k+m \mid k} := cov(\sum_{j=0}^{m-1} A^{m-1-j} G w_{k+j})$. This covariance can be readily evaluated using the independence of the w_k:

$$\Sigma_{k+m \mid k} = cov(\sum_{j=0}^{m-2} A^{m-1-j} G w_{k+j}) + cov(G w_{k+m-1})$$

$$= cov(A \sum_{j=0}^{m-2} A^{m-2-j} G \ w_{k+j}) + GQG^T$$

$$= A[cov(\sum_{j=0}^{m-2} A^{m-2-j} G \ w_{k+j})]A^T + GQG^T$$

$$= A\Sigma_{k+m-1|k}A^T + GQG^T. \tag{1.9}$$

This recursive linear equation can be solved for $\Sigma_{k+m|k}$, $m > 0$, starting with the initial condition

$$\Sigma_{k|k} = 0. \tag{1.10}$$

From (1.9) and (1.10) we see that $\Sigma_{k+m|k}$ depends only upon m and not on k.

Note that in the discussion above the feedback functions $g_1, \ldots, g_k$ (in contrast to g_i for $i \geq k$) were not required to be constant functions. Indeed g_i, for $i \leq k$, could have been any arbitrary nonlinear function of the available information $\{y^i\}$. If the feedback functions are nonlinear, then the control process will not be Gaussian, and so the state process (which is determined by the *unconditional* probabilities) will not be Gaussian either. Nevertheless, as we have seen, the transition probabilities (which are *conditional* probabilities) *are* Gaussian. We summarize this result as a Lemma.

Lemma (1.11)
The m-step transition probability for the system (1.3), (1.4), with g any feedback law such that $g_{k+1} \equiv u_{k+1}, \ldots, g_{k+m-1} \equiv u_{k+m-1}$, is given by the Gaussian density

$$p(x_{k+m} \mid x_k, u_k, \ldots, u_{k+m-1}) \sim$$

$$N(A^m x_k + \sum_{j=0}^{m-1} A^{m-1-j} B \ u_{k+j}, \Sigma_{k+m|k}).$$

Moreover, the covariance matrix $\Sigma_{k+m|k}$ can be calculated from the linear difference equation (1.9) and (1.10).

Suppose now that the control law is open loop for all time; i.e., the functions $g_i \equiv u_i$, are constant functions for all $i \geq 0$ (in particular, for $i \leq k$ also). Then the state process $\{x_k\}$ is Gaussian and so its (unconditional) pd is completely determined by the (unconditional) means Ex_k^g and the (unconditional) covariance function $\Sigma_{k+m,k} := cov(x_{k+m}, x_k)$.

To obtain $p_{x_k}^g$, which determines Ex_k^g and $\Sigma_{k,k} := cov(x_k, x_k) =: \Sigma_k$, note that

$$x_k = A^k x_0 + \sum_{j=0}^{k-1} A^{k-1-j} B \ u_j + \sum_{j=0}^{k-1} A^{k-1-j} G \ w_j.$$

The three terms on the right are all independent and Gaussian. Hence

$$p_{x_k}^g \sim N(\bar{x}_k, \Sigma_k),$$

where

$$\bar{x}_k = A^k \bar{x}_0 + \sum_{j=0}^{k-1} A^{k-1-j} B \, u_j,$$

$$\Sigma_k = cov(A^k x_0 + \sum_{j=0}^{k-1} A^{k-1-j} G \, w_j)$$

$$= cov(A(A^{k-1}x_0 + \sum_{j=0}^{k-2} A^{k-2-j} G \, w_j)) + cov(G \, w_{k-1})$$

$$= A \Sigma_{k-1} A^T + GQG^T, \quad k \ge 1, \tag{1.12}$$

$$\Sigma_0 = cov(x_0). \tag{1.13}$$

Thus the pd of $\{x_k\}$ can be calculated recursively.

To obtain the covariance function $\Sigma_{k+m,k}$ recall that

$$\Sigma_{k+m,k} := E \, (x_{k+m} - \bar{x}_{k+m})(x_k - \bar{x}_k)^T.$$

From (1.7)

$$x_{k+m} - \bar{x}_{k+m} = A^m(x_k - \bar{x}_k) + \sum_{j=0}^{m-1} A^{m-1-j} G \, w_{k+j},$$

and since x_k and w_{k+j}, $j \ge 0$, are independent,

$$\Sigma_{k+m,k} = A^m \Sigma_k. \tag{1.14}$$

Observe that the mean $\bar{x}_k$ depends on the control process $\{u_k\}$, but the covariance function $\Sigma_{k+m,k}$ does not.

When g is open loop, the process y_k is also Gaussian. From (1.3b) we see that

$$p_{y_k}^g \sim N(\bar{y}_k, \Sigma_k^y),$$

where the mean is

$$\bar{y}_k := E \, y_k = C \, E \, x_k = C \, (A^k \bar{x}_0 + \sum_{j=0}^{k-1} A^{k-1-j} B \, u_j),$$

and the covariance is

$$\Sigma_k^y = cov(C \, x_k) + cov(Hv_k) = C \Sigma_k C^T + HRH^T. \tag{1.15}$$

The covariance of y_{k+m} and y_k is

$$\Sigma^y_{k+m,k} := E\ (y_{k+m} - \bar{y}_{k+m})(y_k - \bar{y}_k)^T.$$

From (1.3b) and (1.14), and using the independence of $\{x_k\}$ and $\{v_k\}$,

$$\Sigma^y_{k+m,k} = C\ \Sigma_{k+m,k}\ C^T,\ m \geq 1,$$
$$= C\ A^m\ \Sigma_k\ C^T,\ m \geq 1. \tag{1.16}$$

The next exercise examines the time varying linear stochastic system.

Exercise (1.17)

Suppose that instead of the time-invariant system (1.3), (1.4), we have the time-varying linear system

$$x_{k+1} = A_k\ x_k + B_k\ u_k + G_k\ w_k,$$
$$y_k = C_k\ x_k + H_k\ v_k,$$

and the independent basic variables are Gaussian:

$$x_0 \sim N(\bar{x}_0, \Sigma_0),\quad w_k \sim N(0, Q_k),\quad v_k \sim N(0, R_k).$$

Show that the m-step transition probability is Gaussian,

$$p(x_{k+m}\ |\ x_k, u_k, \ldots, u_{k+m-1}) \sim$$
$$N(A_{k+m-1} \cdots A_k\ x_k + \sum_{j=0}^{m-1} A_{k+m-1} \cdots A_{k+j+1} B_{k+j}\ u_{k+j}, \Sigma_{k+m\,|\,k}).$$

Moreover, the covariance can be calculated from the time-varying linear difference equation

$$\Sigma_{k+m\,|\,k} = A_{k+m-1}\ \Sigma_{k+m-1\,|\,k}\ A^T_{k+m-1} + G_{k+m-1} Q_{k+m-1} G^T_{k+m-1},$$
$$\Sigma_{k\,|\,k} = 0.$$

Throughout the above we have assumed that the noise $\{w_k\}$ entering the system is an independent process. The next exercise shows how to stretch this assumption by enlarging the state vector.

Exercise (1.18)

Suppose the noise sequence $\{w_0, w_1, \ldots\}$ is not independent, so that the system (1.3), (1.4) is not a linear stochastic system as defined. Suppose however that $\{w_k\}$ is itself the output of a linear system,

$$\xi_{k+1} = F\xi_k + \epsilon_k;\ w_k = D\xi_k + \delta_k,$$

where $\{x_0, \xi_0, \ldots, v_0, \ldots \epsilon_0, \ldots \delta_0, \ldots\}$ are independent and Gaussian. Show that $\{y_k\}$ can then be written as the output of a linear stochastic system.

[Hint: Consider the state vector $\zeta_k := (x^T_k, \xi^T_k)^T$.]

Thus our definition of linear stochastic systems encompasses situations where the input noise can be regarded as being generated by filtering white Gaussian noise through some linear stochastic system.

2. Linear non-Gaussian systems

Consider the linear system (1.3), but suppose the independent basic random variables $x_0, w_0, w_1, \ldots, v_0, v_1, \ldots$ are not Gaussian. Assume, however, that their means and covariances are the same as before:

$$Ex_0 = \bar{x}_0, \quad Ew_k = 0, \quad Ev_k = 0,$$

$$cov(x_0) = \Sigma_0, \quad cov(w_k) = Q, \quad cov(v_k) = R.$$

Consider an open loop control so that $g_k = u_k$, $k \geq 0$, are constant functions.

Let

$$\bar{x}_{k+m \mid k} := E\{x_{k+m} \mid x_k, u_k, \ldots, u_{k+m-1}\},$$

$$\Sigma_{k+m \mid k} := E\{(x_{k+m} - \bar{x}_{k+m \mid k})(x_{k+m} - \bar{x}_{k+m \mid k})^T$$
$$\mid x_k, u_k, \ldots, u_{k+m-1}\},$$

be the conditional mean and conditional covariance of x_{k+m} given x_k, respectively.

From (1.7) it follows that

$$\bar{x}_{k+m \mid k} = A^m x_k + \sum_{j=0}^{m-1} A^{m-1-j} B u_{k+j},$$

$$\Sigma_{k+m \mid k} = cov\left(\sum_{j=0}^{m-1} A^{m-1-j} G w_{k+j}\right),$$

which gives

$$\Sigma_{k+m \mid k} = A \Sigma_{k+m-1 \mid k} A^T + GQG^T.$$

Comparing with (1.8) we see that the first two conditional moments are the same as in the Gaussian case. However the conditional distribution need not be Gaussian.

Similarly the (unconditional) mean and covariance

$$\bar{x}_k := Ex_k, \quad \Sigma_k := E(x_k - \bar{x}_k)(x_k - \bar{x}_k)^T,$$

are also given by (1.12) when the control is open loop.

This coincidence of the first two moments also extends to the time-varying case of Exercise (1.17).

3. Asymptotic properties

We study the behavior as $k \rightarrow \infty$ of the covariance matrices Σ_k given by (1.12) and (1.13), which we reproduce here:

$$\Sigma_k = A\Sigma_{k-1}A^T + GQG^T, \quad k \geq 1, \tag{3.1}$$

$$\Sigma_0 = cov(x_0). \tag{3.2}$$

Recursive substitution leads to

$$\Sigma_k = A^k\Sigma_0(A^k)^T + \sum_{j=0}^{k-1} A^j GQG^T(A^j)^T. \tag{3.3}$$

To obtain convergence of Σ_k we need the series to be summable, and so we should require the matrix A^j to converge to 0.

Recall that a matrix A is said to be **stable** if all its eigenvalues are strictly smaller than 1 in magnitude.

Theorem   (3.4)

Suppose A is stable. Then there is a positive semidefinite matrix Σ_∞ such that $\lim_{k \rightarrow \infty} \Sigma_k = \Sigma_\infty$. Moreover, Σ_∞ is the unique solution of the linear equation

$$\Sigma_\infty = A \Sigma_\infty A^T + GQG^T. \tag{3.5}$$

Proof

Since A is stable, there are numbers $K < \infty$ and $0 < \alpha < 1$ such that every coefficient of the matrix A^j is smaller in magnitude than $K\alpha^j$. Hence in (3.3) the first term vanishes and the sum converges as $k \rightarrow \infty$, so

$$\lim_{k \rightarrow \infty} \Sigma_k = \Sigma_\infty := \sum_{j=0}^{\infty} A^j GQG^T (A^j)^T.$$

Next, since Σ_k converges to Σ_∞, we can let $k \rightarrow \infty$ in (3.1) and deduce that Σ_∞ obeys (3.5). It remains to show that (3.5) has a unique solution. Suppose that Σ_∞^1 and Σ_∞^2 are two solutions to (3.5). Then $\Delta := \Sigma_\infty^1 - \Sigma_\infty^2$ satisfies

$$\Delta = A \Delta A^T.$$

Recursive substitution leads to

$$\Delta = A^k \Delta (A^k)^T,$$

and letting $k \rightarrow \infty$ shows that $\Delta = 0$, hence $\Sigma_\infty^1 = \Sigma_\infty^2$.   □

Equation (3.5) is called the *discrete-time Lyapunov equation.*

Exercise   (3.6)
Suppose A is stable. If $cov(x_0) = \Sigma_\infty$, then the covariance function of $\{x_k\}$ is stationary, i.e., $\Sigma_k = \Sigma_\infty$, and $\Sigma_{k+m,k} = A^m \Sigma_\infty$ for all $k > 0$. In general, $\{x_k\}$ can be decomposed as

$$x_k = x'_k + x''_k$$

where the covariance function of $\{x'_k\}$ is stationary, and $E \mid x''_k \mid^2 \to 0$ as $k \to \infty$. (The decomposition may not be unique.)

Remark   (3.7)
It is possible that A is not stable and yet Σ_k converges. A trivial example is given by $\Sigma_0 = 0$, $Q = 0$, in which case $\Sigma_k \equiv 0$ independent of A. It can also happen that Σ_∞ is not strictly positive definite. Another trivial example: take $A = 0$, and rank $GQG^T < n$. These two examples suggest that if the input disturbance $\{w_k\}$ affects all the components of the state vector, then the stability of A would be necessary for the convergence of Σ_k, and the limiting covariance Σ_∞ would be positive definite. This intuition turns out to be well-founded provided the notion of affecting all components of the state is appropriately defined.

A pair of matrices A, S of dimensions $n \times n$ and $n \times g$ respectively, is said to be **reachable** if the $n \times ng$ matrix $[S \ AS \ .. \ A^{n-1}S]$ has rank n. The adjective reachable is justified in the following exercise.

Exercise   (3.8)
Show that the following statements are equivalent.
(1) A, S is a reachable pair.
(2) The $n \times n$ matrix $\sum_{j=0}^{n-1} A^j SS^T (A^j)^T$ is positive definite.
(3) For every $x \in R^n$, there exists an input sequence $w_0, \ldots, w_{n-1}$ with values in R^g, which steers the state of the deterministic linear system

$$x_{k+1} = A \ x_k + S \ w_k, \quad k \ge 0,$$

from the state $x_0 = 0$ to the state $x_n = x$.

Theorem   (3.9)
Suppose A, S is reachable. Then the following statements are equivalent.
(1) A is stable.
(2) The equation

$$\Sigma = A \ \Sigma \ A^T + SS^T \tag{3.10}$$

has a positive definite solution Σ.

Proof
If A is stable then, from Theorem (3.4), there is a unique positive

semidefinite solution, namely,

$$\Sigma = \sum_{j=0}^{\infty} A^j \, S \, S^T \, (A^j)^T.$$

Since A, S is reachable, Σ is positive definite by (3.8b).

Now suppose Σ is a positive definite solution of (3.10). Then

$$\Sigma = A^k \, \Sigma \, (A^k)^T + \sum_{j=0}^{k-1} A^j \, S \, S^T \, (A^j)^T, \quad k \geq 0. \tag{3.11}$$

Let λ be an eigenvalue of A and let $x \neq 0$ be a vector such that $A^T x = \lambda x$. Multiplying (3.11) on the left by x^* (* denotes complex conjugate transpose), and on the right by x gives, for $k = n - 1$,

$$x^* \, \Sigma \, x = |\lambda|^{2(n-1)} x^* \, \Sigma \, x + x^* \, [\sum_{j=0}^{n-1} A^j SS^T (A^j)^T] \, x.$$

The matrix [. .] is positive definite by (3.8b) and so the last term is strictly positive. Hence $|\lambda| < 1$, and A is stable. $\square$

We return to the covariance equation (3.1). Let S be any matrix such that

$$GQG^T = SS^T. \tag{3.12}$$

Such a matrix S is called a **square root** of GQG^T.

Theorem (3.13)

Suppose A, S is reachable. Then the following statements are equivalent.
(1) A is stable.
(2) Σ_k converges to a positive definite matrix Σ_∞ as $k \to \infty$.

Proof

Immediate from Theorem (3.9). $\square$

Remark

The square root S in (3.12) is not unique. However, from (3.8b), we see that A, S is reachable if and only if $\sum_{j=0}^{n-1} A^j SS^T (A^j)^T = \sum_{j=0}^{n-1} A^j GQG^T (A^j)^T$ is positive definite. Hence the reachability of A, S does not depend upon the choice of the square root in (3.12).

4. Gauss-Markov processes

A stochastic process $\{x_k, k \geq 0\}$ with values in R^n is said to be **Gauss-Markov** (GM) if it is Gaussian—i.e., for every k, the pd of the random variables $x_0, \ldots, x_k$ is Gaussian—and Markov—i.e.,

$$p(x_{k+1} \mid x_k, \ldots, x_0) = p(x_{k+1} \mid x_k). \tag{4.1}$$

We have seen that the state process $\{x_k\}$ given by

$$x_{k+1} = A_k x_k + w_k,$$

where $x_0, w_0, w_1, \ldots$ are independent and Gaussian is GM. There is a converse to this fact which we leave as an exercise.

Let $\{x_k\}$ be a GM process. Suppose $E x_k \equiv 0$, and let $\Sigma_k := cov(x_k)$, $\Sigma_{k+1,k} := cov(x_{k+1}, x_k)$.

Exercise   (4.2)

Show that

$$\hat{x}_{k+1|k} := E\{x_{k+1} \mid x_k, \ldots, x_0\} = E\{x_{k+1} \mid x_k\}$$

$$= \Sigma_{k+1,k} \Sigma_k^{-1} x_k. \tag{4.3}$$

Now let

$$w_k := x_{k+1} - \hat{x}_{k+1|k}.$$

Show that the random variables $x_0, w_0, w_1, \ldots$ are Gaussian and independent. Thus the GM process $\{x_k\}$ has a representation of the form

$$x_{k+1} = A_k x_k + w_k, \tag{4.4}$$

with $A_k := \Sigma_{k+1,k} \Sigma_k^{-1}$.

[Hint: The first equality in (4.3) follows from the Markov property (4.1). For the second equality and the remainder of the exercise, proceed as follows. Show that $x_{k+1} - \Sigma_{k+1,k} \Sigma_k^{-1} x_k$ and x_k are uncorrelated by computing their covariance. Recall that two jointly Gaussian random variables are independent if they are uncorrelated. Hence $E\{\hat{x}_{k+1} - \Sigma_{k+1,k} \Sigma_k^{-1} x_k \mid x_k\} = 0$, and so the second equality in (4.3) is also verified.]

In the exercise $w_k = x_{k+1} - \hat{x}_{k+1|k}$ and x_k together determine x_{k+1}. The latter is known at time k; however the former is independent of the past, and hence unknown at time k. For this reason $w_k := \hat{x}_{k+1|k} - x_{k+1}$ is called the **innovation**, or new information about x_{k+1} that is not present in the past. The representation (4.4) is called an *innovations representation* of the process $\{x_k\}$.

5. Notes

1. Properties of linear stochastic systems are discussed extensively in Anderson and Moore (1979). That book also studies reachability and associated concepts including controllability and observability. Further properties of linear stochastic systems are presented in Chapter 7.

6. References

[1]   B.D.O. Anderson and J.B. Moore (1979), *Optimal filtering,* Prentice-Hall, Englewood Cliffs, NJ.

CHAPTER 4
CONTROLLED MARKOV CHAIN MODELS

This chapter presents basic results for stochastic systems modeled as finite state controlled Markov chains. In the case of complete observations and feedback laws depending only on the current state, the state process is a Markov chain. Asymptotic properties of Markov chains are reviewed. Infinite state Markov chains are studied briefly. Finally, a technique is presented that reduces the study of certain continuous time Markov processes to discrete time Markov chains. The technique is illustrated by an example of a queuing system.

1. An example

Until now we have supposed that the state takes values in R^n, but in many situations it is more appropriate to permit the state to take on only a finite number of values. We introduce such an example.

Consider a machine whose condition at time k is described by the state x_k which can take the values 1 or 2 with the interpretation that $x_k = 1$ or $x_k = 2$ depending on whether the machine is in an operational or failed condition. For the moment there are no control actions allowed so that the machine behavior is autonomous. Suppose the machine is operational at time k, so $x_k = 1$, and there is a probability $q > 0$ that it will fail in the next period, so $x_{k+1} = 2$; with probability $1 - q$ it will continue to remain operational, so $x_{k+1} = 1$. Suppose further that q does not depend upon previous values $x_{k-1}, \ldots, x_0$. Finally, suppose that a failed machine continues to remain failed, so that $x_{k+1} = 2$ with probability 1, if $x_k = 2$. Then $\{x_k, \ k \geq 0\}$ is a Markov chain whose transition probabilities are described by the matrix $P = \{P_{ij}\}$:

$$P = \begin{bmatrix} 1-q & q \\ 0 & 1 \end{bmatrix}. \tag{1.1}$$

The transition probability matrix P has the property that all its elements are non-negative and the sum of the elements in every row is 1. Such a matrix is said to be a **stochastic matrix.**

The Markov property is expressed by

$$Prob\{x_{k+1} = j \mid x_k = i, x_{k-1}, \ldots, x_0\}$$

35

$$= P_{ij}, \quad i, j \in \{1, 2\}. \tag{1.2}$$

We now introduce two control actions. Let u_k^1 denote the intensity of machine use at time k. It takes on values $u_k^1 = 0$, 1 or 2 accordingly as the machine is not used, is in light use, or is in heavy use. Suppose that the greater the intensity of use, the larger is the likelihood of machine failure. Let u_k^2 denote the intensity of machine maintenance effort. Suppose it takes only two values 0 or 1, the higher value denoting greater maintenance. The idea is that maintenance reduces the likelihood of machine failure and permits a failed machine to become operational.

The effects of these two control actions, intensity of machine use and maintenance, can be modeled as a controlled transition probability as follows. Let $u_k := (u_k^1, u_k^2)$. Then

$$P\{\, x_{k+1} = 2 \mid x_k = 1, x_{k-1}, \ldots, u_k, u_{k-1}, \ldots\}$$
$$= q_1(u_k^1) - q_2(u_k^2),$$

$$P\{\, x_{k+1} = 1 \mid x_k = 1, x_{k-1}, \ldots, u_k, u_{k-1}, \ldots\}$$
$$= 1 - [q_1(u_k^1) - q_2(u_k^2)],$$

$$P\{\, x_{k+1} = 1 \mid x_k = 2, x_{k-1}, \ldots, u_k, u_{k-1}, \ldots\}$$
$$= q_2(u_k^2),$$

$$P\{\, x_{k+1} = 2 \mid x_k = 2, x_{k-1}, \ldots, u_k, u_{k-1}, \ldots\}$$
$$= 1 - q_2(u_k^2). \tag{1.3}$$

These transition probabilities can be put in matrix form similar to (1.1), except that they will be functions of the control u:

$$P(u^1, u^2) = \begin{bmatrix} 1 - q_1(u^1) + q_2(u^2) & q_1(u^1) - q_2(u^2) \\ q_2(u^2) & 1 - q_2(u^2) \end{bmatrix}. \tag{1.4}$$

Equation (1.3) is illustrated in the state transition diagram below.

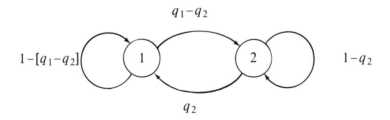

Of course the values of q are such that $q_1(0) < q_1(1) < q_1(2)$ and $q_2(0) < q_2(1)$ because a lightly used or better maintained machine is less likely to fail than a heavily used or less well-maintained machine. We also expect that the last two probabilities in (1.3) should not depend on u_k^1 because when the machine has failed, it cannot be used.

Suppose the state is observed and consider a feedback policy $\{g_0, g_1, \ldots\}$ which is time-invariant, that is, $g_k \equiv g$, and let $u_k = g(x_k)$. This results in the transition probability matrix $P^g = \{P_{ij}^g\}$ where

$$P_{ij}^g := P_{ij}(g(i)), \quad i, j \in \{1, 2\}. \tag{1.5}$$

For example, if $g(1) = (2, 0)$, and $g(2) = (0, 1)$, then

$$P^g = \begin{bmatrix} 1 - q_1(2) + q_2(0) & q_1(2) - q_2(0) \\ q_2(1) & 1 - q_2(1) \end{bmatrix}. \tag{1.6}$$

The resulting process $\{x_k\}$ is a Markov chain with stationary transition probability P^g. The pd of x_k can be written as the *row* vector

$$p_k := (Prob\{x_k = 1\}, Prob\{x_k = 2\}).$$

By the Markov property (1.3)

$$p_{k+m} = p_k[P^g]^m, \quad m \geq 0, \tag{1.7}$$

and, in particular,

$$p_k = p_0[P^g]^k, \tag{1.8}$$

where p_0 is the initial distribution of x_0.

Often, as $k \to \infty$, p_k converges to a pd $p = (p(1), p(2))$ that does *not* depend on the initial distribution p_0. We then say that it is an **ergodic** chain. The limiting pd is called the **steady state** or equilibrium or invariant distribution. It is the solution of the linear equations

$$p = pP^g, \quad p(1) + p(2) = 1. \tag{1.9}$$

Equation (1.9) always has a solution. In the ergodic case the solution is unique and the limiting distribution $p = (p(1), p(2))$ has the following interpretation:

$$p(i) = \lim_{n \to \infty} \frac{1}{n} \sum_{k=1}^{n} I(x_k = i) \text{ wp 1}, \tag{1.10}$$

where I is the *indicator* function—i.e., $I(x_k = i) = 1$ if $x_k = i$, and $I(x_k = i) = 0$ if $x_k \neq i$. Thus $p(1)$ is the average proportion of time that the machine is operational and $p(2)$ the average proportion of time it spends in the failed state.

From (1.9) it is evident that the steady state probability p depends on the feedback law g. So by changing the policy g, that is, by changing the use and maintenance of the machine, we can alter the number of times it fails. One may then ask the following question: Which policy g leads to the best p? We will examine this and related questions in later chapters.

2. Finite state controlled Markov chains

The preceding example generalizes to the case of an arbitrary finite state controlled Markov chain whose state x_k takes values in $\{1, .., I\}$. The control u_k takes values in a prespecified set U. U may be finite or infinite. The transition probabilities are specified by the $I \times I$ matrix valued function on U,

$$u \rightarrow P(u) := \{ P_{ij}(u), \ 1 \le i, j \le I\}, \tag{2.1}$$

with the interpretation that

$$Prob \ \{ \ x_{k+1} = j \ | \ x_k = i, x_{k-1}, .., x_0, u_k, .., u_0\}$$
$$= P_{ij}(u_k). \tag{2.2}$$

The matrix $P(u)$ has the property that every element is non-negative, and the sum of the elements in every row is 1—i.e., it is a stochastic matrix. We must also specify the pd of the initial state x_0. In case the observation $y_k \not\equiv x_k$, one must also specify the observation probability

$$P(y \ | \ i) := Prob\{y_k = y \ | \ x_k = i\}. \tag{2.3}$$

3. Complete observations and Markov policies

Consider the controlled Markov chain model (2.1). Suppose the state is observed, $y_k \equiv x_k$. Let $g = \{g_0, g_1, .. \}$ be a feedback policy such that g_k depends only on the current state x_k (and not on $x_{k-1}, x_{k-2}, ..$). We call such g a **Markov policy.**

Let g be a Markov policy, and let $\{x_k\}$ be the resulting state process. Denote the pd of x_k by the I-dimensional row vector

$$p_k^g := (Prob \{x_k = 1\}, .., Prob\{x_k = I\}),$$

the superscript g emphasizes the dependence on g. However, in the sequel we drop the superscript since the g-dependence will be clear from the context.

Lemma   (3.1)

When a Markov policy g is employed, the resulting state process $\{x_k\}$ is a Markov process. Its one-step transition probability at time k is given by the matrix

$$P_k^g := \{(P_k^g)_{ij} := P_{ij}(g_k(i)), \ 1 \le i, j \le I\},$$

Its m-step transition probability at time k is given by the matrix

$$P_k^g \ .. \ P_{k+m-1}^g,$$

so its ijth element is the probability that the state will be j at time $k+m$ given that it is i at time k. Hence

$$p_{k+m} = p_k \ P_k^g \ .. \ P_{k+m-1}^g.$$

In particular,

$$p_k = p_0 \ P_0^g \ .. \ P_{k-1}^g, \tag{3.2}$$

where p_0 is the pd of the initial state x_0.

Proof
The proof is immediate from the Markov property (2.2). $\qquad\square$

Since the transition probability matrix P_k^g depends on the time k, we say that $\{x_k\}$ is a Markov chain with nonstationary transition probability.

4. The cost of a Markov policy

A Markov policy g determines the pd of the state process $\{x_k\}$ and the control process $\{u_k = g_k(x_k)\}$. Different policies will lead to different probability distributions. In optimal control problems one is interested in finding the best or optimal policy. To do this one needs to compare different policies. This is done by specifying a **cost function**. This is a sequence of real valued functions of the state and control,

$$c_k(i, u), \ 1 \le i \le I, u \in U, k \ge 0.$$

The interpretation is that $c_k(i, u)$ is the cost to be paid if at time k, $x_k = i$ and $u_k = u$.

Fix a Markov policy g. The cost incurred by g up to the time horizon N is $\sum_{k=0}^{N} c_k(x_k, u_k)$. This is a random variable since x_k and u_k are random. Hence the expected cost is

$$J(g) := E^g \sum_{k=0}^{N} c_k(x_k, u_k) = E^g \sum_{k=0}^{N} c_k(x_k, g_k(x_k)). \tag{4.1}$$

Here E^g denotes expectation with respect to the pd of $\{x_k\}$, $\{u_k\}$ determined by g.

$J(g)$ can be readily evaluated in terms of the transition probability matrices P_k^g as follows. From (4.1) we get

$$J(g) = \sum_{k=0}^{N} \sum_{i=1}^{I} Prob\{x_k = i\} \ c_k(i, g_k(i))$$

$$= \sum_{k=0}^{N} p_k \, c_k^g = \sum_{k=0}^{N} p_0 \, (P_0^g \ldots P_{k-1}^g) \, c_k^g, \tag{4.2}$$

where c_k^g is the I-dimensional *column* vector

$$c_k^g := (c_k(1, g(1)), \ldots, c_k(I, g(I)))^T. \tag{4.3}$$

The last equality in (4.2) follows from (3.2).

Thus the best Markov policy g is the one that minimizes $J(g) = \sum_{k=0}^{N} p_0 \, (P_0^g \ldots P_{k-1}^g) \, c_k^g$. In Chapter 5 we study Dynamic Programming as an approach for computing the best g.

Central to dynamic programming is a recursive technique for calculating the cost of a Markov policy g. Since the technique depends only on the fact that the state process corresponding to g is Markov, we introduce it here. For each time $1 \le k \le N$, and state $1 \le i \le I$, let $V_k^g(i)$ denote the expected cost incurred during $k, \ldots, N$ when $x_k = i$. That is,

$$V_k^g(i) := E^g \{ \sum_{l=k}^{N} c_l(x_l, g_l(x_l)) \mid x_k = i \}. \tag{4.4}$$

Observe that with this notation the total cost (4.2) is

$$J(g) = \sum_{i=1}^{I} (p_0)_i \, V_0^g(i). \tag{4.5}$$

Lemma $\qquad\qquad\qquad\qquad\qquad\qquad\qquad\qquad\qquad\qquad\qquad\qquad$ (4.6)

The functions $V_k^g(i)$ can be calculated by backward recursion,

$$V_k^g(i) = c_k(i, g_k(i)) + \sum_{j=1}^{I} (P_k^g)_{ij} \, V_{k+1}^g(j), \quad 0 \le k < N, \tag{4.7}$$

starting with the final condition

$$V_N^g(i) = c_N(i, g_N(i)). \tag{4.8}$$

Proof

From the definition we immediately get (4.8). Next

$$V_k^g(i) = E^g \{ \sum_{l=k}^{N} c_l(x_l, g_l(x_l)) \mid x_k = i \}$$

$$= c_k(i, g_k(i))$$

$$+ E^g \{ E^g \{ \sum_{l=k+1}^{N} c_l(x_l, g_l(x_l)) \mid x_{k+1}, x_k = i \} \mid x_k = i \}$$

$$= c_k(i, g_k(i)) + E^g \{V^g_{k+1}(x_{k+1}) \mid x_k = i\}, \text{ by } (2.2)$$

$$= c_k(i, g_k(i)) + \sum_{j=1}^{I} V^g_{k+1}(j) \, Prob\{x_{k+1} = j \mid x_k = i\}$$

which is (4.7) once we recall the definition of $(P^g_k)_{ij}$.   □

The previous equations can be expressed in a convenient vector notation. Denote the I-dimensional column vector

$$V^g_k := (V^g_k(1), \ldots, V^g_k(I))^T. \tag{4.9}$$

Then, using (4.3), we can express (4.7), (4.8), and (4.5), respectively, as

$$V^g_k = c^g_k + P^g_k V^g_{k+1}, \ 0 \le k < N, \tag{4.10}$$

$$V^g_N = c^g_N, \tag{4.11}$$

$$J(g) = p_0 V^g_0. \tag{4.12}$$

Here, the time horizon N is finite. Often one is interested in the infinite horizon. This is not an immediate extension, since if one simply sets $N = \infty$ in (4.1), in most cases one gets $J(g) = \infty$ for every g. The notion of best g then becomes meaningless. There are two ways to treat the infinite horizon problem.

The first approach is to introduce a discount factor β, $0 < \beta < 1$, and to consider the expected discounted cost

$$J(g) = E^g \sum_{k=0}^{\infty} \beta^k c_k(x_k, u_k).$$

Observe that if c_k is bounded, then $J(g)$ will be finite. Since the cost incurred at time k is weighted by β^k, present costs are more important than future costs. In an economic context, $\beta = (1 + r)^{-1}$, where $r > 0$ is the interest rate. With this interpretation, $J(g)$ is the present value of the cost. From (4.2) it follows that

$$J(g) = \sum_{k=0}^{\infty} \beta^k p_0 (P^g_0 \ldots P^g_{k-1}) c^g_k.$$

Define

$$V^g_k(i) := E^g \{\sum_{l=k}^{\infty} \beta^l c_l(x_l, g_l(x_l)) \mid x_k = i\};$$

then, using the notation (4.9), the counterparts of (4.10) and (4.12) are

$$V^g_k = \beta^k c^g_k + P^g_k V^g_{k+1}, \ k \ge 0, \tag{4.13}$$

$$J(g) = p_0 \, V_0^g.$$

However, in the infinite horizon problem, there is no counterpart of the final condition (4.11).

The second approach is followed when discounting is inappropriate. A policy is then evaluated according to its average cost per unit time,

$$J(g) = \lim_{N \to \infty} \frac{1}{N} E^g \sum_{k=0}^{N-1} c_k(x_k, g_k(x_k)). \tag{4.14}$$

Using (4.2) this cost equals

$$\lim_{N \to \infty} \frac{1}{N} \sum_{k=0}^{N-1} p_0 \, (P_0^g \, . \, . \, P_{k-1}^g) \, c_k^g. \tag{4.15}$$

From this expression we see that if P_k^g varies with k, then the limit above need not exist. If the transition matrix does not depend on k, then the limit always exists as we study in the next section.

So far, the total cost (4.1) is the sum of the costs incurred in each time period. The next exercise shows that other cost functions can be put into this additive form.

Exercise (4.16)

Suppose the cost incurred by a Markov policy $g := \{g_0, . . . , g_N\}$ is

$$J(g) := E^g \, I(\max_{0 \le k \le N} h(x_k) \ge \alpha), \tag{4.17}$$

where I is the indicator function, and h and α are prespecified. [Thus $J(g)$ is the probability that $h(x_k)$ exceeds α at some time k.] Show that (4.17) can be put into the additive form (4.1).
[Hint: Define the new chain $z_k := (x_k, y_k)$, with $y_k \in \{0, 1\}$. The transition probability of x_k is exactly as before, whereas

$$Prob\{y_{k+1} = 1 \mid y_k = 0, x_k, z_{k-1}, . . . , z_0, u_k, . . . , u_0\}$$

$$= \begin{cases} 1 & \text{if } h(x_k) \ge \alpha \\ 0 & \text{if } h(x_k) < \alpha \end{cases}$$

$$Prob\{y_{k+1} = 1 \mid y_k = 1, x_k, z_{k-1}, . . . , z_0, u_k, . . . , u_0\} \equiv 1.$$

Now let $c_k(z, u) \equiv 0$ for $k < N$, and $c_N(x, y, u) \equiv y$.]

5. Stationary Markov policy

A Markov policy $g = \{g_0, g_1, . . \}$ is **stationary** or **time-invariant** if $g_0 \equiv g_1 \equiv . . \equiv g$, with a slight abuse of notation.

Let g be stationary; then the transition probability matrix is stationary, $P_k^g \equiv P^g$. Suppose the cost functions are also time-invariant, $c_k \equiv c$. Fix a discount $0 < \beta < 1$. Then

$$V_k^g(i) = E^g \{\sum_{l=k}^{\infty} \beta^l c(x_l, g(x_l)) \mid x_k = i\}$$

$$= \beta^k E^g \{\sum_{l=0}^{\infty} \beta^l c(x_l, g(x_l)) \mid x_0 = i\}$$

$$= \beta^k V_0^g(i).$$

Using this in (4.13) gives

$$\beta^k V_0^g = \beta^k c^g + \beta^{k+1} P^g V_0^g,$$

or, in matrix notation,

$$[I - \beta P^g] V_0^g = c^g.$$

This is a set of I linear equations in the I unknowns $V_0^g(i)$, for $i = 1, 2 .., I$. Exercise (5.3) shows that $[I - \beta P^g]$ is invertible so that this system of linear equations has a unique solution V_0^g.

Exercise   (5.1)
Define the norm of $x \in R^I$ by $|x| := \max \{|x_1|, .., |x_I|\}$. Define the norm of an $I \times I$ matrix P by $||P|| := \max_i \sum_j |P_{ij}|$. Show that the matrix norm is induced by the vector norm in the sense that

$$||P|| = \max_{x \neq 0} \frac{|Px|}{|x|}.$$

As a consequence show that $||PQ|| \leq ||P|| \, ||Q||$.

Exercise   (5.2)
An $I \times I$ matrix P is a transition probability (or stochastic) matrix if $P_{ij} \geq 0$ and $\sum_j P_{ij} = 1$. Show that $||P|| = 1$.

Exercise   (5.3)
Let P be a transition matrix and $0 < \beta < 1$. Show that $I - \beta P$ is invertible; in fact,

$$[I - \beta P]^{-1} = \lim_{n \to \infty} \sum_{k=0}^{n} [\beta P]^k.$$

(The infinite sum on the right hand side exists since $||P|| = 1$). Fix c and $V(0)$ and generate the sequence $V(n)$ by

$$V(n+1) := c + \beta P V(n).$$

Show that $V(n) \to [I - \beta P]^{-1} c$ as $n \to \infty$.
[Hint : Recursively substitute for $V(n)$.]

Next we study the average cost (4.14), (4.15) when g and c are stationary. The next three lemmas are stated without proof.

Lemma   (5.4)
If P is a transition probability matrix, then the Cesaro limit

$$\lim_{N \to \infty} \frac{1}{N} \sum_{k=0}^{N-1} P^k =: \Pi,$$

always exists. The matrix Π is a stochastic matrix and it satisfies the equation

$$\Pi = \Pi P.$$

Thus for stationary g and time-invariant cost, the average cost per unit time (4.14) and (4.15), is

$$J(g) = \lim_{N \to \infty} \frac{1}{N} E^g \sum_{k=0}^{N-1} c(x_k, g(x_k)) = p_0 \Pi c^g. \tag{5.5}$$

Let π be one of the rows of Π. Then

$$\pi = \pi P.$$

Moreover since Π is a stochastic matrix, π can be regarded as a pd. This has the interpretation that if the Markov chain has initial pd given by π, then the pd of the state remains at π for all time. Thus π is said to be an invariant probability distribution. Clearly if the rows of Π are not all the same, then there is more than one invariant probability distribution.

Exercise   (5.6)
Give an example such that $J(g)$ in (5.5) depends on p_0.
[Hint: Choose $P = I$. Note that there are several invariant probability distributions.]

In many cases (5.5) is independent of the initial distribution p_0. An $I \times I$ transition probability matrix P is reducible or decomposable if there is a renumbering of the states $\{1, .., I\}$ for which P takes the form

$$P = \begin{bmatrix} P_1 & P_2 \\ 0 & P_3 \end{bmatrix},$$

where P_1 and P_3 are square matrices. This means that it is not possible to make a transition from a state indexing a row of P_3 to a state corresponding to P_1. Hence if the initial state happens to lie in the set of states indexing rows of P_3, then the Markov chain stays forever in the same set, with transition probabilities given by P_3. Thus there is a Markov chain with this smaller state space.

A transition matrix P which cannot be put in this form by any renumbering of states is called **irreducible** or **indecomposable**.

Lemma   (5.7)

If P is an irreducible transition probability matrix, then there is a unique row vector π such that

$$\pi P = \pi, \quad \sum_{i=1}^{I} \pi_i = 1.$$

Moreover $\pi_i > 0$, all i. Finally, the matrix Π in (5.5) has all rows equal to π.

[π is called the **steady state** or invariant probability distribution of the Markov chain $\{x_k\}$.]

Exercise   (5.8)

Construct an example of a Markov chain that is reducible and that has several different invariant probability distributions. Is every component $\pi_i > 0$ for every invariant probability distribution π?

[Hint: See Exercise (5.6).]

Exercise   (5.9)

Show that P is irreducible if and only if for every i and j there is a sequence of states $i =: i_0 , i_1 , \ldots , i_{k-1} , i_k := j$ such that $P_{i_l i_{l+1}} > 0$ for $l = 0, 1 .., k-1$. Hence there is a path in the state space from every state to every other state that can be traversed by the Markov chain with positive probability.

[Hint: If not, then group all the states which cannot be reached from a certain state into one set. Identify these states with the matrix P_1.]

Thus if P^g is irreducible, then

$$J(g) = p_0 \, \Pi \, c^g = \pi \, c^g \tag{5.10}$$

since all the rows of Π are identical. The cost $J(g)$ is therefore independent of the initial distribution.

A probability transition matrix P is said to be periodic if there is a renumbering of the states for which P takes the form

$$P = \begin{bmatrix} 0 & P_1 & 0 & \cdot & \cdot & 0 \\ 0 & 0 & P_2 & \cdot & \cdot & 0 \\ \cdot & & \cdot & \cdot & \cdot & \cdot \\ \cdot & & \cdot & \cdot & \cdot & \cdot \\ 0 & 0 & 0 & \cdot & \cdot & P_{n-1} \\ P_n & 0 & 0 & \cdot & \cdot & 0 \end{bmatrix}.$$

This means that one can partition the states into disjoint subsets $I_1, \ldots, I_n$ such that from states in I_{m-1} transitions are possible only to states in I_m. If P cannot be put into this form it is said to be **aperiodic**.

Lemma   (5.11)
If P is irreducible and aperiodic, then

$$\lim_{k \to \infty} P^k = \Pi,$$

where Π is the matrix with all rows equal to π.

Exercise   (5.12)
Show that if P is periodic, then P^k cannot converge.

There is an interesting case of a cost over the infinite time horizon which requires neither a discount factor nor consideration of the average cost per unit time. We introduce this in the next exercise.

Exercise   (5.13)
Let $F \subset \{1, .., I\}$ be a subset of states, and for a stationary Markov policy g let τ be the first time that x_k enters F. That is

$$\tau = \begin{cases} \min \{k \geq 0 \mid x_k \in F\} \\ \infty, \text{ if } x_k \notin F \text{ for all } k. \end{cases}$$

Suppose

$$E\{\tau \mid x_0 = i\} < \infty \quad \text{for all } i.$$

Let $c(i, u)$ be a stationary cost function, and define

$$V^g(i) := E^g \{ \sum_{k=0}^{\tau-1} c(x_k, g(x_k)) \mid x_0 = i \}.$$

Show that $V^g(i) < \infty$ and

$$V^g = c^g + R^g V^g,$$

where $(R^g)_{ij} := (P^g)_{ij}$ if $i \notin F$, and $(R^g)_{ij} := 0$ if $i \in F$. [The random time τ is called a stopping time, and problems of this type are called stopping time problems.]

6. Infinite state Markov chains

The discussion in Sections 2, 3, and 4 carries over with obvious changes to the case of controlled Markov chains whose state x_k takes values in the infinite set $\{1, 2, 3, ..\}$. The control takes values in U, and for $u \in U$, the transition probabilities are specified by an infinite dimensional matrix

$$P(u) := \{P_{ij}(u), \ 1 \leq i, j, < \infty\}$$

with the same interpretation as (2.2).

A Markov policy g is defined exactly as before, and the corresponding one-step and m-step transition probability matrix at time k are

$$P_k^g := \{((P_k^g)_{ij}) := P_{ij}(g_k(i)),\ 1 \le i, j < \infty\},$$

$$P_k^g \cdot \cdot P_{k+m-1}^g.$$

The pd of x_k is now the infinite row vector

$$p_k^g = p_k := (Prob\{x_k = 1\},\ Prob\{x_k = 2\}, \dots),$$

and by the Markov property,

$$p_{k+m} = p_k\ P_k^g \cdot \cdot P_{k+m-1}^g, \quad \text{and} \quad p_k = p_0\ P_0^g \cdot \cdot P_{k-1}^g.$$

For cost functions $c_k(i, u)$, $1 \le i < \infty$, $u \in U$, the expected cost over a finite horizon is given by (4.1). If V_k^g is the infinite-dimensional column vector with components $V_k^g(i)$ defined by (4.5), then the recursion analogous to (4.7) and (4.8) is

$$V_k^g(i) = c_k(i, g_k(i)) + \sum_{j=1}^{\infty} (P_k^g)_{ij}\ V_{k+1}^g(j),\ 0 \le k < N,$$

$$V_N^g(i) = c_N(i, g_N(i)).$$

Similarly, the discounted cost over the infinite horizon is given by the recursion (4.13). If the policy g and the cost function $c(i, u)$ are time-invariant, then the infinite-dimensional vector V_0^g satisfies the linear equation

$$[I - \beta P^g]\ V_0^g = c^g. \tag{6.1}$$

The next exercise shows that (6.1) has a unique solution.

Exercise (6.2)

Let l_∞ be the set of all infinite-dimensional column vectors $x = (x_1, x_2, \dots)^T$. For $x \in l_\infty$, define its norm $|x| := \sup\{|x_1|, |x_2|, \dots\}$. Show that $||P^g|| := \sup_{x \neq 0} \dfrac{|P^g x|}{|x|} = 1$. Now show that the linear map defined on l_∞ by $x \to [I - \beta P^g]x$ is invertible; in fact,

$$[I - \beta P^g]^{-1}x = \sum_{k=1}^{\infty} [\beta P^g]^k x.$$

In particular, if $c^g \in l_\infty$—i.e., if $\sup_i |c^g(i)|$ is finite—then (6.1) has a unique solution.

Similarly, we can define the average cost per unit time as

$$J(g) := \lim_{N \to \infty} \frac{1}{N} E^g \sum_{k=0}^{N} c(x_k, g(x_k))$$

$$= p_0 [\lim_{N \to \infty} \frac{1}{N} \sum_{k=0}^{N} (P^g)^k] c^g. \tag{6.3}$$

The only significant difference between the finite and infinite state cases arises at this point, because Lemmas (5.4), (5.7) and (5.11) do not hold in the infinite case as the next exercise shows.

Exercise (6.4)

Consider a Markov chain with the following state transition diagram.

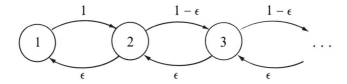

This gives the transition probability matrix

$$P = \begin{bmatrix} 0 & 1 & . & . & . \\ \epsilon & 0 & 1-\epsilon & . & . \\ . & \epsilon & 0 & 1-\epsilon & . \\ . & . & \epsilon & 0 & . \\ . & . & . & \epsilon & . \\ . & . & . & . & . \end{bmatrix}.$$

Show that P is irreducible and periodic for $\epsilon > 0$. Show that there exists a steady-state distribution π with

$$\pi P = \pi, \text{ and } \sum_{i=0}^{\infty} \pi(i) = 1,$$

if and only if $\epsilon > \frac{1}{2}$.

[Hint: First show that a positive solution to $\pi P = \pi$ must be proportional to $\eta = (\eta_1, \eta_2, ..)$ with

$$\eta(1) = 1, \text{ and } \eta(i) = \frac{(1-\epsilon)^{i-2}}{\epsilon^{i-1}}, \ i > 1.$$

Now show that $\sum_i \pi(i) = 1$ is possible iff $\sum_i \eta(i) < \infty$ iff $\epsilon > \frac{1}{2}$.]

If, however, the Cesaro limit in (6.3) exists (as it always does in the finite case),

$$\lim_{N \to \infty} \frac{1}{N} \sum_{k=0}^{N} (P^g)^k = \Pi,$$

then $J(g) = p_0 \Pi c^g$. If, furthermore, the chain is ergodic so that all rows of Π are equal to the steady state distribution π, then the average cost is independent of the initial distribution, $J(g) = \pi c^g$.

7. Continuous time Markov chains

The example of Section 1 also makes sense if the transition of the machine state can occur at any continuous time t. The state process is then given by $\{x(t), t \geq 0\}$ with values in $\{1, 2\}$, and the control is given by $\{u(t), t \geq 0\}$, $u(t) = (u^1(t), u^2(t)) \in \{0,1,2\} \times \{0,1\}$. $x(t)$ is described statistically by giving the probability with which the state will change in an infinitesimal interval $(t, t + dt)$. In the example the specification could take the form [compare (1.3)]:

$$P\{x(t + dt) = 2 \mid x(t) = 1, x(s), u(s), s \leq t\}$$
$$= [q_1(u^1(t)) - q_2(u^2(t))]dt$$
$$P\{x(t + dt) = 1 \mid x(t) = 1, x(s), u(s), s \leq t\}$$
$$= 1 - [q_1(u^1(t)) - q_2(u^2(t))]dt$$
$$P\{x(t + dt) = 1 \mid x(t) = 2, x(s), u(s), s \leq t\}$$
$$= q_2(u^2(t))dt$$
$$P\{x(t + dt) = 2 \mid x(t) = 2, x(s), u(s), s \leq t\}$$
$$= 1 - q_2(u^2(t))dt \qquad (7.1)$$

The first two relations mean that if $x(t) = 1$, then $x(t + dt) = 2$ with probability $[q_1 - q_2]dt$ and $x(t + dt) = 1$ with the complementary probability $1 - [q_1 - q_2]dt$; furthermore, these transition probabilities do not depend upon previous values of the state and control, $x(s)$, $u(s)$, $s < t$—this is the Markov property. A similar interpretation holds for the other two relations.

Let $p(t) = (p_1(t), p_2(t)) := (P\{x(t) = 1\}, P\{x(t) = 2\})$ be the pd of $x(t)$. By Bayes' rule,

$$p_j(t + dt) = \sum_{i=1}^{2} P\{x(t + dt) = j \mid x(t) = i, u(t) = u\}p_i(t).$$

In particular, using (7.1),

$$p_1(t + dt) = p_1(t)\{1 - [q_1(u^1(t)) - q_2(u^2(t))]dt\}$$
$$+ p_2(t)q_2(u^2(t))dt,$$

from which we get the differential equation,

$$\dot{p}_1(t) = -p_1(t)[q_1(u^1(t)) - q_2(u^2(t))] + p_2(t)q_2(u^2(t)).$$

Similarly,

$$\dot{p}_2(t) = p_1(t)[q_1(u^1(t)) - q_2(u^2(t))] - p_2(t)q_2(u^2(t)).$$

Combining this in matrix form gives

$$\dot{p}(t) = p(t) \times$$

$$\begin{bmatrix} -[q_1(u^1(t)) - q_2(u^2(t))] & [q_1(u^1(t)) - q_2(u^2(t))] \\ q_2(u^2(t)) & -q_2(u^2(t)) \end{bmatrix}$$

$$=: p(t)Q(u(t)).$$

$Q(u(t))$ is called the (transition probability) **rate** matrix of the continuous time controlled Markov chain $\{x(t)\}$. Note that the off-diagonal terms of Q are nonnegative, and the row sum is 0. (This is the analog of the fact that the row sum of a transition probability matrix is 1.)

More generally, a continuous time controlled Markov chain $\{x(t)\}$ with values in the finite set $\{1, .., I\}$ or the countable set $\{1, 2, ..\}$ is given by specifying, for each $u \in U$, the rate matrix

$$Q(u) = \{Q_{ij}(u)\},$$

$Q_{ij}(u) \geq 0$ for $i \neq j$, $Q_{ii}(u) = -\sum_j Q_{ij}(u)$, with the interpretation that

$$P\{x(t + dt) = j \mid x(t) = i, x(s), u(s), s \leq t\}$$
$$= Q_{ij}(u(t))dt, \quad i \neq j$$
$$P\{x(t + dt) = i \mid x(t) = i, x(s), u(s), s \leq t\}$$
$$= 1 - \sum_{j \neq i} Q_{ij}(u(t))dt.$$

If $p(t) = (P\{x(t) = 1\}, P\{x(t) = 2\}, ..)$ is the pd of $x(t)$, then it satisfies the differential equation

$$\dot{p}(t) = p(t)Q(u(t)).$$

A Markov policy is a family $g = \{g(t, i), t \geq 0\}$ that at each t gives the control to be used as a function of the state, $u(t) = g(t, x(t))$. This defines the rate matrix at t,

$$Q^g(t) := \{Q_{ij}(g(t, i))\},$$

and the pd of the state evolves according to the linear differential equation

$$\dot{p}(t) = p(t) Q^g(t). \tag{7.2}$$

If $P^g(s, t)$ denotes the fundamental matrix solution of this differential equation, i.e.,

$$\dot{P}^g(s, t) = P^g(s, t) Q^g(t), \quad t \geq s,$$

$$P^g(t, t) = I \text{ (identity matrix)},$$

then

$$p(t) = p(s) P^g(s, t) = p(0) P^g(0, t). \tag{7.3}$$

Equation (7.2) is the **Kolmogorov forward equation**, and (7.3) is the counterpart of (3.2). In case $g(t, i) \equiv g(i)$ is time-invariant, Q^g is stationary, and

$$P^g(s, t) = \exp[(t - s)Q^g], \quad p(t) = p(0) \exp tQ^g.$$

Exercise (7.4)

Let Q be a rate matrix—i.e., the off-diagonal elements are nonnegative, and the row sums are 0. Suppose $x(t) =: (x_1(t), \ldots, x_n(t))$ satisfies the differential equation $\dot{x}(t) = x(t) Q$. Show that $\sum_{i=1}^{n} x_i(t) = $ constant for all t. Thus if $x(0)$ is a pd, then the sum of the components of $x(t)$ remains 1 for all t.
[Hint: Multiply both sides of the differential equation by the column vector $(1, \ldots, 1)^T$ on the right.]

A useful example of an infinite state continuous time Markov chain is the **Poisson** process of rate λ described in the next exercise.

Exercise (7.5)

Consider a Markov chain with state space $\{0, 1, \ldots\}$. Let $Q_{i,i+1} := \lambda > 0$ and $Q_{ii} := -\lambda$. Thus the Markov chain moves to the next higher integer at a rate λ. Let the initial pd be $p(0) = (1, 0, 0, 0, \ldots)$, so the chain starts in state 0. Show that $p_i(t) = \exp(-\lambda t)(\lambda t)^i / i!$ for $i \geq 0$.
[Hint: Recursively solve the differential equation (7.2) for $p_i(t)$ starting with $i = 0$.]

It can be shown that in each visit to a state i, the amount of time the continuous time Markov chain spends in that state before moving to a new state is an exponentially distributed random variable with mean $(-Q_{ii})^{-1}$. That is, if τ denotes this random time, then

$$Prob(\tau > t) = \exp(Q_{ii} t).$$

Exercise (7.6)

If τ is an exponentially distributed random variable, show that

$$Prob(\tau > t + s \mid \tau > s) = Prob(\tau > t), \quad \text{for all } s \geq 0.$$

Hence the time between jumps of a Markov chain is *memoryless*: the time that the Markov chain will continue to remain in state i does not depend on how long it has already spent in state i.

As in the case of discrete time Markov chains, a pd π is invariant if $p(s) = \pi$ implies that $p(t) = \pi$ for all $t \geq s$—i.e., $\dot{p}(t) \equiv 0$. From (7.2) it follows that

$$\pi Q = 0. \tag{7.7}$$

The next exercise shows that this is also a sufficient condition for π being invariant.

Exercise   (7.8)
If π is a pd with $\pi Q = 0$, show that $p(s) = \pi$ implies $p(t) = \pi$ for all $t \geq s$.
[Hint: Show that $\pi \exp(tQ) = \pi$ by expanding the matrix exponential in a power series.]

Equation (7.7) has a simple interpretation. Writing out this vector equation component by component gives

$$\sum_i \pi_i Q_{ij} = 0 \quad \text{for all } j.$$

Noting $Q_{jj} = -\sum_{i \neq j} Q_{ji}$, this can be rewritten as,

$$\sum_{i \neq j} \pi_i Q_{ij} = \pi_j \sum_{i \neq j} Q_{ji}. \tag{7.9}$$

The left hand side is seen to be the total rate of flow into state j, while the right hand side is the total rate of flow out of state j. The condition (7.9) is therefore called a **state balance** equation.

In discrete time Markov chains state transitions can only occur at discrete time instants, but a continuous time Markov chain can change state at any time. Hence there is no notion of periodicity. One consequence is that for a finite state continuous time Markov chain, the limit

$$p(\infty) := \lim_{t \to \infty} p(0)\exp(tQ)$$

always exists, and is, moreover, an invariant pd satisfying $p(\infty)Q = 0$. (Thus there is no need to take Cesaro limits as in (5.4)). However, just as in the discrete case, $p(\infty)$ may depend on the initial pd $p(0)$ if the Markov chain is reducible. For infinite state Markov chains the limit does not always exist.

The next two exercises give applications to queuing systems.

Exercise   (7.10)
Consider a queue to which customers arrive as a Poisson process of rate λ. The queue is served by a single server acting on one customer at a time. When the server acts on a customer, it completes service in an amount of time that is an exponentially distributed random variable with mean μ^{-1}. Let $x(t)$ be the number of customers in the queue at time t.

The process $x(t)$ is a continuous time Markov chain with rate matrix Q given by

$$Q_{01} := \lambda, \quad Q_{00} := -\lambda,$$

$$Q_{i,i+1} := \lambda, \quad Q_{i,i-1} := \mu, \quad Q_{ii} := -(\lambda + \mu) \text{ for } i \geq 1.$$

Show that there is an invariant pd π if and only if $\rho := \dfrac{\lambda}{\mu} < 1$, and it is then given by

$$\pi_i = \frac{\rho^i}{1 - \rho}, i \geq 0.$$

[Hint: Write out the state balance equations and recursively solve for $\pi_1, \pi_2, ..$ in terms of π_0. Then show that $\sum\limits_{i=0}^{\infty} \pi_i = 1$ is possible if and only if $\rho < 1$.]

The queuing system above is called an M/M/1 queue. The first "M" signifies that the arrival process to the queue is a Poisson process, which is a Markov process with memoryless interarrival times; the second "M" signifies that the service times are exponentially distributed and consequently also memoryless; and the "1" signifies that there is one server.

The next exercise uses a queuing model of a telephone exchange.

Exercise   (7.11)
Consider a telephone exchange with a total of n lines. Calls arrive as a Poisson process of rate λ. When a call arrives, a line is assigned to it if one is idle. If all lines are busy, the call is turned away or blocked. Once a line is assigned to a call, the line is kept busy for an exponentially distributed random time with mean μ^{-1}, after which the line again becomes idle. Let $x(t)$ be the number of busy lines at time t. Then $x(t)$ is a continuous time Markov chain with rate matrix Q given by,

$$Q_{i,i-1} := i\mu \text{ for } i = 1, .., n,$$

$$Q_{i,i+1} := \lambda \text{ for } i = 0, .., n-1,$$

$$Q_{00} := -\lambda, \quad Q_{nn} := -n\mu,$$

$$Q_{ii} := -(\lambda + i\mu) \text{ for } i = 1, .., n-1.$$

Show that there is an invariant probability distribution π with

$$\pi_i = \frac{\rho^i / i!}{\sum\limits_{j=0}^{n} \rho^j / j!}, i \geq 0.$$

π_n is the steady state probability of finding all lines in the exchange busy; it is the fraction of calls which are blocked or lost, and the formula for it is called the *Erlang loss formula.* This is an M/M/n/n queue. The last "n" denotes the capacity of the system because arrivals that occur when the queue size is n are turned away.

8. Relationship between continuous and discrete time Markov chains

We now relate continuous time chains to discrete time chains. Let $\{x(t)\}$ be a chain with time-invariant rate matrix Q. Assume that

$$Q_0 := \sup_i (-Q_{ii}) = \sup_i \sum_{j \neq i} Q_{ij} < \infty.$$

The chain is then said to be **uniformizable**. (Note that $Q_0 < \infty$ if the state space is finite.) Fix $q \geq Q_0$. The pd of $x(t)$ satisfies

$$\dot{p}(t) = -p(t)Q = -p(t)q[I - R],$$

where

$$R := [I + \frac{Q}{q}].$$

So

$$p(t) = p(0) \exp -tq[I - R]$$

$$= p(0) \sum_{k=0}^{\infty} [\exp -tq] \frac{(tq)^k}{k!} R^k. \tag{8.1}$$

Note that R is a stochastic matrix. Now consider the discrete time chain $\{x_k\}$ with state transition probability matrix R and initial pd $p(0)$, so the pd of x_k is

$$p_k = p(0)R^k.$$

Comparing this with (8.1) gives

$$p(t) = \sum_{k=0}^{\infty} [\exp -tq] \frac{(qt)^k}{k!} p_k. \tag{8.2}$$

Next let $\{K(t)\}$ be a Poisson process with rate q and independent of the chain $\{x_k\}$, so (see Exercise (7.5))

$$Prob\{K(t) = k\} = [\exp -tq] \frac{(qt)^k}{k!}, \quad k = 0, 1, 2, \ldots, \tag{8.3}$$

and consider the continuous time process $\{\hat{x}(t)\}$ obtained from $\{K(t)\}$ and $\{x_k\}$ by

$$\hat{x}(t) := x_{K(t)}.$$

That is, the value of $\hat{x}$ at time t is the value of the discrete chain x_k at time $k = K(t)$. Alternatively, let $T_0 := 0$, and for $k > 0$, let T_k be the first time that $K(t) = k$. Then

$$\hat{x}(t) = x_k, \quad \text{for } T_k \leq t < T_{k+1}. \tag{8.4}$$

[Note that $T_{k+1} - T_k$, $k = 0, 1, ..$ are iid random variables with probability density $p(T_{k+1} - T_k = t) = q\exp -qt$ since the time between jumps of a Poisson process is exponentially distributed.] By Bayes' rule

$$P\{\hat{x}(t) = i\} = \sum_{k=0}^{\infty} P\{x_k = i\}P\{K(t) = k\}. \tag{8.5}$$

From (8.2), (8.3) and (8.5) we conclude that

$$P\{\hat{x}(t) = i\} = P\{x(t) = i\},$$

so that $x(t)$ and $\hat{x}(t)$ have the same pd.

We illustrate the uniformization procedure by an example of a control problem involving two queuing stations in tandem. Customers arrive in station 1 as a Poisson process with rate λ. The rate at time t of the server in station 1 can be selected to be any number $u(t) \in [0, a]$. Upon completing service at station 1, a customer joins the queue at 2 served by another server at constant rate μ. Let $x_i(t)$ be the number of customers at time t in station i, including the customer being served, and let $x(t) = (x_1(t), x_2(t)) \in \{0, 1, 2, ..\} \times \{0, 1, 2, ..\}$ denote the state at t.

Let $p(t) = \{p(x(t) = (i, j), 0 \leq i, j < \infty\}$ be the pd of $x(t)$. Then the Kolmogorov equation is

$$\dot{p}(0, j) = \mu p(0, j+1) - \lambda p(0, j), \quad j \geq 0,$$

$$\dot{p}(i, 0) = \mu p(i, 1) + \lambda p(i-1, 0) - [\lambda + u]p(i, 0), \quad i > 0,$$

$$\dot{p}(i, j) = \lambda p(i-1, j) + \mu p(i, j+1) + up(i+1, j-1)$$

$$- [\lambda + \mu + u]p(i, j), \quad i > 0, j > 0.$$

Clearly every rate is smaller in magnitude than $q := \lambda + \mu + a$, so that we get the discrete chain whose nonzero transition probabilities are given by:

$$P(x_{k+1} = (0, j) \mid x_k = (0, j+1), u_k = u) = \mu q^{-1}, \quad j \geq 0,$$

$$P(x_{k+1} = (0, j) \mid x_k = (0, j), u_k = u) = 1 - \lambda q^{-1}, \quad j \geq 0,$$

$$P(x_{k+1} = (i, 0) \mid x_k = (i, 1), u_k = u) = \mu q^{-1}, \quad i > 0,$$

$$P(x_{k+1} = (i, 0) \mid x_k = (i-1, 1), u_k = u) = \lambda q^{-1}, \quad i > 0,$$

$$P(x_{k+1} = (i, 0) \mid x_k = (i, 0), u_k = u)$$

$$= 1 - (\lambda + \mu)q^{-1}, \quad i > 0,$$

$$P(x_{k+1} = (i, j) \mid x_k = (i-1, j), u_k = u)$$
$$= \lambda q^{-1}, \ i > 0, j > 0,$$
$$P(x_{k+1} = (i, j) \mid x_k = (i, j+1), u_k = u)$$
$$= \mu q^{-1}, \ i > 0, j > 0,$$
$$P(x_{k+1} = (i, j) \mid x_k = (i+1, j-1), u_k = u)$$
$$= u q^{-1}, \ i > 0, j > 0,$$
$$P(x_{k+1} = (i, j) \mid x_k = (i, j), u_k = u)$$
$$= 1 - (\lambda + \mu + u)q^{-1}, \ i > 0, j > 0,$$

The discrete chain $\{x_k = (x_{1k}, x_{2k})\}$ corresponds to the following diagram.

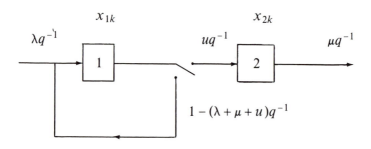

We now relate the costs incurred by the continuous and discrete time Markov chains $\{x(t)\}$—or $\{\hat{x}(t)\}$—and $\{x_k\}$. Suppose $c(i)$ is the instantaneous cost incurred at time t if $x(t) = i$. Then the expected cost incurred by $\{x(t)\}$ over the infinite time horizon, with discount rate $\alpha > 0$, is [using $c := (c(1), c(2), \dots)^T$]

$$E \int_0^\infty [\exp -\alpha t] c(x(t)) dt$$

$$= \int_0^\infty [\exp -\alpha t] p(t) c \, dt$$

$$= \sum_{k=0}^\infty \{ \int_0^\infty [\exp -(\alpha + q)t] \frac{(qt)^k}{k!} \} p_k c, \text{ by (8.2)}$$

$$= (\alpha + q)^{-1} \sum_{k=0}^\infty \beta^k p_k c,$$

where $\beta := q(\alpha + q)^{-1}$. Thus, ignoring the constant factor $(\alpha + q)^{-1}$, this is the same cost as incurred by the discrete time chain $\{x_k\}$ with the discount factor β.

Similarly, the average cost per unit time incurred by $\{x(t)\}$ is

$$\lim_{T \to \infty} \frac{1}{T} E \int_0^T c(x(t))dt$$

$$= \lim_{T \to \infty} \frac{1}{T} E \int_0^T c(\hat{x}(t))dt$$

$$= \lim_{N \to \infty} \frac{1}{ET_N} E \sum_{k=0}^{N-1} (T_{k+1} - T_k)c(x_k), \text{ by (8.4)}$$

$$= \lim_{N \to \infty} \frac{1}{ET_N} \sum_{k=0}^{N-1} E(T_{k+1} - T_k)Ec(x_k),$$

by independence of $\{x_k\}$ and $\{T_k\}$

$$= \lim_{N \to \infty} \frac{1}{N} \sum_{k=0}^{N-1} Ec(x_k),$$

which is also the average cost per unit time incurred by the discrete chain $\{x_k\}$.

9. Notes

1. For an introduction to controlled Markov chain models see Howard (1971). Finite Markov chains are thoroughly treated in Kemeny and Snell (1960) and Feller (1957), where one may find proofs of Lemmas (5.4), (5.7) and (5.11). Continuous time Markov chains are rigorously treated in Doob (1953).

2. According to Lemma (5.11), aperiodic and irreducible finite chains are ergodic. There is no such simple characterization for infinite chains; see Kemeny, Snell and Knapp (1976) and Revuz (1975).

3. If the infinite continuous time chain $\{x(t)\}$ is not uniformizable—i.e., if $\sup_i \sum_{j \neq i} Q_{ij} = \infty$—then $x(t)$ may change values infinitely many times in a finite time interval and there would be no equivalent discrete time chain. A continuous chain $\{x(t)\}$ can be approximated via time discretization, $x_k = x(\Delta k)$, where $\Delta > 0$ is a small number, see Kushner (1977) and van Dijk (1984).

4. Poisson processes can be used as building blocks to construct more general jump processes, including continuous time Markov chains; see Bremaud (1981) and Snyder (1975).

5. Queuing models are discussed in Kleinrock (1975), Bremaud (1981) and Gross and Harris (1974). More general queuing networks are discussed in Kelly (1979). A recent survey of control problems for queuing systems is given in Stidham (1985). The example in Section 8 is from Rosberg, Varaiya and Walrand (1981).

10. References

[1]   P. Bremaud (1981), *Point processes and queues, Martingale Dynamics,* Springer Verlag, New York.

[2]   J. Doob (1953), *Stochastic processes,* John Wiley, New York.

[3]   W. Feller (1957), *An introduction to probability theory and its applications, Vol 1,* John Wiley, New York.

[4]   D. Gross and C. Harris (1974), *Fundamentals of queueing theory,* John Wiley, New York.

[5]   R.A. Howard (1971), *Dynamic probabilistic systems, Vol 1, Markov models,* John Wiley, New York.

[6]   F. Kelly (1979), *Reversibility and stochastic networks,* John Wiley, New York.

[7]   J.G. Kemeny and J.L. Snell (1960), *Finite Markov chains,* Van Nostrand, Princeton, NJ.

[8]   J.G. Kemeny, J.L. Snell and A.W. Knapp (1976), *Denumerable Markov chains,* Springer Verlag, New York.

[9]   L. Kleinrock (1975), *Queueing systems, Vol 1, Theory, Vol 2, Computer applications,* Wiley-Interscience, New York.

[10] H. Kushner (1977), *Probability methods for approximations in stochastic control and for elliptic equations,* Academic Press, New York.

[11] D. Revuz (1975), *Markov chains,* American Elsevier, New York.

[12] Z. Rosberg, P.P. Varaiya and J.C. Walrand, "Optimal control of service in tandem queues," *IEEE Transactions on Automatic Control,* Vol. AC-27(3), 600-609.

[13] D. Snyder (1975), *Random point processes,* Wiley-Interscience, New York.

[14] S. Stidham, Jr. (1985), "Optimal control of admission to a queueing system," *IEEE Transactions on Automatic Control,* Vol. AC-30(8), 705-713.

[15] N.H. van Dijk (1984), *Controlled Markov processes: Time discretization,* CWI Tract 11, Center for Mathematics and Computer Science, Amsterdam.

CHAPTER 5
INPUT OUTPUT MODELS

In this chapter we introduce linear models that relate input and output processes with no reference to the internal states of the systems. These models are particularly useful in identification.

1. Arbitrariness of state variables

Consider the linear stochastic system

$$x_{k+1} = Ax_k + Bu_k + Gw_k, \tag{1.1a}$$

$$y_k = Cx_k + Hv_k, \tag{1.1b}$$

where $x_k \in R^n$. As before, the basic random variables $\{x_0, w_0, w_1, \ldots, v_0, v_1, \ldots\}$ are assumed independent. Let T be a n-dimensional nonsingular matrix, and define new state variables x'_k by

$$x'_k := Tx_k, \quad x_k = T^{-1}x'_k. \tag{1.2}$$

In terms of these new variables (1.1) is transformed into

$$x'_{k+1} = A'x'_k + B'u_k + G'w_k, \tag{1.3a}$$

$$y_k = C'x'_k + Hv_k, \tag{1.3b}$$

where

$$A' = TAT^{-1}, \ B' = TB, \ G' = TG, \ C' = CT^{-1}.$$

If we regard u_k, v_k, w_k as the input (both control and disturbance), and y_k as the output, the models (1.1) and (1.3) are input-output equivalent since both give the same relation between input and output. Indeed, since any nonsingular matrix T could have been used in (1.2), there are infinitely many equivalent systems. If one adopts the viewpoint that the only real variables pertaining to the system are its inputs and outputs (in this view the system is a black box), then the state is seen as an arbitrary artifact of the analyst.

This arbitrariness is particularly annoying in an identification problem. Suppose we know we are dealing with a system with the state space model (1.1), but the matrices $A, B, C, G,$ and H are not known. We want to identify or estimate these matrices on the basis of the observations of the control and output u_k, y_k, over an interval, say $k = 0, 1, \ldots, K$. It is clear that the best we can hope to achieve is to identify these matrices up to equivalence. In particular, the observations cannot provide any justification for choosing between models (1.1) and (1.3). The fact that several specifications of the matrices account equally well for the observations leads to instabilities in identification procedures. There are two ways of preventing this. First, we may choose a single model in each equivalence class, and insist that this is the correct model for the class. (This is often called the *canonical* model.) Second, we may try and eliminate the state variables and directly relate the output to the input variables. The first approach will be studied later. Here we investigate the second approach.

2. Elimination of state variables

Introduce the (forward) **shift operator** q. For any sequence or stochastic process $\xi = \{\xi_k, \ k = 0, 1, \ldots\}$, $q\xi$ is the sequence $(q\xi)_k := \xi_{k+1}$, $k \geq 0$. Then (1.1) can be written as

$$qx = Ax + Bu + Gw, \tag{2.1a}$$

$$y = Cx + Hv, \tag{2.1b}$$

where x, y, u, v, and w denote sequences. Solve (2.1a) for x,

$$x = (qI - A)^{-1}[Bu + Gw],$$

and substitute in (2.1b) to get

$$y = C(qI - A)^{-1}[Bu + Gw] + Hv. \tag{2.2}$$

Let α_i be matrices such that

$$(qI - A)^{-1} =: [det(qI - A)]^{-1}[\alpha_n + q\alpha_{n-1} + \ldots + q^{n-1}\alpha_1].$$

Multiply (2.2) by $q^{-n}[det(qI - A)]$. Use the preceding relation and $e =: (v, w)$ to obtain

$$[\sum_{i=0}^{n} A_i q^{-i}] y = [\sum_{i=1}^{n} B_i q^{-i}] u + [\sum_{i=0}^{n} C_i q^{-i}] e, \tag{2.3}$$

where A_i, B_i, and C_i are matrices with $A_0 = I$. In the time domain this is

$$y_k = -\sum_{i=1}^{n} A_i y_{k-i} + \sum_{i=1}^{n} B_i u_{k-i} + \sum_{i=0}^{n} C_i e_{k-i}. \tag{2.4}$$

This is the **ARMAX** model. The first term on the right in (2.4) expressing the dependence of the current output on its own past values is the **autoregressive**, or AR, term. The last term is the **moving average**, or MA, term since it is a moving combination of independent random variables e_k. The middle, or X, term gives the effect of external or exogenous input. Special cases are the AR ($B_i = 0$, $C_i = 0$, $i \geq 1$), MA ($A_i = 0$, $B_i = 0$, $i \geq 1$), ARX ($C_i = 0$, $i \geq 1$), and the ARMA ($B_i = 0$, $i \geq 1$) models.

Remarks   (2.5)

1. Any ARMAX model (2.4) can be put into the state space model (2.1) so that the two linear models are input-output equivalent. The linear system model (2.1) however generalizes in a natural way to nonlinear models as in Definition (2.2.6), but there is no fully equivalent way to generalize the linear ARMAX model. A straightforward extension of (2.4) often used in practice is

$$f(y_k) = - \sum A_i f(y_{k-i}) + \sum B_i g(u_{k-i}) + \sum C_i e_{k-i},$$

where f and g are known functions. A common example in the scalar case is

$$f(x) = g(x) = \ln x.$$

2. The state model is particularly useful when the state has some empirical significance. It is also important for problems in estimation and control since the Markov property is key. The ARMAX model is useful for identification problems, especially when the true system is complex or poorly understood. In this case all one can assert is the dependence of the current output on past outputs and inputs. Some examples are given next.

3. Examples

Suppose you wish to predict next year's rainfall, y_{k+1}, based on the record $y_k, y_{k-1}, \ldots$ of the rainfall in previous periods. Lacking any model for rainfall generation based on the laws of physics, one way to proceed is to say that the rainfall follows some statistical pattern and use the available record to identify this pattern. The simplest such pattern would be the one where the rainfall in different years is independent, i.e.,

$$y_k = w_k,$$ (3.1)

where $\{w_k\}$ is a sequence of independent identically distributed random variables. From the record one can estimate the mean $\overline{w}$ and variance σ^2 of w_k. The predicted value for y_{k+1} would be simply $\overline{w}$.

A more detailed study of the data may reveal that the rainfall in consecutive years is not uncorrelated as the model (3.1) suggests. For instance, the record may reveal that a year of high rainfall is often followed by a year of low rainfall. In this case you would wish to modify

(3.1) to

$$y_{k+1} = ay_k + w_k,$$ (3.2)

and you would have the prior belief that $a < 0$. Again the record may be used to estimate a, $\overline{w}$, and σ^2. Now your prediction for y_{k+1} will be $ay_k + \overline{w}$. If you feel that the correlation in the pattern of rainfall extends over 4 years, then the model to propose is

$$y_{k+1} = a_1 y_k + a_2 y_{k-1} + a_3 y_{k-2} + w_k.$$ (3.3)

This is an AR model. It is, of course, possible to derive a state space model from (3.3), and this may be necessary for some purposes. But clearly in this example it is more natural to begin with an input-output model.

Consider an example from economics. Suppose you want to predict the level of investment, y_{k+1}, that firms in the electronics industry will undertake next year. The investment reflects the expectation these firms have about the demand for electronic products next year. It is reasonable to argue that the firms base their expectations on past demand, $d_k, d_{k-1}, \ldots$; so as a first specification one might propose the model

$$y_{k+1} = f(d_k, \ldots, d_{k-n}).$$ (3.4)

It is possible that these expectations are not realized, that is, the firms may have made large investments in the past based on optimistic projections of demand which did not materialize. Or the demand may have turned out to be greater than foreseen and so, in retrospect, previous investment appears to have been too low. What this indicates is that the level of future investment also depends on past investment and suggests that (3.4) should be changed to

$$y_{k+1} = f_1(y_k, \ldots, y_{k-n}) + f_2(d_k, \ldots, d_{k-n}).$$ (3.5)

This is a deterministic model. As such it seems too simplistic, since many considerations that go into the formulation of investment plans are left out. Lacking any explicit model for these factors, we can represent their effect as a pure disturbance term and modify (3.5) to

$$y_{k+1} = f_1(y_k, \ldots, y_{k-n}) + f_2(d_k, \ldots, d_{k-n}) + w_{k+1}.$$ (3.6)

If the functions f_i in (3.6) are nonlinear, we get a nonlinear input-output equation. For identification it is usually easier to deal with linear models. This gives an ARX model

$$y_{k+1} = \sum_{i=0}^{n} a_i y_{k-i} + \sum_{i=0}^{n} b_i d_{k-i} + w_{k+1}.$$ (3.7)

Assuming that the w_k are iid, one can use past investment and sales data to estimate the unknown coefficients a_i, and b_i and the mean and variance of the disturbance term, and then use these estimates to make a prediction of y_{k+1}. The assumption that the disturbance is iid may be unreasonable in this context however. We know, for instance, that there are business cycles in investment. In an upturn in the economy investment is high, and it is lower than normal during a downturn. Since we have not explicitly accounted for these business cycle effects, they are one of the factors in the disturbance term. This term must therefore reflect a cyclical character. The easiest way of doing this is to specify it is as a moving average,

$$w_k = \sum_{i=0}^{n} c_i z_{k-i}, \tag{3.8}$$

where $\{z_k\}$ is as an iid process. Equations (3.7) and (3.8) constitute an ARMAX model of the investment process.

A more production-oriented view of the same situation may make a state space model seem more appropriate. The value of electronics goods produced in year k, q_k, is proportional to the installed production capacity, c_k,

$$q_k = ac_k. \tag{3.9}$$

Installed capacity decreases over time due to obsolescence and increases with new investment, suggesting a relationship of the form

$$c_{k+1} = \delta c_k + y_k, \tag{3.10}$$

where $0 < \delta < 1$ is the fraction of capacity in year k that survives the deterioration process and, as before, y_k is the investment in year k. The annual production augments the stock of goods in inventory which in turn is depleted by the amount sold. This fact leads to an equation

$$v_{k+1} = v_k + q_k - d_k, \tag{3.11}$$

where, as before, d_k is the annual sales and v_k is the stock of inventory at the beginning of year k.

Investment plans attempt to meet forecasted demand and to maintain a constant level of inventory. This suggests an equation of the form

$$y_k = b_1(v_k - \bar{v}) + b_2(d_k - d_{k-1}). \tag{3.12}$$

Here $\bar{v}$ is the desired inventory level and the last term gives the effect of forecasted demand.

Equations (3.9)-(3.12) describe a model with the state (c_k, v_k). In an identification problem the coefficients a, b_1, b_2, δ, and $\bar{v}$ will be unknown. The problem is to estimate them on the basis of the observations q_k, v_k, d_k, y_k and possibly c_k, although in practice the installed capacity is not directly observable. The demand d_k is a random variable whose statistics are usually unknown, and these have to be estimated as well. In this formulation the state variables are clearly not arbitrary artifacts since they have a physical significance.

4.Impulse response model

Another formulation equivalent to (2.1) and (2.4) is the impulse response model, where y_k is expressed as a function of past values of inputs and noise. Write (2.3) as

$$y = [A(q^{-1})]^{-1} [B(q^{-1})] u + [A(q^{-1})]^{-1} [C(q^{-1})] e \qquad (4.1)$$
$$= H_u(q^{-1}) u + H_e(q^{-1}) e.$$

The matrices H_u, H_e are **transfer functions** from u to y and e to y, respectively. Their entries are rational functions of q^{-1}. These can be expanded formally in a power series in q^{-1}. Expressing the result in the time domain gives the **impulse response** (IR) model

$$y_k = \sum_{i=0}^{\infty} h_i^u u_{k-i} + \sum_{i=0}^{\infty} h_i^e e_{k-i}. \qquad (4.2)$$

Since the dependence of y_k extends over the infinite past of input and noise, (4.2) is sometimes called the **infinite impulse response** (IIR) model. If the h_i^u, h_i^e are 0 for $i \geq n$, for some finite n, one gets the **finite impulse response** (FIR) model. FIR is a very special case of ARMAX. FIR models have become popular in digital signal processing.

Suppose $\{u_k, e_k, -\infty < k < \infty\}$ is stationary and has finite variance; then from (4.2), $\{y_k\}$ will also be stationary and with finite variance provided that

$$\sum_{i=0}^{\infty} (||h_i^u||^2 + ||h_i^e||^2) < \infty. \qquad (4.3)$$

Suppose that the matrix $A(z)$ is **stable**, i.e.,

$$\det A(z) = 0 \text{ implies } |z| > 1, \text{ or} \qquad (4.4)$$
$$\det A(q^{-1}) = 0 \text{ implies } |q| < 1.$$

Then there is a power series expansion,

$$[\det A(q^{-1})]^{-1} = \sum_{i=0}^{\infty} a_i q^{-i},$$

with $|a_i| < \rho\lambda^i$ for some $\lambda < 1$ and ρ. From (4.1) we see that then $||h_i^u||$ and $||h_i^e||$ decline geometrically and so (4.3) will hold. In the stationary case one can use frequency domain models.

5. Frequency domain model

A stochastic process $\xi = \{\xi_k, -\infty < k < \infty\}$ with values in R^n is said to be **wide sense** (or second-order) stationary if $E\xi_k = E\xi_l$ and $E\xi_k\xi_l^T = E\xi_{k+m}\xi_{l+m}^T$ for all k, l, m. Denote the mean and covariance function of ξ by

$$\bar{\xi}_k := E\xi_k, \quad \Sigma(m) := E(\xi_{m+k} - \bar{\xi}_{m+k})(\xi_k - \bar{\xi}_k)^T.$$

The process ξ has a (power) *spectral density* $\Phi(\omega)$, which is obtained from $\Sigma(m)$ by

$$\Phi(\omega) := \frac{1}{2\pi} \sum_{-\infty}^{\infty} \Sigma(m)e^{-im\omega}, \quad -\pi \le \omega \le \pi. \tag{5.1}$$

Conversely, $\Sigma(m)$ can be recovered from $\Phi(\omega)$ by

$$\Sigma(m) = \int_{-\pi}^{\pi} e^{im\omega} \Phi(\omega)d\omega, \quad -\infty < m < \infty. \tag{5.2}$$

Moreover there exists a n-dimensional stochastic process with complex values $\{z(\omega), -\pi \le \omega \le \pi\}$ with zero mean and

$$E\, z(\omega)z(\omega')^* = \Phi(\omega)\delta(\omega - \omega'), \tag{5.3}$$

such that

$$\xi_m = \bar{\xi}_m + \int_{-\pi}^{\pi} e^{im\omega} z(\omega)d\omega, \quad -\infty < m < \infty. \tag{5.4}$$

In (5.3), * denotes complex conjugate transpose and δ is the (impulse) delta function. The interpretation of (5.4) is that the process ξ can be viewed as being produced by a combination of sinusoids, with the random amplitude and phase given by $z(\omega)$. The power spectral density $\Phi(\omega)$ represents the power in the frequency ω.

We note some properties of the spectral density.

Exercise (5.5)
Show the following:
(1) $\Phi(\omega)$ is Hermitian—i.e. $\Phi(\omega) = \Phi^*(\omega)$)—and positive definite.
(2) $\Phi(-\omega) = \Phi^T(\omega)$.
(3) $\int_{-\pi}^{\pi} \Phi(\omega)d\omega = \Sigma(0)$ is the total power or variance.

The process ξ is said to be wide sense **white** noise if its spectral density is constant, $\Phi(\omega) \equiv \Phi(0)$. From (5.1) and (5.2), this is equivalent to $\Sigma(0) = 2\pi\Phi(0)$ and $\Sigma(m) = 0$, $m \ne 0$, i.e., ξ is uncorrelated.

Now consider the model (4.2) and for simplicity take $e_k \equiv 0$,

$$y_k = \sum_0^\infty h_i u_{k-i}. \tag{5.6}$$

Suppose $u_k \in R^n$, $y_k \in R^p$, so that the h_i are $p \times n$ matrices. Assume stability:

$$\sum_{i=0}^\infty || h_i ||^2 < \infty. \tag{5.7}$$

Let H be the corresponding $p \times n$ matrix transfer function,

$$H(q^{-1}) := \sum_{i=0}^\infty h_i q^{-i}.$$

If the IIR model (5.6) corresponds to a finite dimensional stochastic system as in (4.1), then the entries of $H(q^{-1})$ will be rational functions of q^{-1}; otherwise they will not be. Finite dimensionality is not assumed for the moment.

Suppose $\{u_k\}$ is zero mean wide sense stationary with covariance and spectral density functions $\Sigma^u(m)$ and $\Phi^u(\omega)$, respectively. Because of (5.7), $\{y_k\}$ is also wide sense stationary. Its covariance and spectral density functions can be calculated from (5.6), as in the following exercise.

Exercise (5.8)

Show that the covariance and spectral density functions of $\{y_k\}$ are given by

$$\Sigma^y(m) = \sum_{i=0}^\infty \sum_{j=0}^\infty h_i \, \Sigma^u(m-i+j) \, h_j^T,$$

$$\Phi^y(\omega) = H(e^{-i\omega}) \, \Phi^u(\omega) \, H^*(e^{-i\omega}).$$

In fact, for the wide sense stationary process $\{z_k^T := (u_k^T, y_k^T)\}$, one gets

$$\Sigma^z(m) = \begin{bmatrix} \Sigma^u(m) & \Sigma^{uy}(m) \\ [\Sigma^{uy}(m)]^T & \Sigma^y(m) \end{bmatrix}, \quad \text{and}$$

$$\Phi^z(\omega) = \begin{bmatrix} \Phi^u(\omega) & \Phi^{uy}(\omega) \\ \Phi^{uy}(\omega)^* & \Phi^y(\omega) \end{bmatrix}. \tag{5.9}$$

It follows that the (cross) covariance and the (cross) spectral density functions between u and y are given by

$$\Sigma^{uy}(m) := E \, u_{k+m} y_k^T = \sum_{i=0}^\infty \Sigma^u(m+i) \, h_i^T,$$

$$\Phi^{uy}(\omega) = \Phi^u(\omega) H^*(e^{-i\omega}).$$

Suppose that u is white, $\Phi^u(\omega) \equiv I$; then the spectral density of y is

$$\Phi^y(\omega) = H(e^{-i\omega}) H^*(e^{-i\omega}). \tag{5.10}$$

Conversely, given $\Phi^y(\omega)$, one can find a transfer function $H(q^{-1})$ which satisfies (5.10) by factoring Φ^y. (This is known as the *spectral factorization* problem.) Then $\{y_k\}$ can be represented as the output of a linear system with transfer function $H(q^{-1})$ driven by white noise. When y is generated by a finite-dimensional system as in (4.1), $H(q^{-1})$ is a rational function; hence $\Phi^y(\omega)$ is also rational in $e^{i\omega}$, and one obtains the rational spectral factorization problem.

In the single-input single-output (SISO) rational case, $n = p = 1$, and so (5.10) simplifies to

$$\Phi^y(\omega) = |H(e^{i\omega})|^2. \tag{5.11}$$

As an example, consider

$$\Phi^y(\omega) = \frac{1.09 + 0.6\cos\omega}{1.16 + 0.8\cos\omega} = \frac{(e^{i\omega} + 0.3)(e^{-i\omega} + 0.3)}{(e^{i\omega} + 0.4)(e^{-i\omega} + 0.4)}.$$

It can be factored in four ways:

$$H_1(q^{-1}) = \frac{q + 0.3}{q + 0.4},$$

$$H_2(q^{-1}) = \frac{q^{-1} + 0.3}{q + 0.4},$$

$$H_3(q^{-1}) = \frac{q + 0.3}{q^{-1} + 0.4},$$

$$H_4(q^{-1}) = \frac{q^{-1} + 0.3}{q^{-1} + 0.4}.$$

Recall that the true transfer function must be stable—otherwise y will not be stationary. By (4.4) this implies that the poles of $H(q^{-1})$ [i.e., the values of q where the denominator of $H(q^{-1})$ vanishes] must lie inside the unit circle, which eliminates H_3, and H_4 from consideration. Both H_1, and H_2 are permissible, however. It is customary to choose H_1 because its zeros [i.e., the values of q where the numerator of $H(q^{-1})$ vanishes] also lie inside the unit circle. (A transfer function whose zeros are inside the unit circle is called **minimum phase**.) The general result is stated next.

Theorem   (5.12)
Let $\Phi(\omega)$ be a scalar rational spectral density function. Then $\Phi(\omega)$ can be uniquely factored as $\Phi(\omega) = |H(e^{i\omega})|^2$ where $H(q^{-1}) = \dfrac{N(q^{-1})}{D(q^{-1})}$ is such

that
(1) $N(q^{-1})$, $D(q^{-1})$ have no common factors,
(2) $D(q^{-1}) = 0$ implies $|q| < 1$,
(3) $N(q^{-1}) = 0$ implies $|q| \leq 1$.

Proof
Uniqueness is guaranteed by (1) and the requirement that the zeros of $D(q^{-1})$ and $N(q^{-1})$ must lie on or inside the unit circle. Since, by (5.5),

$$\int_{-\pi}^{\pi} \Phi(\omega) d\omega = \int_{-\pi}^{\pi} \frac{|N(e^{i\omega})|^2}{|D(e^{i\omega})|^2} d\omega = Ey^2 < \infty,$$

$D(q^{-1})$ cannot have zeros on the unit circle. $\square$

Thus any wide sense scalar stationary process $\{y_k\}$ with rational spectral density can be represented as the output of a stable linear system with transfer function $H(q^{-1})$ whose input $\{u_k\}$ is white. In (5.12), the zeros of $N(q^{-1})$ are often strictly inside the unit circle. In this case the transfer function $\dfrac{1}{H(q^{-1})}$ is also stable and one can recover the input from the output by $u = \dfrac{1}{H(q^{-1})} y$. The filter with transfer function $\dfrac{1}{H(q^{-1})}$ is called a **whitening filter.**

The spectral factorization theorem generalizes to the vector case.

Theorem   (5.13)
Let $\Phi(\omega)$ be a rational spectral density function of a process $\{y_k\}$ with values in R^p. Suppose $\det \Phi(\omega)$ is not identically zero, then Φ can be uniquely factored as

$$\Phi(\omega) = H(e^{-i\omega}) \, \Sigma \, H^*(e^{-i\omega}),$$

where Σ and H are $p \times p$ matrices such that

(1) $\Sigma = \Sigma^T$ is positive definite, $\lim\limits_{q \to \infty} H(q^{-1}) = I$,

(2)   the poles of $H(q^{-1})$ lie inside the unit circle,

(3)   the poles of $[H(q^{-1})]^{-1}$ lie on or inside the unit circle.

6. Notes

1. ARMAX models are widely used in identification of linear models; see Astrom (1970), and Box and Jenkins (1970). For applications in control see Astrom and Wittenmark (1984); and for economic applications see Chow (1981).

2. Impulse response models are popular in digital signal processing, where the FIR model is known as the *tapped delay line* or *transversal filter*. An introduction to this subject is given in Haykin (1984). A more

advanced treatment can be found in Messerschmitt and Honig (1984).

3. The results (5.1)-(5.4) are proved in Doob (1953). For a discussion of frequency domain models, including a proof of Theorem (5.13), see Hannan (1970) or Anderson and Moore (1979).

7. References

[1] B.D.O. Anderson and J.B. Moore (1979), *Optimal filtering*, Prentice-Hall, New York.

[2] K. Astrom (1970), *Introduction to stochastic control theory*, Academic Press, New York.

[3] K.J. Astrom and B. Wittenmark (1984), *Computer controlled systems*, Prentice-Hall, New York.

[4] G.E.P. Box and G.M. Jenkins (1970), *Time series analysis, forecasting, and control*, Holden Day, San Francisco.

[5] G. Chow (1981), *Econometric analysis by control methods*, John Wiley, New York.

[6] J. L. Doob (1953), *Stochastic processes*, John Wiley, New York.

[7] E.J. Hannan (1970), *Multiple time series*, John Wiley, New York.

[8] S. Haykin (1984), *Introduction to adaptive filters*, Macmillan, New York.

[9] D.G. Messerschmitt and M. Honig (1984), *Adaptive filters: structures, algorithms and applications*, Kluwer, Hingham, MA.

CHAPTER 6
DYNAMIC PROGRAMMING

We consider the problem of selecting a feedback control so as to minimize the expected cost. The optimality conditions are obtained by dynamic programming. The case of complete observations is quite straightforward, but the case of partial observations requires the concept of information state. When the state takes on only a finite number of values, the optimal control is easy to compute.

1. Optimal control laws

Consider the stochastic system described by the state space model

$$x_{k+1} = f_k(x_k, u_k, w_k),$$

$$y_k = h_k(x_k, v_k), k = 0, 1, \ldots$$

Suppose that for each k, the control value u_k is to be selected from a prespecified control set $U \subset R^m$. A **feasible** control law is then any sequence $g = \{g_0, g_1, \ldots\}$ such that

$$u_k = g_k(y^k) \in U$$

for all y^k. Let G denote the set of all feasible control laws.

We wish to find the best law in G. To make this precise it is necessary to have a criterion for comparing different laws. We follow the development in deterministic control theory and associate a cost (or loss) with each law and then the best law is simply the one that minimizes the cost.

Suppose we are given the cost function

$$\sum_{k=0}^{N-1} c_k(x_k, u_k) + c_N(x_N). \tag{1.1}$$

It is assumed that the time horizon is $N < \infty$. This means that we are only concerned with the system behavior over the interval $0 \leq k \leq N$. Later we shall deal with the infinite horizon problem. The term $c_k(x_k, u_k)$ is called the **immediate** or **one-period** cost, and $c_N(x_N)$ is the **terminal** cost. Note that since x_N determines the way in which the system can evolve after N, a judicious choice of the terminal cost can incorporate the system behavior beyond N. Since $N < \infty$, a control law is now specified by a finite sequence $g = \{g_0, \ldots, g_{N-1}\}$.

Let g be a feasible law. Let x^g, y^g, and u^g denote the processes corresponding to it. By definition the cost associated with g is

$$C^g = \sum_0^{N-1} c_k(x_k^g, u_k^g) + c_N(x_N^g). \tag{1.2}$$

Since x_k^g and u_k^g are random variables, C^g is also random. Denote its realization at a sample point ω by $C^g(\omega)$. Let g' be another feasible law and let $C^{g'}$ be its random cost. It is very likely that for some sample points ω, we have $C^g(\omega) > C^{g'}(\omega)$, whereas for other sample points ω', $C^g(\omega') < C^{g'}(\omega')$. In this case it is not possible to say which policy gives the lower cost. (This does not happen in the deterministic case, since there the cost is a real number and not random, so for any two policies one must give a lower cost.) To get out of this indeterminacy we consider instead the *expected* cost

$$J(g) := EC^g. \tag{1.3}$$

Here E denotes expectation. This permits us to define an optimal law.

Definition   (1.4)
g^* in G is **optimal** if $J(g^*) = J^* = Inf\{J(g) \mid g \in G\}$. J^* is called the minimum (expected) cost.

Our aim is to characterize the optimal laws using dynamic programming.

It is convenient to drop the superscript g when there is no possibility of confusion. In most cases, in order to preserve the distinction between different policies g and g', it will be sufficient to use the symbols E^g and $E^{g'}$ to distinguish between the pd induced by g and g' respectively. Thus, for example, $E^g c_k(x_k, u_k)$ and $E\, c_k(x_k^g, u_k^g)$ denote the same expected value.

2. Case of complete information I

In this section we assume that the state is observed, $y_k \equiv x_k$. Let $g = \{g_0, \ldots, g_{N-1}\}$ be a feasible policy. Then $u_k = g_k(x_0, \ldots, x_k)$ is allowed to depend on previous values of the state. From Lemma (2.6.6) we know that the conditional probability of x_{k+1} given $x_k, \ldots, x_0, u_k, \ldots, u_0$ depends only on x_k, u_k and not on preceding values. This suggests that in controlling the system, it is enough to let u_k depend only on x_k. We make this precise.

Definition   (2.1)
A feasible law $g = \{g_k\}$ is said to be **Markovian** or a **Markov policy** if g_k depends only on $y_k = x_k$. Let $G_M \subset G$ be the set of Markovian laws.

Intuition suggests that optimal laws belong to G_M.

Exercise (2.2)

Show that if $g \in G_M$, then $\{x^g\}$ is a Markov process.
[Hint: See Lemma (4.3.1).]

We investigate some properties of the cost associated with Markovian policies. The following notation is helpful.

Notation (2.3)

Let $h: R^n \times R^m \to R$ be any function. Let $z \in R^n$ be a fixed vector and let w be a m-dimensional random variable. Define

$$(E_w h)(z) := \int h(z,w) P_w(dw).$$

Exercise (2.4)

Suppose z, w are independent random variables. Then

$$E\{ h(z,w) \mid z \} = (E_w h)(z).$$

Exercise (2.5)

Let $\{x_k, 0 \le k \le N\}$ be a Markov chain with transition probabilities $P_{x_{k+1} \mid x_k}$. Let $\{h_k \mid 0 \le k \le N\}$ be real-valued functions. Define the functions

$$H_N(x) := h_N(x),$$

$$H_k(x) := h_k(x) + \int H_{k+1}(x') P_{x_{k+1} \mid x_k}(dx' \mid x).$$

Then the random variables $H_k(x_k)$ satisfy

$$H_k(x_k) = E\{ \sum_k^N h_l(x_l) \mid x_k \} = E\{ \sum_k^N h_l(x_l) \mid x_k, \ldots, x_0 \}.$$

Here, it is important to distinguish between the function $H_k(x)$ and the random variable $H_k(x_k)$.
[Hint: See Lemma (4.4.6).]

Lemma (2.6)

Let $g = \{g_0, \ldots, g_{N-1}\}$ be a Markov policy. Define recursively the functions

$$V_N^g(x) := c_N(x),$$

$$V_k^g(x) := c_k(x, g_k(x)) + (E_{w_k} V_{k+1}^g)[f_k(x, g_k(x), w_k)].$$

Then the random variable $V_k^g(x_k^g)$ satisfies

$$V_k^g(x_k^g) = E\{ \sum_{l=k}^{N-1} c_l(x_l^g, u_l^g) + c_N(x_N^g) \mid x_k^g \}, \quad k = 0, \ldots, N.$$

Proof
The proof follows from Exercises (2.2) and (2.5).   □

Note that since $\{x_k\}$ is Markov, we also have

$$V_k^g(x_k^g) = E\{ \sum_{k}^{N-1} c_l(x_l^g, u_l^g) + c_N(x_N^g) \mid x_k^g, \ldots ,x_0^g \}. \tag{2.7}$$

Definition   (2.8)
Let $g \in G$. The random variable

$$J_k^g = E\{ \sum_{k}^{N-1} c_l(x_l^g, u_l^g) + c_N(x_N^g) \mid x_0^g, \ldots , x_k^g \}$$

is called the **cost-to-go** at k corresponding to g.

From Definitions (2.8) and (1.2) we see that

$$J_0^g = E\{C^g \mid x_0\}, \quad J(g) = E\, J_0^g. \tag{2.9}$$

While Lemma (2.6) is valid only for Markov policies, the comparison principle below holds for arbitrary feedback policies.

Lemma (Comparison Principle)   (2.10)
Let $V_k(x)$, $0 \le k \le N$, be any functions such that

$$V_N(x) \le c_N(x), \tag{2.11}$$

$$V_k(x) \le c_k(x, u) + E_{w_k} V_{k+1}[f_k(x, u, w_k)], \tag{2.12}$$

for all x and for all u in U. Let $g \in G$ be arbitrary. Then, w.p. 1,

$$V_k(x_k^g) \le J_k^g, \quad k = 0, \ldots ,N. \tag{2.13}$$

Proof
We proceed by induction. From (2.8) and (2.11)

$$J_N^g = E\{c_N(x_N^g) \mid x_0^g, \ldots , x_N^g\} = c_N(x_N^g) \ge V_N(x_N^g),$$

so that (2.13) is true for $k = N$. Suppose it is true for $k + 1$. Then by (2.12) and (2.4),

$$V_k(x_k^g) \le E\{c_k(x_k^g, u_k^g) + V_{k+1}^g[f_k(x_k^g, u_k^g, w_k)] \mid x_0^g, \ldots , x_k^g\}$$

$$\le E\{c_k(x_k^g, u_k^g) +$$

$$E\{ \sum_{k+1}^{N-1} c_l(x_l^g, u_l^g) + c_N(x_N^g) \mid x_0^g, \ldots , x_{k+1}^g \} \mid x_0^g, \ldots , x_k^g\}$$

$$= E\{ \sum_{k}^{N-1} c_l(x_l^g, u_l^g) + c_N(x_N^g) \mid x_0^g, \ldots , x_k^g\} = J_k^g,$$

and so (2.13) holds for k.   □

We get an immediate corollary.

Corollary   (2.14)
Let $V_k(x)$ be functions satisfying (2.11), (2.12). Then $J^* \geq E V_0(x_0)$. Hence if $g \in G$ is such that $J_0^g = V_0(x_0)$, then g is optimal.

Proof
For any g in G we have $J_0^g(x_0) \geq V_0(x_0)$ by (2.13). Taking expectations and using (2.9) gives $J(g) \geq E V_0(x_0)$ and so $J^* \geq E V_0(x_0)$. Finally, if $J_0^g = V_0(x_0)$, then $J(g) = E V_0(x_0) \leq J^*$, so that g must be optimal and $J(g) = J^*$.   □

The two preceding lemmas can be combined to obtain the fundamental result of dynamic programming.

Theorem   (2.15)
Define recursively the functions

$$V_N(x) := c_N(x) \tag{2.16}$$

$$V_k(x) := \operatorname*{Inf}_{u \in U} [c_k(x, u) + E_{w_k} V_{k+1}(f_k(x, u, w_k))]. \tag{2.17}$$

(1) Let g in G be arbitrary. Then $V_k(x_k^g) \leq J_k^g$ w.p.1; in particular, $J(g) \geq EV_0(x_0)$.
(2) A Markov policy $g = \{g_0, \ldots, g_{N-1}\}$ in G_M is optimal if the infimum in (2.17) is achieved at $g_k(x)$, and then $V_k(x_k^g) = J_k^g$ w.p.1 and $J^* = J(g) = E V_0(x_0)$.
(3) A Markov policy $g = \{g_0, \ldots, g_{N-1}\}$ in G_M is optimal only if for each k, the infimum at x_k^g in (2.17) is achieved by $g_k(x_k^g)$, i.e.

$$V_k(x_k^g) = c_k(x_k^g, g_k(x_k^g)) + E_{w_k} V_{k+1}[f_k(x_k^g, g_k(x_k^g), w_k)],$$

w.p. 1.

Proof
The functions $V_k(x)$ defined by (2.16) and (2.17) clearly satisfy (2.11) and (2.12) and so part (1) follows from Lemma (2.10). To prove the sufficiency in part (2), let $g = \{g_k\}$ be a Markov policy that achieves the infimum in (2.17), so

$$V_k(x) = c_k(x , g_k(x)) + E_{w_k} V_{k+1}[f_k(x, g_k(x), w_k)].$$

By Lemma (2.6) it follows that $V_k(x_k^g) = J_k^g$ for all k and in particular $J_0^g = V_0(x_0)$. By Corollary (2.14), g is optimal and $J^* = J(g) = EV_0(x_0)$.

To prove the necessity in part (3) suppose the Markovian policy g is optimal. We prove by induction that $g_k(x_k^g)$ achieves the infimum in (2.17) at x_k^g with probability 1. Consider $k = N - 1$. Suppose the assertion is false. Then there exists another function $g'_{N-1}: R^n \rightarrow U$ such that

$$c_{N-1}(x^g_{N-1}, g_{N-1}(x^g_{N-1})) +$$

$$E_{w_{N-1}} V_N[f_{N-1}(x^g_{N-1}, g_{N-1}(x^g_{N-1}), w_{N-1})]$$

$$\geq c_{N-1}(x^g_{N-1}, g'_{N-1}(x^g_{N-1})) +$$

$$E_{w_{N-1}} V_N[f_{N-1}(x^g_{N-1}, g'_{N-1}(x^g_{N-1}), w_{N-1})]$$

w.p.1; moreover, the inequality is strict with positive probability. Hence taking expectations on both sides and using (2.16) gives

$$E[c_{N-1}(x^g_{N-1}, g_{N-1}(x^g_{N-1})) +$$

$$c_N(f_{N-1}(x^g_{N-1}, g_{N-1}(x^g_{N-1}), w_{N-1}))]$$

$$> E[c_{N-1}(x^g_{N-1}, g'_{N-1}(x^g_{N-1})) +$$

$$c_N(f_{N-1}(x^g_{N-1}, g'_{N-1}(x^g_{N-1}), w_{N-1}))]. \tag{2.18}$$

Consider the Markov policy $g' = \{g_0, \ldots, g_{N-2}, g'_{N-1}\}$. Evidently $x^g_k = x^{g'}_k$, $0 \leq k \leq N-1$ and so $u^g_k = u^{g'}_k$, $0 \leq k \leq N-1$, $u^{g'}_{N-1} = g'_{N-1}(x^g_{N-1})$. It follows that

$$E\, c_k(x^g_k, u^g_k) = E\, c_k(x^{g'}_k, u^{g'}_k), \;\; 0 \leq k \leq N-2. \tag{2.19}$$

Adding (2.18) and (2.19) gives $J(g) > J(g')$ and so g cannot be optimal contrary to the hypothesis. Thus $g_{N-1}(x^g_{N-1})$ does achieve the infimum in (2.17) for $N-1$, and so $J^g_{N-1} = V_{N-1}(x^g_{N-1})$.

Now suppose by induction that $g_{k+1}(x^g_{k+1})$ achieves the infimum and that $J^g_{k+1} = V_{k+1}(x^g_{k+1})$. We prove this for k. Indeed, otherwise there is a function $g'_k: R^n \to U$ such that

$$c_k(x^g_k, g_k(x^g_k)) + E_{w_k} V_{k+1}[f_k(x^g_k, g_k(x^g_k), w_k)]$$

$$\geq c_k(x^g_k, g'_k(x^g_k)) + E_{w_k} V_{k+1}[f_k(x^g_k, g'_k(x^g_k), w_k)]$$

w.p.1. This inequality is strict with positive probability so that taking expectations gives

$$E\, c_k(x^g_k, g_k(x^g_k)) + E\, V_{k+1}(x^g_{k+1})$$

$$> E\, c_k(x^g_k, g'_k(x^g_k)) + E\, V_{k+1}(f_k(x^g_k, g'_k(x^g_k), w_k)). \tag{2.20}$$

Consider the policy $g' = \{g_0, \ldots, g_{k-1}, g'_k, g_{k+1}, \ldots, g_{N-1}\}$. Then certainly

$$E\, c_l(x^g_l, u^g_l) = E\, c_l(x^{g'}_l, u^{g'}_l), \;\; l = 0, \ldots, k-1. \tag{2.21}$$

Also, by the induction hypothesis, $g_{k+1}, \ldots, g_{N-1}$ achieve the infimum in (2.20), and so by Lemma (2.6)

$$E\, J^g_{k+1}(x^g_{k+1}) = E\, V_{k+1}(x^g_{k+1}), \;\; \text{and}$$

$$E\, J^{g'}_{k+1}(x^{g'}_{k+1}) = E\, V_{k+1}(x^{g'}_{k+1}). \tag{2.22}$$

From (2.20), (2.21), and (2.22) it follows that

$$J(g) = E \sum_{l=0}^{k-1} c_l(x_l^g, u_l^g) + E \, c_k(x_k^g, u_k^g) + V_{k+1}(x_{k+1}^g)$$

$$> E \sum_{l=0}^{k-1} c_l(x_l^g, u_l^g) + E \, c_k(x_k^g, u_k^{g'}) + V_{k+1}(x_{k+1}^{g'})$$

$$= J(g'),$$

and so g cannot be optimal contrary to hypothesis. Thus $g_k(x_k^g)$ must achieve the infimum in (2.20) and the result follows by induction. □

The result suggests the following dynamic programming algorithm for finding the optimal policy.

Algorithm (2.23)

First, define $V_N(x) := c_N(x)$; then find the function $g_{N-1} : R^n \to U$ by

$$g_{N-1}(x) = Arg \, Inf \, [c_{N-1}(x, u) +$$

$$E_{w_{N-1}} V_N(f_{N-1}(x, u, w_{N-1})) \mid u \in U],$$

and call the resulting value $V_{N-1}(x)$.

Second, find the function $g_{N-2} : R^n \to U$ by

$$g_{N-2}(x) = Arg \, Inf \, [c_{N-2}(x, u) +$$

$$E_{w_{N-2}} V_{N-1}(f_{N-2}(x, u, w_{N-2})) \mid u \in U],$$

and call this resulting value $V_{N-2}(x)$.

Proceeding in this way we obtain $g_{N-1}, V_{N-1}, \ldots, g_0, V_0$. It is evident from Theorem (2.15) that the Markov policy $\{g_k\}$ obtained in this way is optimal and $V_k(x_k^g) = J(x_k^g)$.

Definition (2.24)

The functions $V_k(x)$, $0 \le k \le N$, obtained here are called the **value**, or optimal value, functions.

Exercise (2.25)

Show that the feedback control g obtained by the algorithm is optimal for all starting times and starting states, i.e., for any initial time k and (random) state x_k, the policy $\{g_k, \ldots, g_{N-1}\}$ is optimal. Moreover, the minimum cost is $E \, V_k(x_k)$.

Remark (2.26)

The result reveals the important property that the optimal policy at time k depends only on the current state x_k and is independent of the previous values $x_{k-1}, \ldots, x_0, u_{k-1}, \ldots, u_0$ of the state and control. For this property to hold the state must be observed. This does not occur in the partial observation case, which makes the search for an optimal policy in that case more difficult.

The algorithm (2.23) is not easy to implement since at each step we have to carry out the minimization (2.24) for every value of x. Therefore, if x takes values in R^n, for example, then we may have to use some approximation procedure. Sometimes it is possible to get a closed form expression for $V_k(x)$ and solve (2.23) exactly. The most important instance of this is when the f_k are linear, the basic random variables are all Gaussian and the cost functions c_k are all quadratic. This so-called LQG case will be studied later. Another instance where no approximation may be needed is when the state takes on only a finite number of different values. We study this next, assuming a controlled Markov chain model.

3. Case of complete information II

Consider a stochastic system whose state x_k takes on only finitely many values $i = 1, \ldots, I$. The control u_k is selected from a control constraint set U. The system is described by the state transition probability which can be represented by the matrix $P(u) := \{P_{ij}(u), 1 \le i, j \le I\}$, where

$$P_{ij}(u) := P_{x_{k+1} \mid x_k, u_k}(j \mid x_k = i, u_k = u)$$

$$= Prob\{x_{k+1} = j \mid x_k = i, u_k = u\}. \tag{3.1}$$

As before, for any feedback policy $g = \{g_0, \ldots, g_{N-1}\}$ the cost is

$$J(g) = E\left\{\sum_{k=0}^{N-1} c_k(x_k^g, u_k^g) + c_N(x_N^g)\right\}.$$

In terms of the notation (3.1), Theorem (2.18) can be expressed as follows.

Theorem $\hspace{12cm}$ (3.2)
Define recursively the functions $V_k(i)$, $0 \le k \le N$, $1 \le i \le I$ by

$$V_N(i) := c_N(i), \tag{3.3}$$

$$V_k(i) := \underset{u \in U}{Inf}\left\{c_k(i, u) + \sum_{j=1}^{I} V_{k+1}(j)P_{ij}(u)\right\} \tag{3.4}$$

(1) Let $g \in G$. Then $V_k(x_k^g) \le J_k^g$, w.p.1.
(2) A Markov policy $g = \{g_0, \ldots, g_{N-1}\}$ is optimal if, for each i, the infimum in (3.4) is achieved at $g_k(i)$, and then $V_k(x_k^g) = J_k^g$ w.p.1 and $J^* = J(g) = E\, V_0(x_0)$.

We can rewrite (3.3) and (3.4) more conveniently. Define the I-dimensional column vectors $V_k := (V_k(1), \ldots, V_k(I))^T$, $c_N := (c_N(1), \ldots, c_N(I))^T$, $c_k(u) := (c_k(1, u), \ldots, c_k(I, u))^T$. Then (3.3), (3.4) can be written in vector notation as

$$V_N = c_N \tag{3.5}$$

$$V_k = \underset{u \in U}{Inf} \{c_k(u) + P(u)V_{k+1}\}, \tag{3.6}$$

where the infimum in u is taken separately for each component of the vector equation.

4. Case of partial information

For the stochastic system model

$$x_{k+1} = f_k(x_k, u_k, w_k), \tag{4.1a}$$

$$y_k = h_k(x_k, v_k), \tag{4.1b}$$

we know that for any control law g

$$P_{x_{k+1} \mid x_k, \ldots, x_0, u_k, \ldots, u_0} = P_{x_{k+1} \mid x_k, u_k}.$$

This means that given the present values x_k and u_k, the future value of x_{k+1} is independent of the past values of the state and control. In other words, in order to predict the value of the state at $k + 1$ it is enough to know the value of the state and control at k and the previous values are irrelevant. Indeed, this is a property that we demand of something called state. It is clear that to use this property the state must be observed. But this is not possible if the observation available $y_k \neq x_k$. In this case we would look for another property, namely, that the state at time k should be computable from the information available at time k which is $y^k = (y_0, \ldots, y_k)$, and $u^{k-1} := (u_0, \ldots, u_{k-1})$. This discussion suggests the following definition.

Definition (4.2)

ζ_k is an **information state** for the stochastic system (4.1) if (1) ζ_k is a function of y^k, u^{k-1}, and (2) ζ_{k+1} can be determined from ζ_k, y_{k+1}, and u_k.

The trivial case, $\zeta_k \equiv$ constant, satisfies this definition. Our interest, however, is to find information states that contain all relevant information in the observations. We show next that the conditional distribution of x_k given y^k, u^{k-1} is such an information state.

5. Conditional probability as information state

The notation is simpler if all the probability distributions have densities, so we assume this. Let $g = \{g_0, g_1, \ldots\}$ be a feedback policy for the system (4.1). Let $\{x_k\}$, $\{y_k\}$, and $\{u_k\}$ be the resulting processes. Recall that $u_k = g_k(y^k)$. Denote $z^k := (y^k, u^{k-1})$, so z^k consists of all the observations available at time k, before the control u_k is selected. For simplicity denote

$$p^g_{k \mid k}(x_k \mid z^k) := p_{x_k \mid z^k}(x_k \mid z^k), \tag{5.1}$$

$$p^g_{k+1 \mid k}(x_{k+1} \mid z^k, u_k) := p_{x_{k+1} \mid z^k, u_k}(x_{k+1} \mid z^k, u_k). \tag{5.2}$$

It is an important fact that these densities do not depend on the policy g.

Exercise (5.3)

Show in a way similar to Lemma (2.6.6) that

$$p^g(x_{k+1} \mid x_k, z^k, u_k) = p(x_{k+1} \mid x_k, u_k)$$

where the right-hand side is just the state transition probability. In particular, it does not depend on g.

Exercise (5.4)

Show that

$$p^g(y_{k+1} \mid x_{k+1}, z^k, u_k) = p(y_{k+1} \mid x_{k+1})$$

where the right-hand side is the output transition probability.
[Hint: Use $y_{k+1} = h_{k+1}(x_{k+1}, v_{k+1})$ and v_{k+1} independent of $\{x_{k+1}, z^k, u_k\}$.]

We now develop a recursive relation for calculating the densities (5.1) and (5.2).

$$p^g_{k+1 \mid k+1}(x_{k+1} \mid z^{k+1})$$

$$= \frac{p^g(y_{k+1} \mid x_{k+1}, z^k, u_k) p^g(x_{k+1}, z^k, u_k)}{p^g(z^{k+1})},$$

by Bayes' rule. By (5.4) this equals

$$= \frac{p(y_{k+1} \mid x_{k+1}) p^g(x_{k+1}, z^k, u_k)}{\int p^g(x_{k+1}, z^{k+1}) dx_{k+1}}. \tag{5.5}$$

Next

$$p^g(x_{k+1}, z^{k+1})$$

$$= p^g(y_{k+1} \mid x_{k+1}, z^k, u_k) \, p^g(x_{k+1}, z^k, u_k)$$

$$= p(y_{k+1} \mid x_{k+1}) \, p^g(x_{k+1} \mid z^k, u_k) \, p^g(z^k, u_k).$$

Substituting into (5.5) gives

$$p^g_{k+1 \mid k+1}(x_{k+1} \mid z^{k+1})$$

$$= \frac{p(y_{k+1} \mid x_{k+1}) \, p^g_{k+1 \mid k}(x_{k+1} \mid z^k, u_k)}{\int p(y_{k+1} \mid x_{k+1}) \, p^g_{k+1 \mid k}(x_{k+1} \mid z^k, u_k) dx_{k+1}}, \tag{5.6}$$

which we write as

$$p_{k+1|k+1}^g(x_{k+1} \mid z^{k+1})$$
$$= \Phi_k \ [p_{k+1|k}^g(\cdot \mid z^k, u_k), y_{k+1}](x_{k+1}), \qquad (5.7)$$

with the function Φ_k chosen appropriately. Note that Φ_k depends on the entire function $p_{k+1|k}^g(\cdot \mid z^k, u_k)$ and not just its value at any particular x_{k+1}.

On the other hand,

$$p_{k+1|k}^g(x_{k+1} \mid z^k, u_k)$$
$$= \int p^g(x_{k+1} \mid x_k, z^k, u_k) \, p^g(x_k \mid z^k, u_k) dx_k$$
$$= \int p(x_{k+1} \mid x_k, u_k) \, p_{k|k}^g(x_k \mid z^k) dx_k,$$

by (5.3), which we write as

$$p_{k+1|k}^g(x_{k+1} \mid z^k, u_k) = \Psi_k \ [p_{k|k}^g(\cdot \mid z^k), u_k](x_{k+1}), \qquad (5.8)$$

with the function Ψ_k chosen appropriately. Equations (5.7) and (5.8) can be solved recursively starting with either of the following initial conditions

$$p_{0|0}^g(x_0 \mid y_0) = \frac{p(y_0 \mid x_0) \, p(x_0)}{\int p(y_0 \mid x_0) \, p(x_0) dx_0}, \qquad (5.9a)$$

$$p_{0|-1}^g(x_0) = p(x_0). \qquad (5.9b)$$

Lemma (5.10)

The conditional probability density $p_{k|k}^g(\cdot \mid z^k)$ does not depend on g. It is an information state in that it can be evaluated from z^k and there is a function T_k such that

$$p_{k+1|k+1}(\cdot \mid z^{k+1}) = T_k \ [p_{k|k}(\cdot \mid z^k), y_{k+1}, u_k]. \qquad (5.11)$$

The initial value $p_{0|0}(\cdot \mid z^0) = p_{0|0}(\cdot \mid y_0)$ is given by (5.9a).

Proof

Combining (5.7) and (5.8) we obtain (see diagram on p.82)

$$p_{k+1|k+1}^g(\cdot \mid z^{k+1}) = \Phi_k \ [\Psi_k(p_{k|k}^g(\cdot \mid z^k), u_k), y_{k+1}]$$
$$= T_k \ [p_{k|k}^g(\cdot \mid z^k), y_{k+1}, u_k],$$

which defines the function T_k. This function does not depend on g and from (5.9a) it is clear that the initial condition does not depend on g either. It follows that $p_{k|k}^g$ cannot depend on g.   □

The proof of the next lemma is left as an exercise.

Lemma (5.12)

The conditional probability density $p_{k+1|k}^g(\cdot \mid z^k, u_k)$ does not depend on g. There is a function S_k such that

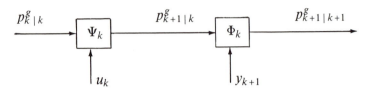

$$p_{k+1|k}(\,.\ |\ z^k, u_k) = S_k\,[p_{k|k-1}(\,.\ |\ z^{k-1}, u_{k-1}), y_k, u_k].$$

The initial value $p_{0|-1}(\,.\ )$ is given by (5.9b).

Remark (5.13)

Note that the information state is the entire probability density *function* $p_{k|k}(\,.\ |\ z^k)$ and not its value at any particular x_k. This is so because to calculate the left–hand side of (5.7) or (5.8) for any particular x_k we need the entire functions $p_{k+1|k}(\,.\ |\ z^k, u_k)$, $p_{k|k}(\,.\ |\ z^k)$. Thus this information state takes its values in the space of all probability densities on R^n (if x is n-dimensional). This is generally an infinite-dimensional space. We shall see later that in the linear, Gaussian case, the information state becomes finite-dimensional because $p_{k|k}(\,.\ |\ z^k)$ turns out to be Gaussian and so it can be summarized by its mean and covariance. Another instance where we get a finite-dimensional information state is when x_k takes its values in a finite set. We study this next.

6. Information state for a controlled Markov chain

Consider a stochastic system whose state x_k takes values $i = 1, \ldots, I$. The state transition is described by the transition probability matrix $P(u) = \{P_{ij}(u), 1 \le i, j \le I\}$. The output at time k is y_k, and for simplicity denote the output transition probability by

$$p(y_k \mid i) = p_{y_k \mid x_k}(y_k \mid x_k = i).$$

From (5.7) and (5.8) we have

$$p_{k+1|k+1}(i \mid z^{k+1}) = \frac{p(y_{k+1} \mid i)\, p_{k+1|k}(i \mid z^k, u_k)}{\sum_i p(y_{k+1} \mid i)\, p_{k+1|k}(i \mid z^k, u_k)},$$

$$p_{k+1|k}(i \mid z^k, u_k) = \sum_j p_{k|k}(j \mid z^k)\, P_{ji}(u_k).$$

Combining these yields

$$p_{k+1|k+1}(i \mid z^{k+1}) = \frac{\displaystyle\sum_j p_{k|k}(j \mid z^k)\, P_{ji}(u_k)\, p(y_{k+1} \mid i)}{\displaystyle\sum_i \sum_j p_{k|k}(j \mid z^k)\, P_{ji}(u_k)\, p(y_{k+1} \mid i)}. \qquad (6.1)$$

This can be written in a more revealing and compact manner. Define the
I-dimensional *row* vector

$$\pi_{k \mid k}(z^k) := (p_{k \mid k}(1 \mid z^k), \ldots, p_{k \mid k}(I \mid z^k)),$$

the *column* vector $\underline{1} := (1, \ldots, 1)^T$, and the $I \times I$ diagonal matrix $D(y_k)$
whose ith entry is $p(y_k \mid i)$. Then (6.1) is equivalent to

$$\pi_{k+1 \mid k+1}(z^{k+1}) = \frac{\pi_{k \mid k}(z^k) \, P(u_k) \, D(y_{k+1})}{\pi_{k \mid k}(z^k) \, P(u_k) \, D(y_{k+1}) \, \underline{1}} \tag{6.2}$$

$$=: T_k \, [\pi_{k \mid k}(z^k), \, y_{k+1}, \, u_k],$$

and the initial condition is

$$\pi_{0 \mid 0}(z^0) = \frac{\pi_0 \, D(y_0)}{\pi_0 \, D(y_0) \, \underline{1}}, \tag{6.3}$$

where $\pi_0 := (p(x_0 = 1), \ldots, p(x_0 = I))$ is the initial distribution of the
state.

Remark (6.4)
The denominator in the right-hand side of (6.2) and (6.3) is simply a normalizing factor to guarantee that we have a probability. If we did not
have this factor then (6.3) would be a linear equation in the state. This
suggests the next result.

Lemma (6.5)
Consider the linear system in the state variable $\rho_{k \mid k}(z^k)$ given by

$$\rho_{k+1 \mid k+1}(z^{k+1}) = \rho_{k \mid k}(z^k) \, P(u_k) \, D(y_{k+1}) \tag{6.6}$$

with the initial condition

$$\rho_{0 \mid 0}(z^0) = \pi_0 \, D(y_0). \tag{6.7}$$

Then $\rho_{k \mid k}(z^k)$ is an information state. Moreover,

$$\pi_{k \mid k}(z^k) = \frac{\rho_{k \mid k}(z^k)}{\rho_{k \mid k}(z^k) \, \underline{1}}. \tag{6.8}$$

Proof
For convenience let $r_k(z^k) := \rho_{k \mid k}(z^k) \, \underline{1}$. Certainly we have

$$\pi_{0 \mid 0}(z^0) = \frac{\rho_{0 \mid 0}(z^0)}{r_0(z^0)}.$$

Suppose next that (6.8) holds for k; then, from (6.2),

$$\pi_{k+1 \mid k+1}(z^{k+1}) = \frac{r_k(z^k) \, \rho_{k \mid k}(z^k) \, P(u_k) \, D(y_{k+1})}{r_k(z^k) \, \rho_{k \mid k}(z^k) \, P(u_k) \, D(y_{k+1}) \, \underline{1}}$$

$$= \frac{\rho_{k+1|k+1}(z^{k+1})}{\rho_{k+1|k+1}(z^{k+1}) \, \underline{1}}$$

$$= \frac{\rho_{k+1|k+1}(z^{k+1})}{r_{k+1}(z^{k+1})} \, ,$$

and so (6.8) holds for $k + 1$.   □

Exercise   (6.9)
Consider (5.11). Show that there is an unnormalized density $\rho_{k|k-1}(\cdot \mid z^{k-1}, u_{k-1})$ such that it can be obtained from $\rho_{k-1|k-2}(\cdot \mid z^{k-2}, u_{k-2})$ by a *linear* transformation and

$$p_{k|k-1}(x_k \mid z^{k-1}, u_{k-1}) = \frac{\rho_{k|k-1}(x_k \mid z^k, u_{k-1})}{\int \rho_{k|k-1}(x_k \mid z^k, u_{k-1}) dx_k} \, .$$

Remark   (6.10)
For many purposes the unnormalized probability $\rho_{k|k}$ is as useful as $p_{k|k}$ or $\pi_{k|k}$. Of course it is much more convenient to deal with the unnormalized version since its evolution is described by a linear equation.

7. Dynamic programming with partial information

We study the problem of optimal control of a stochastic system whose state x_k takes values in the set $\{1, \ldots, I\}$. The system is described by the state transition matrix $P(u)$ and the output transition probability $p(y_k \mid i)$. A feedback policy $g = \{g_0, \ldots, g_{N-1}\}$ is feasible if

$$u_k = g_k(y^k) = g_k(y_0, \ldots, y_k) \in U$$

where U is the control constraint set. Let G be the set of feasible policies. The aim is to find the policy g that minimizes

$$J(g) := E^g \left\{ \sum_{k=0}^{N-1} c_k(x_k, u_k) + c_N(x_N) \right\}.$$

Since the state is not directly observed we cannot expect Markovian policies to be optimal. But from Lemma (5.10) we expect that an optimal policy will be a function only of the information state $\pi_{k|k}$. We state this precisely.

A policy $g = \{g_0, \ldots, g_N\}$ is said to be **separated** if g_k depends on y^k only through the information state, i.e., $u_k = g_k(\pi_{k|k}(\cdot \mid z^k))$. Let G_S denote the set of separated policies. The adjective "separated" is used to signal the fact that in implementing such a policy one first calculates the conditional probability and then chooses the control -- the task of estimation and control are separated. See the following diagram.

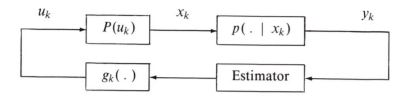

We will show that an optimal control is separated and obtain a recursion relation similar to Theorem (3.2). We need some notation. Let Π be the set of all probability row vectors $\pi = (\pi(1), \ldots, \pi(I))$ where $\pi(i) \geq 0$ and $\sum \pi(i) = 1$.

Theorem   (7.1)
Define recursively the functions $V_k(\pi)$, $0 \leq k \leq N$, $\pi \in \Pi$, by

$$V_N(\pi) := E\{\, c_N(x_N) \mid \pi_{N \mid N} = \pi \,\},\tag{7.2}$$

$$V_k(\pi) := \underset{u \in U}{Inf}\ E\{c_k(x_k, u)$$

$$+ V_{k+1}(T_k[\pi, y_{k+1}, u]) \mid \pi_{k \mid k}(z^k) = \pi \,\}.\tag{7.3}$$

(1) Let $g \in G$. Then

$$V_k(\pi_{k \mid k}(z^{g,k})) \leq J_k^g$$

$$:= E\{\, \sum_{l=k}^{N-1} c_l(x_l^g, u_l^g) + c_N(x_N^g) \mid z^{g,k} \,\}.\tag{7.4}$$

(2) Let g be a separated policy such that for all π in Π, $g_k(\pi)$ achieves the infimum in (7.3); then g is optimal and

$$V_k(\pi_{k \mid k}(z^{g,k})) = J_k^g, \ w.p.\ 1.\tag{7.5}$$

Note
The right-hand side in (7.2) is simply $\sum_i c_N(i)\pi(i)$. The expectation in (7.3) is

$$\sum_i c_k(i,u)\pi(i)$$

$$+ \int V_{k+1}(T_k[\pi, y_{k+1}, u])p(y_{k+1} \mid \pi_{k \mid k} = \pi, u)dy_{k+1}.$$

Here, from (6.2),

$$T_k[\pi, y_{k+1}, u] = \frac{\pi\, P(u)\, D(y_{k+1})}{\pi\, P(u)\, D(y_{k+1})\, \underline{1}},$$

and $p(y_{k+1} \mid \pi, u)$ is the conditional probability density of y_{k+1} given $\pi_{k \mid k}(z^k) = \pi$ and $u_k = u$. Hence

$$p(y_{k+1} \mid \pi, u) = \sum_j p(y_{k+1} \mid j)p(x_{k+1} = j \mid \pi, u)$$

$$= \sum_j \sum_i p(y_{k+1} \mid j)\,\pi(i)\,P_{ij}(u).$$

Proof

We first prove (1) by induction. For $k = N$

$$J_N^g = E^g\{\,c_N(x_N) \mid z^N\,\} = \sum_i c_N(i)\pi_{N \mid N}(z^N)(i), \qquad (7.6)$$

so (7.4) holds with equality. Suppose now that (7.4) holds for $k + 1$. Then

$$J_k^g = E^g\{c_k(x_k, u_k)$$
$$+ E^g\{\sum_{k+1}^{N-1} c_l(x_l, u_l) + c_N(x_N) \mid z^{k+1}\,\} \mid z^k\,\} \qquad (7.7)$$

$$\geq E^g\{c_k(x_k, u_k)$$

$$+ V_{k+1}(\pi_{k+1 \mid k+1}(z^{k+1})) \mid z^k\}, \quad \text{by hypothesis}$$

$$= E^g\{E^g\,\{c_k(x_k, u_k)$$

$$+ V_{k+1}(\pi_{k+1 \mid k+1}(z^{k+1})) \mid z^k, u_k\,\} \mid z^k\,\}$$

$$= E^g\{E^g\{\,c_k(x_k, u_k)$$

$$+ V_{k+1}(T_k[\pi_{k \mid k}(z^k), y_{k+1}, u_k]) \mid \pi_{k \mid k}(z^k), u_k\,\} \mid z^k\,\}$$

$$\geq E^g\{V_k(\pi_{k \mid k}(z^k)) \mid z^k\} = V_k(\pi_{k \mid k}(z^k))$$

using (7.3) and so (7.4) holds for k.

To prove (2) suppose $g_k(\pi)$ achieves the infimum. We first prove (7.5) by induction. From (7.6) we know that (7.5) holds for $k = N$; suppose it holds for $k + 1$. Then in the derivation of (7.7) both inequalities become equalities, the first by the induction hypothesis and the second because u_k achieves the infimum. Thus (7.5) holds for all k.

It remains to show that g is optimal. Taking $k = 0$ in (7.5) gives

$$V_0(\pi_0 \mid 0) = J_0^g.$$

Taking expectations gives

$$J(g) = EJ_0^g = E \ V_0(\pi_0 \mid 0).$$

On the other hand, for any g' in G, setting $k = 0$ in (7.4) and taking expectations gives

$$J(g') \geq E \ V_0(\pi_0 \mid 0).$$

Therefore g is optimal. □

8. The concept of dual control

We saw in Section 5 that when $x_k \in R^n$ and $y_k \neq x_k$, the information state $p_{k \mid k}(\, . \mid z^k)$ is usually infinite-dimensional. The dynamic programming solution of the optimal control problem is then computationally much more difficult than in the complete information case. This difficulty is easily seen when we compare Theorems (3.2) and (7.1). In the former, one needs to compute for each k the value function $V_k(i)$, $1 \leq i \leq I$, that is, I numbers; in the latter, one needs to compute the function $V_k(\pi)$ defined on the I-dimensional set Π.

The additional computational burden has a deeper significance. In the case of complete information, the state x_k is known, and the effect of control is to alter the future evolution of the state. In the partial information case, the effect of control is to alter the information state $p_{k \mid k}(z^k)$. This effect occurs in two ways: (1) the control can alter the future values of the actual state, x_k, and (2) it can alter future values of the available information z^k and hence the knowledge that one has of x_k through $p_{k \mid k}(z^k)$. It is helpful to have an accurate estimate of x_k in order to control it effectively—i.e., with low cost. Therefore the choice of u_k must take into account two factors: how does u_k affect $x_{k+1}, x_{k+2}, \ldots$, and how does it affect knowledge of these values through $z^k, z^{k+1}, \ldots$. This aspect of learning about the system state is called the **dual aspect**, or function, of control. In the case of complete information the control does not have this function.

The dual aspect is revealed more clearly in the following important special case of partial information. Suppose again that x_k takes values in $\{1, \ldots, I\}$, and u_k in U. The state transition probability matrix is $P(u, \theta) = \{P_{ij}(u, \theta), 1 \leq i, j \leq I\}$. Here θ is an unknown parameter, x_k is observed, and $g = \{g_0, \ldots, g_N\}$ has to be chosen so that $u_k = g_k(x^k) = g_k(x_0, \ldots, x_k)$ is in U and minimizes

$$J(g) := E^g \{ \sum_0^{N-1} c_k(x_k, u_k) + c_N(x_n) \}. \tag{8.1}$$

At first sight this appears to fit the case of complete information, but this is not so since θ—hence the transition probability matrix—is unknown. The problem now seems to lie outside the range we have considered. A way of transforming it into the class of problems considered earlier is to treat θ as another state variable, so that the augmented state is now $\{x_k, \theta_k\}$. We want $\theta_{k+1} = \theta_k$, since the parameter does not change. We shall suppose that we know a priori that the true parameter value belongs to the *finite* set $\Theta = \{\theta^1, \ldots, \theta^J\}$.

The transition probability for the augmented state is then

$$Prob\{x_{k+1} = j, \theta_{k+1} = \beta \mid x_k = i, \theta_k = \alpha, u_k = u\}$$

$$= P_{ij}(u, \alpha)\, \delta(\beta, \alpha) := P_{i\alpha, j\beta}(u), \tag{8.2}$$

where $\delta(\beta, \alpha)$ is the Kronecker delta, i.e., $\delta(\beta, \alpha) = 1$ or 0 accordingly as $\beta = \alpha$ or $\beta \neq \alpha$. To complete the specification of the augmented system, we need to give the distribution of the initial state (x_0, θ_0). It is natural to assume that x_0, θ_0 are independent, so

$$p(x_0, \theta_0) = p(x_0)p(\theta_0),$$

where $p(\theta_0 = \theta)$ is the analyst's prior guess of the likelihood that the true parameter value is θ for each θ in Θ. Notice that the problem is one of partial information since the augmented state (x_k, θ_k) is not observed.

We want to apply Theorem (7.1). The information state is now the conditional probability distribution

$$p_{k \mid k}(i, \alpha \mid z^k) := Prob\{x_k = i, \theta_k = \alpha \mid z^k\}.$$

Since x_k is a component of $z^k = x^k$,

$$p_{k \mid k}(i, \alpha \mid z^k) = \delta(i, x_k)\, \pi_{k \mid k}(\alpha \mid z^k), \tag{8.3}$$

where

$$\pi_{k \mid k}(\alpha \mid z^k) := Prob\{\theta_k = \alpha \mid z^k\}. \tag{8.4}$$

We now calculate the transition function T_k, where

$$p_{k+1 \mid k+1}(\cdot \mid z^{k+1}) = T_k[p_{k \mid k}(\cdot \mid z^k), x_{k+1}, u_k].$$

From (8.3) and (6.1),

$$\delta(i, x_{k+1})\, \pi_{k+1 \mid k+1}(\alpha \mid z^{k+1}) = \frac{N}{D}, \tag{8.5}$$

where the numerator is

$$N = \sum_j \sum_\beta p_{k \mid k}(j, \beta \mid z^k)$$

$$\times P_{j\beta, i\alpha}(u_k)\, p(x_{k+1} \mid x_{k+1} = i, \theta_{k+1} = \alpha) \tag{8.6}$$

$$= \sum_j \sum_\beta \delta(j, x_k) \, \pi_{k \mid k}(\beta \mid z^k) \, P_{ji}(u_k, \beta) \delta(\beta, \alpha) \delta(i, x_{k+1})$$

$$= P_{x_k x_{k+1}}(u_k, \alpha) \, \pi_{k \mid k}(\alpha \mid z^k) \, \delta(i, x_{k+1}),$$

and the denominator is

$$D = \sum_i \sum_\alpha N = \sum_\alpha P_{x_k x_{k+1}}(u_k, \alpha) \, \pi_{k \mid k}(\alpha \mid z^k). \tag{8.7}$$

From (8.5), (8.6), and (8.7) we get

$$\pi_{k+1 \mid k+1}(\theta \mid z^{k+1}) = \frac{P_{x_k x_{k+1}}(u_k, \theta) \, \pi_{k \mid k}(\theta \mid z^k)}{\sum_\theta P_{x_k x_{k+1}}(u_k, \theta) \, \pi_{k \mid k}(\theta \mid z^k)} \tag{8.8}$$

$$=: S_k [\pi_{k \mid k}(\,. \mid z^k), x_k, x_{k+1}, u_k](\theta).$$

$\pi_{k \mid k}$ is the prior probability of θ_{k+1} at time $k + 1$ and $\pi_{k+1 \mid k+1}$ is the posterior probability given the new observations u_k and the transition x_k, x_{k+1}.

It is evident from (8.3) that the pair $(x_k, \pi_{k \mid k})$ is equivalent to the information state $p_{k \mid k}$. This new state belongs to the set $\{1, \ldots, I\} \times \Pi$, where Π is the set of row vectors $\{\pi\}$ = $\{\pi(\theta), \theta \in \Theta \mid \pi(\theta) \geq 0, \sum \pi(\theta) = 1\}$. Theorem (7.1) now gives the next result.

Theorem (8.9)
Define recursively the functions $V_k(i, \pi)$, $0 \leq k \leq N$, $1 \leq i \leq I$, and $\pi \in \Pi$, by

$$V_N(i, \pi) := c_N(i), \tag{8.10}$$

$$V_k(i, \pi) = \underset{u}{Inf} \, [c_k(i, u) + E\{V_{k+1}(x_{k+1}, S_k[\pi, i, x_{k+1}, u])$$

$$\mid x_k = i, \pi_{k \mid k}(\,. \mid z^k) = \pi\}]. \tag{8.11}$$

(1) Let g be arbitrary; then

$$V_k(x_k^g, \pi_{k \mid k}(\,. \mid z^{g,k})) \leq J_k^g, \text{ w.p. } 1. \tag{8.12}$$

(2) Let $g = \{g_0, \ldots, g_{N-1}\}$ be any separated policy such that $u_k = g_k(i, \pi)$ achieves the infimum in (8.11); then g is optimal and gives equality in (8.12).

Note
The second term on the right in (8.11) is simply

$$\sum_j V_{k+1}(j, S_k[\pi, i, j, u]) \, p(x_{k+1} = j \mid \pi_{k \mid k} = \pi, x_k = i, u_k = u)$$

$$= \sum_{\theta} \sum_{j} P_{ij}(u, \theta) \, \pi(\theta) \, V_{k+1}(j, S_k[\pi, i, j, u])$$

Remark   (8.13)
The dependence on u_k of the (information state) transition function S_k shows how learning about the unknown parameter θ depends on the control chosen. The cost of uncertainty about θ is reflected in the dependence on π of the value function $V(i, \pi)$. In practice one expects that for fixed i, $V_k(i, \pi)$ decreases as the spread of the distribution π decreases; then (8.11) shows that the choice of u_k must balance the decrease in cost achieved by reducing the spread in π with the possible increase in the cost due to changes in x.

9. Notes

1. Many books are devoted to dynamic programming; see, for example, Bellman and Dreyfus (1962), Bertsekas (1978), Ross (1983), and Whittle (1982). In addition to the basic results derived in this chapter, these books study several applications including scheduling, inventory control, stopping times, and sequential hypothesis testing problems. Applications to macroeconomic policy are studied in Chow (1981) and Kendrick (1981). Most books deal only with the complete information case. We see from Section 5, however, that the case of partial observations reduces to that of complete information if we define a new state to be the conditional distribution of the original state given the available observations.

2. In our discussion we paid no attention to technical measurability problems. In order that the expected cost corresponding to a policy be well-defined, the policy must be an appropriately measurable function; for example, $g_k(x)$ must be a measurable function of x. Thus when taking the infimum in (2.17) for each x, the resulting policy should be chosen to be measurable in x. These issues are treated in Bertsekas and Shreve (1978).

3. Equations (5.11) and (6.1) are sometimes called nonlinear filtering equations—nonlinear refers to the nonlinear state equations, and filtering refers to the fact that in these equations one filters the noisy observations to estimate the distribution of the state. The fact that filtering equations are linear in the unnormalized conditional distribution, Lemma (6.5), has not been previously observed in the literature. The analogous result, known as Zakai's equation, is of major importance in the theory of nonlinear filtering for continuous time stochastic control. A recent review of that theory is given by Kallianpur and Karandikar (1983).

4. The optimality of separated policies, Theorem (7.1), was first proved by Striebel (1965). There is a martingale formulation of the fundamental results of dynamic programming, Theorems (2.15) and (7.1), which has proved to be particularly useful in continuous time problems; see Davis and Varaiya (1973), and Davis (1979).

5. The concept of dual control is generally attributed to Fel'dbaum (1965). When one is controlling a system with unknown parameters, as in the example in Section 8, it seems intuitive that the control action would be more aggressive than if the parameter is known, since one is trying to probe the system to estimate the unknown parameter. For attempts to classify control actions along these lines see Jacobs and Patchell (1972) and Kendrick (1981). For approximations to the dynamic programming equations see Tse, Bar-Shalom and Meier (1973).

6. For a discussion of the applications of dynamic programming, see Bellman (1957).

10. References

[1] R. Bellman (1957), *Dynamic programming,* Princeton University Press, Princeton, NJ.

[2] R.E. Bellman and S.E. Dreyfus (1962), *Applied dynamic programming,* Princeton University Press, Princeton, NJ.

[3] D. Bertsekas (1976), *Dynamic programming and stochastic control,* Academic Press, New York.

[4] D. Bertsekas and S. Shreve (1978), *Stochastic optimal control : the discrete time case,* Academic Press, New York.

[5] G. Chow (1981), *Econometric analysis by control methods,* McGraw-Hill, New York.

[6] M.H.A. Davis (1979), "Martingale methods in stochastic control," in M. Kohlmann and W. Vogel (eds), *Stochastic Control Theory and Stochastic Differential Systems,* Lecture Notes in Control and Information Sciences, Vol 16, 85-117, Springer Verlag, Berlin.

[7] M.H.A. Davis and P. Varaiya (1973), "Dynamic programming conditions for partially observable stochastic systems," *SIAM J. Control,* Vol 11, 226-261.

[8] A.A. Fel'dbaum (1965), *Optimal control systems,* Academic Press, New York.

[9] O.L.R. Jacobs and J.W. Patchell (1972), "Caution and probing in stochastic control," *Int. J. Control.* Vol. 16(1), 189-199.

[10] G. Kallianpur and R.L. Karandikar (1983), "Some recent developments in nonlinear filtering theory," *Acta Applicandae*

Mathematicae, Vol 1(4), 399-434.

[11] D. Kendrick (1981), *Stochastic control for economic models,* McGraw-Hill, New York.

[12] S. Ross (1983), *Introduction to stochastic dynamic programming,* Academic Press, New York.

[13] C. Striebel (1965), "Sufficient statistics in the optimum control of stochastic systems," *J. Mathematical Analysis and Applications,* Vol 12(3), 576-592.

[14] E. Tse, Y. Bar-Shalom and L. Meier (1973), "Wide sense adaptive dual control of stochastic nonlinear systems," *IEEE Transactions on Automatic Control,* Vol. AC-18(2), 98-108.

[15] P. Whittle (1982), *Optimization over time,* John Wiley, New York.

CHAPTER 7
LINEAR SYSTEMS: ESTIMATION AND CONTROL

In this chapter we study the special case of partial information when the state equations are linear and the input and observation noise processes are independent Gaussian processes. The conditional distribution of the state is given by the Kalman filter and is finite-dimensional. The dual function of control is absent. Consequently, the calculation of the optimal control becomes very easy; when the cost is quadratic, a closed form solution is obtained. Minimum variance control laws for ARMAX models are derived. Lastly we present the Levinson algorithm and the lattice filter for estimation problems that arise in signal processing applications.

1. The linear Gaussian model

Consider the stochastic system

$$x_{k+1} = A_k x_k + B_k u_k + G_k w_k,$$
$$y_k = C_k x_k + H_k v_k,$$

where $x_k \in R^n$, $u_k \in R^m$, $y_k \in R^p$, $w_k \in R^g$, $v_k \in R^h$; A_k, B_k, G_k, C_k, and H_k are possibly time-varying, known matrices of appropriate dimension. The basic random variables $\{x_0, w_0, \ldots, v_0, \ldots\}$ are all independent and Gaussian, with

$$x_0 \sim N(0, \Sigma_0), \quad w_k \sim N(0, Q), \quad v_k \sim N(0, R).$$

The covariance matrices Σ_0, Q, and R are all known.

The information available at time k is $z^k = (y^k, u^{k-1})$, and the information state is the conditional density $p_{k|k}(x_k \mid z^k)$. We want to calculate its transition function T_k specified in Lemma (6.5.10).

2. The case $u = 0$

Suppose that $u_k \equiv 0$. The system equations then become

$$x_{k+1} = A_k x_k + G_k w_k, \tag{2.1}$$
$$y_k = C_k x_k + H_k v_k. \tag{2.2}$$

The available information is $z^k = y^k$. The random variables x_k, x_{k+1}, and y^k are jointly Gaussian. Hence the conditional densities are also Gaussian; denote

$$p_{k\,|\,k}(x_k \mid y^k) \sim N(x_{k\,|\,k}, \Sigma_{k\,|\,k}), \quad \text{and}$$

$$p_{k+1\,|\,k}(x_{k+1} \mid y^k) \sim N(x_{k+1\,|\,k}, \Sigma_{k+1\,|\,k}).$$

By definition,

$$x_{k\,|\,k} := E\{x_k \mid y^k\}, \quad \text{and}$$

$$\Sigma_{k\,|\,k} := E\{(x_k - x_{k\,|\,k})(x_k - x_{k\,|\,k})^T \mid y^k\}. \tag{2.3}$$

That is, the conditional mean of x_k given y^k is $x_{k\,|\,k}$, while the conditional covariance is $\Sigma_{k\,|\,k}$. Similarly,

$$x_{k+1\,|\,k} := E\{x_{k+1} \mid y^k\}, \quad \text{and}$$

$$\Sigma_{k+1\,|\,k} := E\{(x_{k+1} - x_{k+1\,|\,k})(x_{k+1} - x_{k+1\,|\,k})^T \mid y^k\}. \tag{2.4}$$

To obtain the recursion rule T_k for the conditional densities one can apply the formulas (6.5.7) and (6.5.8). It is easier and more instructive to proceed directly using the following fact.

Lemma $\hspace{8cm}$ (2.5)

Let x, y be jointly Gaussian random variables, with $x \sim N(\bar{x}, \Sigma_x)$, $y \sim N(\bar{y}, \Sigma_y)$, $\Sigma_{xy} := cov(x, y) = E(x - \bar{x})(y - \bar{y})^T$. Let $\hat{x} := E\{x \mid y\}$, $\tilde{x} := x - \hat{x}$. Then

$$\hat{x} = \bar{x} + \Sigma_{xy} \Sigma_y^{-1} (y - \bar{y}). \tag{2.6}$$

Moreover $\tilde{x}$ is independent of y, consequently of $\hat{x}$, and $\tilde{x} \sim N(0, \Sigma_{\tilde{x}})$ where

$$\Sigma_{\tilde{x}} := \Sigma_x - \Sigma_{xy} \Sigma_y^{-1} \Sigma_{xy}^T. \tag{2.7}$$

Proof

Let $\hat{v} := \bar{x} + \Sigma_{xy} \Sigma_y^{-1}(y - \bar{y})$ and $\tilde{v} := x - \hat{v}$. Clearly $E\tilde{v} = 0$. Also

$$E\tilde{v}(y - \bar{y})^T = E(x - \bar{x})(y - \bar{y})^T - \Sigma_{xy}\Sigma_y^{-1}E(y - \bar{y})(y - \bar{y})^T$$

$$= 0,$$

which, since they are jointly Gaussian, proves that $\tilde{v}$ and y are independent. Hence $\hat{x} = E\{x \mid y\} = E\{\hat{v} + \tilde{v} \mid y\} = \hat{v} + E\{\tilde{v} \mid y\} = \hat{v}$. This proves (2.6). Finally, since $\hat{x}$ and $\tilde{x}$ are independent, $cov(x) = cov(\hat{x}) + cov(\tilde{x})$, from which follows (2.7) upon substituting from (2.6). $\hspace{2cm}$ $\square$

Exercise $\hspace{8cm}$ (2.8)

Let x, y, and z be jointly Gaussian with y and z independent. Let $\hat{x} := E\{x \mid y, z\}$, $\tilde{x} := x - \hat{x}$. Then

$$\hat{x} = \bar{x} + \Sigma_{xy}\Sigma_y^{-1}(y - \bar{y}) + \Sigma_{xz}\Sigma_z^{-1}(z - \bar{z}),$$

$$\Sigma_{\widetilde{x}} = \Sigma_x - \Sigma_{xy}\Sigma_y^{-1}\Sigma_{xy}^T - \Sigma_{xz}\Sigma_z^{-1}\Sigma_{xz}^T.$$

[Hint: Apply Lemma (2.5) to x and (y, z).]

We now obtain the transition function T_k in several steps.

Step 1

From (2.1),

$$E\{x_{k+1} \mid y^k\} = A_k E\{x_k \mid y^k\} + G_k E\{w_k \mid y^k\}.$$

The last term vanishes because w_k and y^k are independent, so

$$x_{k+1\mid k} = A_k x_{k\mid k}. \tag{2.9}$$

For convenience denote

$$\widetilde{x}_{k+1\mid k} := x_{k+1} - x_{k+1\mid k}, \quad \widetilde{x}_{k\mid k} := x_k - x_{k\mid k}. \tag{2.10}$$

From (2.10), (2.1), and (2.9) we have

$$\widetilde{x}_{k+1\mid k} = A_k x_k + G_k w_k - A_k x_{k\mid k} = A_k \widetilde{x}_{k\mid k} + G_k w_k.$$

The two terms on the right being independent,

$$\Sigma_{k+1\mid k} = A_k \Sigma_{k\mid k} A_k^T + G_k Q G_k^T. \tag{2.11}$$

Step 2

Denote

$$y_{k\mid k-1} := E\{y_k \mid y^{k-1}\}, \quad \widetilde{y}_{k\mid k-1} := y_k - y_{k\mid k-1}. \tag{2.12}$$

Recall

$$y_k = C_k x_k + H_k v_k.$$

Since v_k and y^{k-1} are independent, this gives

$$y_{k\mid k-1} = C_k x_{k\mid k-1},$$

$$\widetilde{y}_{k\mid k-1} = C_k(x_k - x_{k\mid k-1}) + H_k v_k = C_k \widetilde{x}_{k\mid k-1} + H_k v_k. \tag{2.13}$$

The two terms on the right being independent,

$$\Sigma_{k\mid k-1}^y := cov(\widetilde{y}_{k\mid k-1}) = C_k \Sigma_{k\mid k-1} C_k^T + H_k R H_k^T. \tag{2.14}$$

From (2.13) we also get

$$E\, x_k \widetilde{y}_{k\mid k-1}^T = E\, x_k \widetilde{x}_{k\mid k-1}^T C_k^T + E\, x_k v_k^T H_k^T. \tag{2.15}$$

Now $x_k = x_{k\mid k-1} + \widetilde{x}_{k\mid k-1}$, with the two terms on the right being independent by Lemma (2.5), so $E\, x_k \widetilde{x}_{k\mid k-1}^T = \Sigma_{k\mid k-1}$. Also x_k and v_k are independent. Using these in (2.15) gives

$$E\, x_k \widetilde{y}_{k\mid k-1}^T = \Sigma_{k\mid k-1} C_k^T. \tag{2.16}$$

Step 3

From (2.12) we see that y^k and $(y^{k-1}, \widetilde{y}_{k|k-1})$ are functions of each other, i.e. y^k can be obtained from knowledge of $(y^{k-1}, \widetilde{y}_{k|k-1})$ and vice-versa. Hence they are equivalent in the amount of information that they contain, and so

$$x_{k|k} = E\{x_k \mid y^{k-1}, \widetilde{y}_{k|k-1}\}.$$

Moreover by (2.5) y^{k-1} and $\widetilde{y}_{k|k-1}$ are independent, so by (2.8)

$$x_{k|k} = E\{x_k \mid y^{k-1}\}$$
$$+ (E\, x_k \widetilde{y}^T_{k|k-1})(\Sigma^y_{k|k-1})^{-1}\, \widetilde{y}_{k|k-1}, \tag{2.17}$$

$$\Sigma_{k|k} = \Sigma_{k|k-1}$$
$$- (E\, x_k \widetilde{y}^T_{k|k-1})(\Sigma^y_{k|k-1})^{-1}(E\, x_k \widetilde{y}^T_{k|k-1})^T. \tag{2.18}$$

Substituting from (2.14) and (2.16),

$$x_{k|k} = x_{k|k-1}$$
$$+ \Sigma_{k|k-1}\, C^T_k\, [C_k \Sigma_{k|k-1} C^T_k + H_k R H^T_k]^{-1}\, \widetilde{y}_{k|k-1}, \tag{2.19}$$

$$\Sigma_{k|k} = \Sigma_{k|k-1}$$
$$- \Sigma_{k|k-1}\, C^T_k\, [C_k \Sigma_{k|k-1} C^T_k + H_k R H^T_k]^{-1}\, C_k \Sigma_{k|k-1}, \tag{2.20}$$

Theorem (2.21)

The conditional density $p_{k|k} \sim N(x_{k|k}, \Sigma_{k|k})$ can be obtained from the following recursion relations.

$$x_{k+1|k+1} = A_k x_{k|k} + L_{k+1}[y_{k+1} - C_{k+1} A_k x_{k|k}], \tag{2.22}$$

$$x_{0|0} = L_0 y_0, \tag{2.23}$$

$$\Sigma_{k+1|k+1} = (I - L_{k+1} C_{k+1}) \Sigma_{k+1|k}, \tag{2.24}$$

$$\Sigma_{k+1|k} = A_k \Sigma_{k|k} A^T_k + G_k Q G^T_k, \tag{2.25}$$

$$\Sigma_{0|0} = (I - L_0 C_0) \Sigma_0. \tag{2.26}$$

Here

$$L_k := \Sigma_{k|k-1} C^T_k\, [C_k \Sigma_{k|k-1} C^T_k + H_k R H^T_k]^{-1}, \tag{2.27}$$

$$L_0 := \Sigma_0 C^T_0\, [C_0 \Sigma_0 C^T_0 + H_0 R H^T_0]^{-1}. \tag{2.28}$$

Proof

For $k \geq 0$,

$$\widetilde{y}_{k+1|k} = y_{k+1} - y_{k+1|k} = y_{k+1} - C_{k+1} x_{k+1|k}$$
$$= y_{k+1} - C_{k+1} A_k x_{k|k},$$

by (2.9). Substituting from this relation and (2.9) into (2.19) gives (2.22). Substitution from (2.27) into (2.20) gives (2.24), whereas (2.25) is the same as (2.11). Finally, by definition,

$$\tilde{y}_{0|-1} = y_0, \; x_{0|-1} = Ex_0 = 0,$$

and so $\Sigma_{0|-1} = \Sigma_0$. Substituting these into (2.19) for $k = 0$, and using (2.28) gives (2.23) and (2.26). $\square$

The recursive algorithm (2.22)-(2.26) specifies the transition function of the information state. It is known as the **Kalman state estimator** or **filter**. The reason for the term estimator is that $x_{k|k}$ is the best estimate (in the sense given below) of the state x_k given the observations y^k.

Exercise   (2.29)
Let x and y be random vectors. Show that for any function f of y, $cov(x - E\{x \mid y\}) - cov(x - f(y)) \le 0$. (Here and later, if Σ_1 and Σ_2 are symmetric matrices, $\Sigma_1 - \Sigma_2 \le 0$, or $\Sigma_1 \le \Sigma_2$ means that $\Sigma_1 - \Sigma_2$ is a negative semidefinite matrix.) In particular deduce that, $E \mid x - E\{x \mid y\} \mid^2 \le E \mid x - f(y) \mid^2$. Moreover show that equality holds if and only if $E\{x \mid y\} = f(y)$ w.p.1—i.e., $E\{x \mid y\}$ is the unique minimum mean square estimate of x given y.
[Hint: First show that $x - E\{x \mid y\}$ and $g(y)$ are uncorrelated for any g, and in particular, for $g(y) := E\{x \mid y\} - f(y)$. Write $x - f(y) = [x - E\{x \mid y\}] + [E\{x \mid y\} - f(y)]$, and evaluate its covariance.]

Thus $x_{k|k}$ is the minimum mean square estimate of x_k given y^k and $x_{k+1|k}$ is the minimum mean square estimate of x_{k+1} given y^k. In the literature $x_{k+1|k}$ is sometimes called the *one-step-ahead state predictor*. From (2.9) and (2.22) we obtain the pd of the predictor.

Theorem   (2.30)
The conditional density $p_{k+1|k} \sim N(x_{k+1|k}, \Sigma_{k+1|k})$ can be obtained from the following recursive relations:

$$x_{k+1|k} = A_k x_{k|k-1} + A_k L_k(y_k - C_k x_{k|k-1}), \; k \ge 0 \qquad (2.31)$$

$$x_{0|-1} = Ex_0 = 0. \qquad (2.32)$$

The covariance matrix $\Sigma_{k+1|k}$ is given by (2.24)-(2.26).

It is reasonable to expect that if the initial covariance Σ_0 of x_0 is decreased, then the covariance at subsequent times also decreases. The next lemma makes this precise.

Lemma   (2.33)
Suppose $\Sigma_{k+1|k}$ and $\bar{\Sigma}_{k+1|k}$ are the covariances corresponding to the initial conditions $\Sigma_{0|-1}$ and $\bar{\Sigma}_{0|-1}$. If $\Sigma_{0|-1} \le \bar{\Sigma}_{0|-1}$, then $\Sigma_{k+1|k} \le \bar{\Sigma}_{k+1|k}$.

Proof

Let $x_0 \sim N(0, \Sigma_{0|-1})$, $\zeta_0 \sim N(0, \overline{\Sigma}_{0|-1} - \Sigma_{0|-1})$ be independent of the disturbances $\{w_k, v_k\}$. Let $\bar{x}_0 := x_0 + \zeta_0$; then $\bar{x}_0 \sim N(0, \overline{\Sigma}_{0|-1})$. Let $\{\bar{x}_k\}$ and $\{\bar{y}_k\}$ be the state and output processes with initial condition $\bar{x}_0$, and $\{x_k\}$, $\{y_k\}$ the state and output processes corresponding to the initial condition x_0. Since $\bar{x}_{k+1} = x_{k+1} + A_k A_{k-1} \ldots A_0 \zeta_0$ and ζ_0 is independent of (x_{k+1}, y^k), it follows that

$$E\{\bar{x}_{k+1} \mid y^k, \zeta_0\} = E\{x_{k+1} \mid y^k, \zeta_0\}$$
$$+ E\{A_k A_{k-1} \ldots A_0 \zeta_0 \mid y^k, \zeta_0\}$$
$$= E\{x_{k+1} \mid y^k\} + A_k A_{k-1} \ldots A_0 \zeta_0.$$

Hence

$$\Sigma_{k+1|k} = cov(\bar{x}_{k+1} - E\{\bar{x}_{k+1} \mid y^k, \zeta_0\}).$$

However,

$$\overline{\Sigma}_{k+1|k} = cov(\bar{x}_{k+1} - E\{\bar{x}_{k+1} \mid \bar{y}^k\}).$$

From Exercise (2.29), identifying

$$x := \bar{x}_{k+1}, \quad y := (y^k, \zeta_0), \quad f(y) := E\{\bar{x}_{k+1} \mid \bar{y}^k\},$$

it follows that $\Sigma_{k+1|k} \leq \overline{\Sigma}_{k+1|k}$.   □

We study next the asymptotic behavior of the Kalman filter when the state equations (2.1) and (2.2) are time-invariant. The predictor error covariance is then governed by

$$\Sigma_{k+1|k} = A\{\Sigma_{k|k-1} - \Sigma_{k|k-1}$$
$$\times C^T[C\Sigma_{k|k-1}C^T + HRH^T]^{-1}C\Sigma_{k|k-1}\}A^T + GQG^T,$$
$$\Sigma_{0|-1} = \Sigma_0.$$

We investigate the boundedness and convergence of $\Sigma_{k+1|k}$ and its dependence on $\Sigma_{0|-1}$. The following exercise suggests that boundedness depends on the pair A, C.

Exercise   (2.34)

Give an example for which $\Sigma_{k+1|k}$ is unbounded.
[Hint: Take $C = 0$ and A unstable.]

A pair of matrices A, C of dimension $n \times n$ and $p \times n$ respectively is said to be **observable** if the $n \times np$ matrix $[C^T, A^T C^T \ldots (A^{n-1})^T C^T]$ has rank n. Observability refers to the fact that the initial state x_0 can be deduced from measurements of the output of the system, as shown in the following exercise.

Exercise   (2.35)

(1) Show that statements (a)-(d) are equivalent.

(a) A, C is observable.

(b) A^T, C^T is reachable.

(c) The $n \times n$ matrix $\sum_{j=0}^{n-1} (A^j)^T C^T C A^j$ is positive definite.

(d) The linear map $x_0 \to (y_0, \ldots, y_{n-1})$ given by

$$x_{k+1} = A x_k, \quad y_k = C x_k,$$

is one-to-one.

(2) Show that A, C is not observable if there is an eigenvector of A in the kernel of C—i.e., if there exists v with $Cv = 0$ and $Av = \lambda v$, for some λ.

[Hint: For (1) see Exercise (3.3.8). For (2), let $x_0 = v$ and show that (d) is violated.]

Assume now that A, C is observable. We will find an estimate of x_{k+1} given y^k whose error covariance is bounded. Denote $w^{k,i} := (w_k, \ldots, w_{k+i-1})$, and similarly for $v^{k,i}$ and $y^{k,i}$. By (2.1), there exist constant matrices $G(i)$ so that

$$x_{k+i} = A^i x_k + G(i) w^{k,i}, \tag{2.36}$$

$$y_{k+i} = C x_{k+i} + H v_{k+i} = C A^i x_k + C G(i) w^{k,i} + H v_{k+i}.$$

Multiply this equation on the left by $(A^i)^T C^T$ and sum to get

$$\sum_{i=0}^{n-1} (A^i)^T C^T y_{k+i} = M x_k + \sum_{i=0}^{n-1} (A^i)^T C^T C G(i) w^{k,i}$$

$$+ \sum_{i=0}^{n-1} (A^i)^T C^T H v_{k+i},$$

where $M := \sum_{i=0}^{n-1} (A^i)^T C^T C A^i$ is positive definite. Multiply by M^{-1} and rewrite the resulting equation as

$$x_k = \alpha y^{k,n} + \beta w^{k,n} + \gamma v^{k,n},$$

where α, β, and γ are appropriate matrices. Substitute this into (2.36) for $i = n$ to get

$$x_{k+n} = A^n \alpha y^{k,n} + [A^n \beta + G(n)] w^{k,n} + A^n \gamma v^{k,n}. \tag{2.37}$$

If we estimate x_{k+n} by $A^n \alpha y^{k,n}$, the resulting error is $[A^n \beta + G(n)] w^{k,n} + A^n \gamma v^{k,n}$ which has covariance—call it Ψ—not depending on k. By Exercise (2.29), $\Sigma_{k+n \mid k+n-1} \leq \Psi$, hence bounded.

The steady state behavior is addressed in the next theorem.

Theorem $\hspace{6cm}$ (2.38)

Let Γ be a square root of $G Q G^T$, ($\Gamma \Gamma^T = G Q G^T$). Suppose A, Γ is

reachable, A, C is observable and $HRH^T > 0$.
(1) Then $\lim_{k \to \infty} \Sigma_{k+1 \mid k} = \bar{\Sigma}$, and $\bar{\Sigma}$ is the unique nonnegative definite solution to the steady state or algebraic Riccati equation,

$$\bar{\Sigma} = A\{\bar{\Sigma} - \bar{\Sigma}C^T[C\bar{\Sigma}C^T + HRH^T]^{-1}C\bar{\Sigma}\}A^T + GQG^T. \qquad (2.39)$$

In particular, $\bar{\Sigma}$ is independent of the initial covariance $\Sigma_{0 \mid -1}$.
(2) Consider the time-invariant predictor,

$$\bar{x}_{k+1 \mid k} = A\bar{x}_{k \mid k-1} + A\bar{L}[y_k - C\bar{x}_{k \mid k-1}], \qquad (2.40)$$

with $\bar{L} := \bar{\Sigma}C^T[C\bar{\Sigma}C^T + HRH^T]^{-1}$, and arbitrary initial condition $\bar{x}_{0 \mid -1}$; then,

$$\lim_{k \to \infty} \text{cov}(x_k - \bar{x}_{k \mid k-1}) = \bar{\Sigma}.$$

The predictor (2.40) uses the steady state limit $\bar{\Sigma}$ in place of the optimal choice $\Sigma_{k \mid k-1}$. Its asymptotic error covariance is the same as that of the optimal predictor. The proof of this theorem is developed next.

It is easy to show that $\Sigma_{k \mid k-1}$ converges when the initial covariance is $\Sigma_{0 \mid -1} = 0$. This is done in the following exercise using the monotonicity property of Lemma (2.33) and boundedness.

Exercise $\qquad\qquad\qquad\qquad\qquad\qquad\qquad\qquad\qquad\qquad\qquad (2.41)$
Show that if A, C is observable and $\Sigma_{0 \mid -1} = 0$, then $\Sigma_{k \mid k-1}$ converges.
[Hint: Let $\Sigma_{k \mid k-1}$ correspond to the initial covariance $\Sigma_{0 \mid -1} = 0$, and $\tilde{\Sigma}_{k \mid k-1}$ to some other initial covariance $\tilde{\Sigma}_{0 \mid -1}$. Show that if $\tilde{\Sigma}_{0 \mid -1} = \Sigma_{1 \mid 0}$, then $\tilde{\Sigma}_{k \mid k-1} = \Sigma_{k+1 \mid k}$. Since $\tilde{\Sigma}_{0 \mid -1} \geq \Sigma_{0 \mid -1}$, it follows from the monotonicity property of Lemma (2.33) that $\Sigma_{k+1 \mid k} = \tilde{\Sigma}_{k \mid k-1} \geq \Sigma_{k \mid k-1}$, and so the sequence $\Sigma_{k \mid k-1}$ is monotone increasing. Moreover, it is bounded above. To establish convergence of the diagonal elements of $\Sigma_{k \mid k-1}$, note that $e_i^T \Sigma_{k \mid k-1} e_i$ is a monotone increasing and bounded sequence, where e_i is the unit vector with a 1 in the i-th position. For the convergence of the off-diagonal elements, examine $(e_i + e_j)^T \Sigma_{k \mid k-1}(e_i + e_j)$.]

Let $\bar{\Sigma}$ be defined as the limit of $\Sigma_{k \mid k-1}$ with $\Sigma_{0 \mid -1} = 0$; then $\bar{\Sigma}$ satisfies (2.39).

The following exercise shows that the proof of Theorem (2.38) is reduced to showing that $\Sigma_{k \mid k-1}$ converges to $\bar{\Sigma}$ when $\Sigma_{0 \mid -1}$ is a multiple of the identity matrix.

Exercise $\qquad\qquad\qquad\qquad\qquad\qquad\qquad\qquad\qquad\qquad\qquad (2.42)$
Suppose that $\lim_k \Sigma_{k \mid k-1} = \bar{\Sigma}$ whenever $\Sigma_{0 \mid -1} = \delta I$ for some $\delta > 0$. Show the following.
(1) $\Sigma_{k \mid k-1}$ converges to $\bar{\Sigma}$ for every $\Sigma_{0 \mid -1}$.

(2) There is only one solution of (2.39) within the class of symmetric non-negative definite matrices. $\overline{\Sigma}$ is a symmetric positive definite solution.
[Hint: Every $\Sigma_{0|-1}$ satisfies $0 \le \Sigma_{0|-1} \le \delta I$ for some $\delta > 0$. The convergence of $\Sigma_{k|k-1}$ follows by the monotonicity property of Lemma (2.33). To prove (2) suppose $\hat{\Sigma}$ is a nonnegative definite solution of (2.39). Show that $\Sigma_{0|-1} = \hat{\Sigma}$ implies $\Sigma_{k|k-1} = \hat{\Sigma}$ for every k.]

Define $\overline{L} := \overline{\Sigma} C^T [C \overline{\Sigma} C^T + HRH^T]^{-1}$ and $\overline{K} := A \overline{L}$. Clearly $\overline{L}$ is the limit of L_k given by (2.17) when $\Sigma_{0|-1} = 0$. The following property of $\overline{K}$ is important.

Exercise (2.43)
Show that $(A - \overline{K}C)$ is a stable matrix, i.e., all its eigenvalues are less than 1 in magnitude.
[Hint: Rewrite $\overline{\Sigma}$ as

$$\overline{\Sigma} = (A - \overline{K}C)\overline{\Sigma}(A - \overline{K}C)^T + \overline{K}HRH^T \overline{K}^T + GQG^T.$$

Suppose $(A - \overline{K}C)^T v = \lambda v$ with $|\lambda| \ge 1$. Show that $v^* \overline{\Sigma} v = v^* \overline{K} HRH^T \overline{K}^T v = v^* GQG^T v = 0$. Hence $\overline{K}^T v = 0$, $\Gamma^T v = 0$ and so $A^T v = \lambda v$. This implies that A, Γ is not reachable, see Exercise (2.35).]

Consider $\Sigma_{0|-1} := \delta I$ for some $\delta > 0$. Then we can rewrite,

$$\Sigma_{k+1|k} - \overline{\Sigma} = (A - \overline{K}C)(\Sigma_{k|k-1} - \overline{\Sigma})(A - AL_k C)^T.$$

We can solve this recursively to get,

$$\Sigma_{k+1|k} - \overline{\Sigma} = (A - \overline{K}C)^{k+1}(\Sigma_{0|-1} - \overline{\Sigma})\Phi^T(k+1, 0), \qquad (2.44)$$

where $\Phi(k, j)$ is the transition matrix

$$\Phi(k, j) := (A - AL_{k-1}C)(A - AL_{k-2}C) \ldots (A - AL_j C).$$

The first term on the right side of (2.44) converges to 0 since $(A - \overline{K}C)$ is stable. Hence if $\Phi(k, 0)$ is bounded, then $\Sigma_{k+1|k} \to \overline{\Sigma}$. So all that remains is to show $\Phi(k, 0)$ is bounded. Write the recursion for $\Sigma_{k+1|k}$ as

$$\Sigma_{k+1|k} = (A - AL_k C)\Sigma_{k|k-1}(A - AL_k C)^T$$
$$+ AL_k HRH^T L_k^T A^T + GQG^T$$
$$\ge (A - AL_k C)\Sigma_{k|k-1}(A - AL_k C)^T$$

and so

$$\Sigma_{k+1|k} \ge \delta \, \Phi(k+1, 0)\Phi(k+1, 0)^T.$$

Since $\Sigma_{k+1|k}$ is bounded, the boundedness of $\Phi(k, 0)$ follows.

The proof of part (2) of Theorem (2.38) follows from Theorems (3.3.9) and (3.3.13) using the stability of $(A - \overline{K}C)$. The proof of Theorem (2.38) is therefore complete.

3. The case with arbitrary u

Now suppose that u_k is selected according to the feedback policy $g = \{g_0, g_1, ..\}$. Corresponding to g are the various processes,

$$x^g_{k+1} = A_k x^g_k + B_k u^g_k + G_k w_k, \tag{3.1}$$

$$y^g_k = C_k x^g_k + H_k v_k, \tag{3.2}$$

$$u^g_k = g_k(y^{g,k}) = g_k(y^g_0, .., y^g_k). \tag{3.3}$$

As before the basic random variables are independent and

$$x_0 \sim N(\bar{x}_0, \Sigma_0), \quad w_k \sim N(0, Q), \quad v_k \sim N(0, R).$$

The feedback functions g_k need not be linear, so the processes, x^g, y^g, and u^g need not be Gaussian. Nevertheless, as we shall see, the conditional densities $p_{k|k}$, $p_{k+1|k}$ are Gaussian. Using the linearity of (3.1), (3.2) we can express

$$x^g_k = \bar{x}^g_k + x_k, \quad y^g_k = \bar{y}^g_k + y_k, \tag{3.4}$$

where

$$\bar{x}^g_{k+1} = A_k \bar{x}^g_k + B_k u^g_k, \quad \bar{x}^g_0 = \bar{x}_0 \tag{3.5}$$

$$\bar{y}^g_k = C_k \bar{x}^g_k, \tag{3.6}$$

$$x_{k+1} = A_k x_k + G_k w_k, \quad x_0 \sim N(0, \Sigma_0), \tag{3.7}$$

$$y_k = C_k x_k + H_k v_k. \tag{3.8}$$

The system (3.7) and (3.8) is the same as that considered in the previous section and does *not* depend on g. Consequently,

$$p_{k|k}(x_k \mid y^k) \sim N(x_{k|k}, \Sigma_{k|k}), \quad \text{and}$$

$$p_{k+1|k}(x_{k+1} \mid y^k) \sim N(x_{k+1|k}, \Sigma_{k+1|k}), \tag{3.9}$$

where the conditional means and covariances are given by Theorems (2.21) and (2.30). Denote

$$z^{g,k} := (y^{g,k}, u^{g,k-1}).$$

Lemma $\hfill$ (3.10)

For each k the random variables $z^{g,k}$ and y^k are functions of each other. (More precisely, they generate the same σ-fields.)

Proof

From (3.4) it follows that $\bar{x}^g_k$ and hence $\bar{y}^g_k$ are functions of $u^{g,k-1}$. Therefore $y_k = y^g_k - \bar{y}^g_k$ is also a function of $z^{g,k}$. It follows that y^k is a function of $z^{g,k}$.

We prove by induction that $z^{g,k}$ is a function of y^k. For $k = 0$, $z^{g,0} = y^{g,0} = \bar{y}_0 + y_0$ which is a function of y_0 since $\bar{y}_0$ is a constant. Suppose that $z^{g,k}$ is a function of y^k. Now $z^{g,k+1} = (z^{g,k}, y^g_{k+1}, u^g_k)$, $u^g_k = g_k(y^{g,k})$ and so, by the induction hypothesis, u^g_k is a function of y^k. Since by (3.5), $\bar{x}^g_{k+1}$ is a function of $u^{g,k}$, it follows that $\bar{x}^g_{k+1}$ and hence $\bar{y}^g_{k+1}$ are functions of y^k. Finally, $y^g_{k+1} = \bar{y}^g_{k+1} + y_{k+1}$, and so y^g_{k+1} is a function of y^{k+1}. The result now follows by induction.   □

Since $\bar{x}^g_k$ is a function of $u^{g,k-1}$, it follows that the conditional density of $x^g_k = \bar{x}^g_k + x_k$ given $z^{g,k}$ is the same as the conditional density of x_k given $z^{g,k}$ shifted by $\bar{x}^g_k$. By (3.10) the conditional density of x_k given $z^{g,k}$ is $p_{k|k}(x_k \mid y^k)$. These two observations and (3.9) imply

$$p_{k|k}(x^g_k \mid z^{g,k}) \sim N(\bar{x}^g_k + x_{k|k}, \Sigma_{k|k}) =: N(x^g_{k|k}, \Sigma_{k|k}).$$

Similarly

$$p_{k+1|k}(x^g_{k+1} \mid z^{g,k}) \sim N(\bar{x}^g_{k+1} + x_{k+1|k}, \Sigma_{k+1|k})$$
$$=: N(x^g_{k+1|k}, \Sigma_{k+1|k}).$$

Theorem   (3.11)
The conditional means $x^g_{k|k}$ and $x^g_{k+1|k}$ can be obtained from the following recursion relations.

$$x^g_{k+1|k+1} = A_k x^g_{k|k} + B_k u^g_k$$
$$+ L_{k+1}[y^g_{k+1} - C_{k+1}(A_k x^g_{k|k} + B_k u^g_k)], \qquad (3.12)$$

$$x^g_{k+1|k} = A_k x^g_{k|k-1} + B_k u^g_k$$
$$+ A_k L_k [y^g_k - C_k x^g_{k|k-1}], \qquad (3.13)$$

$$x^g_{0|0} = \bar{x}_0 + L_0(y_0 - C_0\bar{x}_0), \quad x^g_{0|-1} = \bar{x}_0. \qquad (3.14)$$

The covariance matrices $\Sigma_{k|k}$, $\Sigma_{k+1|k}$, and L_k are given by (2.24)-(2.28).

Proof
Adding (2.22) and (3.5) gives

$$x^g_{k+1|k+1} = A_k x^g_{k|k} + B_k u^g_k$$
$$+ L_{k+1}[y_{k+1} - C_{k+1}A_k x_{k|k}], \qquad (3.15)$$

from which (3.12) follows after substituting

$$y_{k+1} = y^g_{k+1} - \bar{y}^g_{k+1} = y^g_{k+1} - C_{k+1}(A_k \bar{x}^g_k + B_k u^g_k),$$

$$x_{k|k} = x^g_{k|k} - \bar{x}^g_k.$$

Adding (2.31) and (3.5) gives

$$x^g_{k+1|k} = A_k x^g_{k|k-1} + B_k u^g_k + A_k L_k [y_k - C_k x_{k|k-1}],$$

from which we easily get (3.13).

Finally $x_{0|-1}^g = Ex_0 = \bar{x}_0$, and $x_{0|0}^g = E\{x_0 \mid y_0\} = \bar{x}_0 + L_0(y_0 - Ey_0)$, from which we get (3.14).   □

The structure of the state estimator has an intuitive appeal. The estimator contains the system model. At time k it uses the estimate $x_{k|k}^g$ and the known control value u_k^g to predict the next output value $y_{k+1|k}^g$. This value is compared with the actual output value y_{k+1}^g. The output-prediction error, $\tilde{y}_{k+1|k}^g$, is then multiplied by the gain matrix L_{k+1} to modify the estimate of the state at time $k+1$.

Recall that the information state for the system (2.1), (2.2) is the density $p_{k|k}(x_k^g \mid z^{g,k}) \sim N(x_{k|k}^g, \Sigma_{k|k})$. This conditional density is Gaussian even when the feedback functions are nonlinear (in which case the state process is *not* Gaussian.) Moreover, the covariance matrix $\Sigma_{k|k}$ does *not* depend on the feedback law g, in fact it can be precomputed. This implies that the conditional mean $x_{k|k}^g$ is also an equivalent information state. It has the same dimension as x_k^g in contrast with the observation made earlier (see Section 6.5) that the information state is usually infinite-dimensional.

We saw in Section 6.8 that in the case of partial observations the control has a dual, learning function. But in the linear Gaussian or LG case considered here, the control law g affects only the conditional mean $x_{k|k}^g$, whereas the covariance $\Sigma_{k|k}$ does not depend on g. Thus the spread of the distribution $p_{k|k}$ cannot be altered by g. In other words, the learning function is absent in the LG case. This drastically simplifies the optimal control problem as we see next.

The next exercise shows how to reduce the more general case where w_k and v_k are correlated, to the uncorrelated case studied so far.

Exercise   (3.16)

Consider the linear system,

$$x_{k+1} = A_k x_k + B_k u_k + w_k,$$

$$y_k = C_k x_k + v_k,$$

where, with $d_k := (w_k^T, v_k^T)^T$, the random variables $x_0, d_0, d_1, \ldots$ are independent and Gaussian, with

$$x_0 \sim N(\bar{x}_0, \Sigma_0), \quad w_k \sim N(0, Q_k), \quad v_k \sim N(0, R_k),$$

$$\text{cov}(w_k, v_k) =: S_k, \quad \text{and} \quad R_k > 0.$$

Show that the system can be written as

$$x_{k+1} = \bar{A}_k x_k + \bar{u}_k + \bar{w}_k,$$

with an appropriate choice of $\bar{A}, \bar{u}$, and $\bar{w}$ such that $x_0, \bar{w}_0, \ldots, v_0, \ldots$ are independent and Gaussian.

[Hint: Take $\bar{A}_k = A_k - S_k R_k^{-1} C_k$, $\bar{w}_k = w_k - S_k R_k^{-1} v_k$].

4. Optimal control in the LG case

Consider again the linear system

$$x_{k+1} = A_k x_k + B_k u_k + G_k w_k,$$

$$y_k = C_k x_k + H_k v_k,$$

where the basic random variables are independent and

$$x_0 \sim N(\bar{x}_0, \Sigma_0), \ w_k \sim N(0, Q), \ v_k \sim N(0, R).$$

Let U be the control set and G the set of all feasible policies $g = \{g_0, \ldots, g_{N-1}\}$ where $u_k = g_k(y^k) \in U$. We wish to find the policy g that minimizes

$$J(g) := E^g \left\{ \sum_{k=0}^{N-1} c_k(x_k, u_k) + c_N(x_N) \right\}.$$

From Theorem (6.7.1) we know that an optimal policy will be a function of the information state $p_{k|k}(\cdot \mid z^k)$, where $z^k = (y^k, u^{k-1})$. From Theorem (3.11) we know that for any g

$$p_{k|k}(x_k^g \mid z^k) \sim N(x_{k|k}^g, \Sigma_{k|k}).$$

Moreover, $\Sigma_{k|k}$ does not depend on g. Thus we expect that in this linear Gaussian case an optimal policy need only be a function of $x_{k|k}^g$.

Definition (4.1)
$g = \{g_0, \ldots, g_{N-1}\}$ is **separated** if g_k depends on y^k only through $x_{k|k}^g$, i.e., $u_k^g = g_k(x_{k|k}^g)$.

The following notation is useful.

Definition (4.2)
Let $\phi_k: R^n \times U \to R$ be any function. For each fixed x in R^n and u in U define

$$\phi_{k|k}(x, u) := \int \phi_k(\xi, u)[(2\pi)^n \mid \Sigma_{k|k} \mid]^{-\frac{1}{2}}$$

$$\times \exp[-\frac{1}{2}(\xi - x)^T \Sigma_{k|k}^{-1}(\xi - x)]d\xi.$$

The interpretation of $\phi_{k|k}(x, u)$ is given in the next exercise.

Exercise (4.3)
For any g

$$E\{\phi_k(x_k^g, u_k^g) \mid y^{g,k}\} = E\{\phi_k(x_k^g, u_k^g) \mid z^{g,k}\}$$

$$= \phi_{k|k}(x_{k|k}^g, u_k^g).$$

The next result is the analog of Lemma (6.2.6).

Lemma   (4.4)

Let g be a separated policy. Define recursively the functions $V_k^g(x)$, $0 \le k \le N$, $x \in R^n$, by

$$V_N^g(x) := c_{N \mid N}(x) \tag{4.5}$$

$$V_k^g(x) := c_{k \mid k}(x, g_k(x)) + \int V_{k+1}^g(\xi) \left[(2\pi)^n \mid \Delta_{k+1} \mid\right]^{-\frac{1}{2}} \tag{4.6}$$

$$\times \exp\left[-\tfrac{1}{2} (\xi - A_k x - B_k g_k(x))^T \Delta_{k+1}^{-1} (\xi - A_k x - B_k g_k(x))\right] d\xi,$$

where

$$\Delta_{k+1} := \Sigma_{k+1 \mid k} - \Sigma_{k+1 \mid k+1}. \tag{4.7}$$

Then the random variable $V_k^g(x_{k \mid k}^g)$ satisfies w.p.1

$$V_k^g(x_{k \mid k}^g) = J_k^g := E\left\{ \sum_{l=k}^{N-1} c_l(x_l^g, u_l^g) + c_N(x_N^g) \mid y^{g,k} \right\}. \tag{4.8}$$

Proof

By (4.3)

$$J_N^g = E\{ c_N(x_N^g) \mid y^{g,N} \} = c_{N \mid N}(x_{N \mid N}^g) = V_N^g(x_{N \mid N}^g),$$

so (4.8) holds for $k = N$. Suppose it holds for $k + 1$. Then, using (4.3) again,

$$E\left\{ \sum_k^{N-1} c_l(x_l^g, u_l^g) + c_N(x_N^g) \mid y^{g,k} \right\} \tag{4.9}$$

$$= c_{k \mid k}(x_{k \mid k}^g, u_k^g)$$

$$+ E\left\{E\left\{ \sum_{k+1}^{N-1} c_l(x_l^g, u_l^g) + c_N(x_N^g) \mid y^{g,k+1}\right\} \mid y^{g,k} \right\}$$

$$= c_{k \mid k}(x_{k \mid k}^g, u_k^g) + E\{V_{k+1}^g(x_{k+1 \mid k+1}^g) \mid y^{g,k} \}$$

by the induction hypothesis. To evaluate the second term on the right we calculate the conditional density of $x_{k+1 \mid k+1}^g$ given $y^{g,k}$. From (3.12)

$$x_{k+1 \mid k+1}^g = A_k x_{k \mid k}^g + B_k u_k^g + L_{k+1} \bar{y}_{k+1 \mid k}^g.$$

The first two terms on the right are functions of $y^{g,k}$. The third term is independent of $y^{g,k}$; it has zero mean and, by (2.14), its covariance is

$$L_{k+1} \left[C_{k+1}\Sigma_{k+1 \mid k} C_{k+1}^T + H_{k+1}RH_{k+1}^T\right] L_{k+1}^T$$

$$= \Sigma_{k+1 \mid k} C_{k+1}^T \left[C_{k+1}\Sigma_{k+1 \mid k} C_{k+1}^T + H_{k+1}RH_{k+1}^T\right]^{-1} C_{k+1}\Sigma_{k+1 \mid k}$$

$$= \Sigma_{k+1 \mid k} - \Sigma_{k+1 \mid k+1} = \Delta_{k+1},$$

where the first equality follows from (2.27) and the second from (2.20). Thus

$$p(x_{k+1|k+1}^g \mid y^{g,k}) \sim N(A_k x_{k|k}^g + B_k u_k^g, \Delta_{k+1}).$$

From this fact, (4.9) and (4.6) we obtain (4.8) for k. The result now follows by induction.   □

The next result is the analog of the comparison principle (6.2.10), and the proof is similar.

Lemma   (4.10)
Let $V_k(x)$, $0 \le k \le N$, $x \in R^n$, be functions such that

$$V_N(x) \le c_{N|N}(x)$$

$$V_k(x) \le c_{k|k}(x,u) + \int V_{k+1}(\xi) \, [(2\pi)^n \mid \Delta_{k+1} \mid]^{-1/2}$$
$$\times \exp[-\tfrac{1}{2}(\xi - A_k x - B_k u)^T \Delta_{k+1}^{-1} (\xi - A_k x - B_k u)]d\xi,$$

for all $u \in U$. Let g in G be arbitrary. Then,

$$V_k(x_{k|k}^g) \le J_k^g, \quad k = 0, \ldots, N, \; w.p.\,1.$$

Theorem   (4.11)
Define recursively the functions $V_k(x)$ by

$$V_N(x) := c_{N|N}(x) \tag{4.12}$$

$$V_k(x) := \underset{u}{Inf} \; \{c_{k|k}(x,u) + \int V_{k+1}(\xi)[(2\pi)^n \mid \Delta_{k+1} \mid]^{-1/2}$$

$$\times \exp[-\tfrac{1}{2}(\xi - A_k x - B_k u)^T \Delta_{k+1}^{-1} (\xi - A_k x - B_k u)]d\xi\} \tag{4.13}$$

(1) Let g in G be arbitrary; then $V_k(x_{k|k}^g) \le J_k^g$, w.p. 1.
(2) Suppose the infimum in (4.13) is achieved at $g_k(x)$; then the separated policy $g = \{g_0, \ldots, g_N\}$ with $u_k^g = g_k(x_{k|k}^g)$ is optimal. Moreover, $V_k(x_{k|k}^g) = J_k^g$, w.p.1.

Proof
The proof can be easily constructed using Lemmas (4.4) and (4.10).   □

5. The linear, quadratic, Gaussian (LQG) problem

When the cost $c_k(x, u)$ is quadratic, the minimization in (4.13) can be done in closed form. The following fact is useful.

Lemma   (5.1)
Let $x \sim N(\bar{x}, \Sigma)$. Let S be a symmetric matrix. Then $E \, x^T S \, x = \bar{x}^T S \, \bar{x} + Tr(S\Sigma)$ where Tr denotes trace.

Proof
Let $e := x - \bar{x}$. Then $e \sim N(0, \Sigma)$, and

$$E\, x^T S\, x = E(\bar{x} + e)^T S(\bar{x} + e)$$
$$= \bar{x}^T S\, \bar{x} + 2\bar{x}^T S\, Ee + E\, e^T S\, e.$$

The second term on the right vanishes and

$$E\, e^T S\, e = E \sum e_i S_{ij} e_j = \sum S_{ij} E\, e_i e_j$$
$$= Tr(S\, E\, ee^T) = Tr(S\Sigma).$$

$\square$

Corollary (5.2)

Suppose $\phi_k(x, u) = x^T R_k x + u^T S_k u$, where R_k, S_k are symmetric matrices. Then [recall (4.2)]

$$\phi_{k\,|\,k}(x, u) = x^T R_k x + u^T S_k u + Tr(R_k \Sigma_{k\,|\,k}).$$

Assume now that

$$U = R^m, \quad \text{and} \tag{5.3}$$

$$c_N(x) = x^T P_N\, x, \quad c_k(x,u) = x^T P_k\, x + u^T T_k u, \tag{5.4}$$

where P_k and T_k are symmetric positive semidefinite matrices and T_k is positive definite.

We now apply Theorem (4.11) for this quadratic cost.

Lemma (5.5)

Suppose that $V_{k+1}(x) = x^T S_{k+1} x + s_{k+1}$ for some symmetric positive definite matrix S_{k+1} and real number s_{k+1}. Then $V_k(x)$, defined by (4.13), is given by

$$V_k(x) = x^T S_k x + s_k, \tag{5.6}$$

where

$$S_k := P_k + A_k^T \{S_{k+1}$$
$$\quad - S_{k+1} B_k\, [T_k + B_k^T S_{k+1} B_k]^{-1}\, B_k^T S_{k+1}\} A_k, \tag{5.7}$$

$$s_k := s_{k+1} + Tr(P_k \Sigma_{k\,|\,k} + S_{k+1} \Delta_{k+1}). \tag{5.8}$$

Moreover, the infimum in (4.13) is achieved at

$$u = - [T_k + B_k^T S_{k+1} B_k]^{-1} B_k^T\, S_{k+1}\, A_k x. \tag{5.9}$$

Proof

From (5.1), (5.2), and (4.13)

$$V_k(x) = \underset{u}{Inf}\, [x^T P_k x + u^T T_k u + Tr(P_k \Sigma_{k\,|\,k}) + s_{k+1}$$
$$\quad + (A_k x + B_k u)^T S_{k+1}(A_k x + B_k u) + Tr(S_{k+1} \Delta_{k+1})]. \tag{5.10}$$

The term in [..] is a strictly convex function of u since T_k is positive definite. Hence the infimum is achieved when

$$0 = \frac{\partial}{\partial u} [..] = 2T_k u + 2 B_k^T S_{k+1} B_k u + 2 B_k^T S_{k+1} A_k x,$$

which is equivalent to (5.9). Substituting this value of u in (5.10) gives (5.6), (5.7), and (5.8).   □

The preceding lemma and Theorem (4.11) yield the next result.

Theorem   (5.11)
For the LQG problem the value function is given by

$$V_k(x) = x^T S_k x + s_k, \quad 0 \le k \le N,$$ (5.12)

where S_k, s_k are obtained from the recursion relations (5.7) and (5.8) with the boundary conditions

$$S_N = P_N, \quad s_N = Tr(P_N \Sigma_{N|N}).$$ (5.13)

Moreover, the optimal control is given by the linear feedback law $u_k^g = K_k x_{k|k}^g$, where

$$K_k = - [T_k + B_k^T S_{k+1} B_k]^{-1} B_k^T S_{k+1} A_k.$$ (5.14)

Proof
It is only necessary to show (5.13). From (4.12)

$$V_N(x) = c_{N|N}(x) = x^T P_N x + Tr(P_N \Sigma_{N|N})$$

by (5.2).   □

Remark   (5.15)
Note that the statistics of the disturbances $\{w_k\}$, $\{v_k\}$ do not figure in the optimal law (5.14). In particular, this law is optimal even if there is no noise, i.e., $w_k = 0$, $v_k = 0$, w.p.1. Thus the optimal law is the same as if $x_{k|k}^g$ were the true state value instead of x_k^g. Hence this result is known as the **certainty equivalence** principle: The optimal control has the same form as if the estimate were certain, without error. Of course the minimum cost depends on the noise. The increase in cost is given by s_k. As expected, $s_k \equiv 0$ if $\Sigma_{k|k} \equiv 0$, i.e., if x_k^g is known exactly.

6. Numerical considerations in Kalman filter calculations
The Kalman filter is given by the equations

$$\Sigma_{k+1|k} = A_k \Sigma_{k|k} A_k^T + G_k Q G_k^T,$$ (6.1)

$$L_{k+1} = \Sigma_{k+1|k} C_{k+1}^T [C_{k+1} \Sigma_{k+1|k} C_{k+1}^T$$

$$+ H_{k+1} R H_{k+1}^T]^{-1},$$ (6.2)

$$\Sigma_{k+1|k+1} = (I - L_{k+1} C_{k+1}) \Sigma_{k+1|k}.$$ (6.3)

Call (6.1) the *time update*, (6.3) the *measurement update*, and (6.2) the *gain* equation. The state estimate is updated according to

$$x_{k+1|k+1} = A_k x_{k|k} + L_{k+1}[y_{k+1} - C_{k+1} A_k x_{k|k}].\tag{6.4}$$

If these equations are directly used in filter calculations, then roundoff errors may cause $\Sigma_{k+1|k}$ and $\Sigma_{k+1|k+1}$ to become asymmetric or nonpositive definite, resulting in degraded filter performance. Loss of symmetry and positive definiteness can occur, for instance, because (6.3) involves taking the difference between two matrices. The next exercise shows how to replace (6.3) by a relation involving the sum of two positive definite symmetric matrices which is a better representation.

Exercise (6.5)
Show that (6.3) is equivalent to

$$\Sigma_{k+1|k+1} = (I - L_{k+1}C_{k+1})\Sigma_{k+1|k}(I - L_{k+1}C_{k+1})^T$$
$$+ L_{k+1}H_{k+1}RH_{k+1}^T L_{k+1}^T \tag{6.6}$$

We now present a fundamentally different approach to improving numerical accuracy based on factoring $\Sigma_{k|l}$ as

$$\Sigma_{k|l} = U_{k|l} D_{k|l} U_{k|l}^T \tag{6.7}$$

where $U_{k|l}$ is a unit upper triangular matrix (i.e., diagonal elements equal to 1), and $D_{k|l}$ is a positive diagonal matrix. We will find time update equations to obtain $U_{k+1|k}$ and $D_{k+1|k}$ from $U_{k|k}$ and $D_{k|k}$, and measurement update equations to obtain $U_{k+1|k+1}$ and $D_{k+1|k+1}$. Methods based on updating factors of $\Sigma_{k|l}$ are called **factorization** or **square root** methods.

The factorization method presented next has several advantages. First, the representation (6.7) guarantees that $\Sigma_{k|l}$ is symmetric and positive definite. Second, since $\Sigma_{k|l} = (D_{k|l}^{1/2} U_{k|l}^T)^T (D_{k|l}^{1/2} U_{k|l}^T)$, we are effectively dealing with square roots, which reduces the dynamic range of the numbers involved. For instance, numbers between 10^{-n} and 10^n are reduced to numbers in the range $10^{-1/2 n}$ to $10^{1/2 n}$. Moreover, the matrix updates are performed without the need for computing square roots of any scalar numbers (which is computationally more complex in comparison with addition and multiplication). Finally, the computational complexity of the proposed method is almost the same as the computational complexity of the straightforward implementation of the Kalman filter equations, while being numerically more stable.

We separately consider the time and the measurement updates. The following exercise shows that there is no loss in generality if we assume that Q is diagonal, and $H_{k+1}RH_{k+1}^T = r$, a scalar.

Exercise (6.8)
(1) Suppose the state equation is $x_{k+1} = A_k x_k + G_k w_k$, with

$cov\ (w_k) = Q$ (not necessarily diagonal). Define $\bar{w}_k := S_k G_k w_k$. Show that if S_k is chosen as the inverse of the matrix square root of $G_k Q G_k^T$, i.e., $S_k^T S_k = (G_k Q G_k^T)^{-1}$, then $cov\ (\bar{w}_k) = I$ and $G_k w_k = S_k^{-1} \bar{w}_k$. The noise term $G_k w_k$ can be replaced by $S_k^{-1} \bar{w}_k$.
(2) Suppose the observation equation is $y_k = C_k x_k + H_k v_k$, with $cov\ (v_k) = R$. Define $\bar{y}_k = M_k y_k$ where M_k is the inverse of the matrix square root of $H_k R H_k^T$. Show that $\bar{y}_k = \bar{H}_k x_k + \bar{v}_k$ where $cov\ (\bar{v}_k) = I$. We can replace the observations by $\bar{y}_k$. Moreover, by considering each component of $\bar{y}_k$ as a separate observation, we can just as well assume that we receive p separate scalar observations. Thus we simply do p measurement updates before doing a time update.

We begin with the time update equation.

For simplicity of exposition we drop all subscripts. The time update problem is the following. Given matrices Q(diagonal), A, G, a unit upper triangular matrix U, and a positive diagonal matrix D, find a unit upper triangular matrix $\tilde{U}$, and a positive diagonal matrix $\tilde{D}$, such that

$$\tilde{U}\tilde{D}\tilde{U}^T = AUDU^T A^T + GQG^T \tag{6.9}$$

(If $\Sigma_{k\,|\,k} = UDU^T$, then $\Sigma_{k+1\,|\,k} = \tilde{U}\tilde{D}\tilde{U}^T$).

Define the matrices

$$W := [AU, G], \quad \bar{D} := \text{diag}(D, Q),$$

then the problem (6.9) is to find $\tilde{U}$ and $\tilde{D}$ such that

$$\tilde{U}\tilde{D}\tilde{U}^T = W\bar{D}W^T.$$

The basic idea underlying the derivation is this. Let $w_1^T, \ldots, w_n^T$ be the rows of W. Suppose that we find a unit upper triangular matrix $\tilde{U}$ and a matrix V such that

$$W = \tilde{U}V, \tag{6.10}$$

so

$$W\bar{D}W^T = \tilde{U}(V\bar{D}V^T)\tilde{U}^T.$$

Clearly, if $V\bar{D}V^T$ is diagonal, then we are done. Let $v_1^T, \ldots, v_n^T$ be the rows of V. The ijth element of $V\bar{D}V^T$ is

$$(V\bar{D}V^T)_{ij} = v_i^T \bar{D} v_j.$$

Hence this matrix is diagonal if

$$v_i^T \bar{D} v_j = 0 \quad \text{for } i \neq j. \tag{6.11}$$

Let us examine the requirements (6.10) and (6.11) more carefully. Since $\tilde{U}$ is a unit upper triangular matrix, (6.10) implies

$$w_n^T = v_n^T,$$

$$w_{n-1}^T = v_{n-1}^T + \tilde{U}_{n-1,n} v_n^T,$$

$$\vdots$$

$$w_j^T = v_j^T + \sum_{k=j+1}^{n} \tilde{U}_{j,k} v_k^T.$$

Hence for $j = n, \ldots, 1$,

$$w_j \in \text{span } [v_n, v_{n-1}, \ldots, v_j], \tag{6.12}$$

and the jth row of $\tilde{U}$ merely gives the coefficients of the particular linear combination of $[v_n, v_{n-1}, \ldots, v_j]$ which yields w_j. (Here span $[..]$ is the linear subspace spanned by the vectors listed in $[..]$.)

Defining the inner product on R^n by

$$<x, y> := x^T \overline{D} y, \tag{6.13}$$

we see that (6.11) requires the vectors $\{v_1, \ldots, v_n\}$ to form an orthogonal basis for R^n with respect to this inner product.

Hence if we can obtain a basis $\{v_n, \ldots, v_1\}$ of R^n, which is orthogonal with respect to the inner product (6.13), and such that

$$\text{span } [v_n, \ldots, v_j] = \text{span } [w_n, \ldots, w_j] \quad \text{for all } j, \tag{6.14}$$

then (6.10) and (6.11) will be satisfied. This can be simply done by using the well-known Gram-Schmidt orthogonalization method given next.

Exercise (Gram-Schmidt orthogonalization)   (6.15)
Let $\{w_1, \ldots, w_n\}$ be linearly independent vectors in R^n. Recursively define $\{v_n, \ldots, v_1\}$ by

$$v_n := w_n, \tag{6.16}$$

$$v_{n-1} := w_{n-1} - \frac{<w_{n-1}, v_n>}{<v_n, v_n>} v_n,$$

$$\vdots$$

$$v_j := w_j - \sum_{k=j+1}^{n} \frac{<w_j, v_k>}{<v_k, v_k>} v_k. \tag{6.17}$$

Show that $\{v_n, \ldots, v_1\}$ is an orthogonal basis satisfying (6.14).
[Hint: Use as induction hypothesis that for some j, $\{v_n, \ldots, v_j\}$ is an orthogonal set of vectors, and span $[v_n, \ldots, v_j] = \text{span } [w_n, \ldots, w_j]$.]

This implies

$$\widetilde{D}_{jj} := \,<v_j, v_j> \tag{6.18}$$

$$\widetilde{U}_{ij} := \frac{<w_i, v_j>}{<v_j, v_j>} \quad \text{for } i = 1, \ldots, j-1. \tag{6.19}$$

Equations (6.16)-(6.19) define a recursive algorithm for obtaining $\widetilde{U}$ and $\widetilde{D}$.

However for computational reasons, the following modified algorithm is preferred.

Algorithm for U-D factorization time update

Data: $A := A_k$, $G := G_k$, Q, U, D, with Q, D diagonal and $UDU^T := \Sigma_{k|k}$. Define $W := [AU, G] =: [w_1, \ldots, w_n]^T$, $\overline{D} := \text{diag}(D, Q)$, and the inner product $< , >$ by $<x, y> := x^T \overline{D} y$.

Initialization: Set $w_i^{(0)} := w_i$ for $i = 1, \ldots, n$.

Recursive Step: For $j = n, n-1, \ldots, 1$ do

$$\widetilde{D}_{jj} := \,<w_j^{(n-j)}, w_j^{(n-j)}>$$

$$\widetilde{U}_{ij} := \frac{1}{\widetilde{D}_{jj}} <w_i^{(n-j)}, w_j^{(n-j)}> \quad \text{for } i = 1, \ldots, j-1$$

$$w_i^{(n-j+1)} := w_i^{(n-j)} - \widetilde{U}_{ij} w_j^{(n-j)} \quad \text{for } i = 1, 2, \ldots, j-1$$

End Result: $\widetilde{U}, \widetilde{D}$ with $\Sigma_{k+1|k} = \widetilde{U} \widetilde{D} \widetilde{U}^T$.

Exercise (6.20)

Show that the time update algorithm produces $\widetilde{U}$ and $\widetilde{D}$ coinciding with (6.18) and (6.19).

[Hint: Use as induction hypothesis that for some j, $\widetilde{U}_{ij}$ and $\widetilde{D}_{jj}$ are as claimed, and furthermore,

$$w_i^{(n-j+1)} = w_i - \sum_{k=j}^{n} \frac{<w_i, v_k>}{<v_k, v_k>} v_k$$

for $i = 1, \ldots, j-1$. Use also the fact that $w_j^{(n-j)} = v_j$. Start the induction with $j = n$.]

We now turn to the measurement update equation.

Again, for simplicity of exposition we drop all subscripts. We are given C, r, a unit upper triangular matrix $\widetilde{U}$, and a positive diagonal matrix $\widetilde{D}$ (with $\Sigma_{k+1|k} = \widetilde{U} \widetilde{D} \widetilde{U}^T$). Our goal is to find a unit upper triangular matrix $\overline{U}$ and a positive diagonal matrix $\overline{D}$, such that

$$\overline{U}\,\overline{D}\,\overline{U}^T = \widetilde{U} \left\{ \widetilde{D} - \frac{\widetilde{D}\,\widetilde{U}^T C^T C \widetilde{U} \widetilde{D}}{[C \widetilde{U} \widetilde{D} \widetilde{U}^T C^T + r]} \right\} \widetilde{U}^T \tag{6.21}$$

[compare with (6.2) and (6.3)], and also to calculate L defined by

$$L := \frac{\tilde{U}\tilde{D}\tilde{U}^T C^T}{C\tilde{U}\tilde{D}\tilde{U}^T C^T + r} \tag{6.22}$$

in an efficient manner. (Note that we will then have $\Sigma_{k+1|k+1} = \overline{U}\,\overline{D}\,\overline{U}^T$ and $L_{k+1} = L.$)

If we factor the term $\{..\}$ in (6.21) as

$$\tilde{D} - \frac{\tilde{D}\tilde{U}^T C^T C\tilde{U}\tilde{D}}{[C\tilde{U}\tilde{D}\tilde{U}^T C^T + r]} = \hat{U}\hat{D}\hat{U}^T, \tag{6.23}$$

where $\hat{U}$ is unit upper triangular and $\hat{D}$ is positive diagonal, then we can take $\overline{U}$ and $\overline{D}$ as

$$\overline{U} := \tilde{U}\hat{U}, \quad \overline{D} := \hat{D}. \tag{6.24}$$

To obtain (6.23) set

$$\alpha := [C\tilde{U}\tilde{D}\tilde{U}^T C^T + r], \tag{6.25}$$

$$F := \hat{U}^T C^T := (f_1, f_2, \ldots, f_n)^T, \tag{6.26}$$

$$v := (v_1, v_2, \ldots, v_n)^T := \tilde{D}F, \tag{6.27}$$

and rewrite (6.23) as

$$\tilde{D} - \frac{vv^T}{\alpha} = \hat{U}\hat{D}\hat{U}^T. \tag{6.28}$$

If

$$x^T [\tilde{D} - \frac{vv^T}{\alpha}] x = x^T \hat{U}\hat{D}\hat{U}^T x$$

for all $x \in R^n$, then (6.28) is satisfied. Let $x = (x_1, \ldots, x_n)^T$, and write this as

$$\sum_{j=1}^{n} \hat{D}_{jj}(x_j + \sum_{i=1}^{j-1} \hat{U}_{ij}x_i)^2 = \sum_{j=1}^{n} \tilde{D}_{jj}x_j^2 - \alpha^{-1}(\sum_{j=1}^{n} x_j v_j)^2. \tag{6.29}$$

Our goal is to choose $\hat{U}$ and $\hat{D}$ so that (6.29) holds for all $(x_1, \ldots, x_n)$. To do this we equate the coefficients of all the terms containing x_n, then we equate all the terms containing x_{n-1}, and so on. This will give a recursive procedure. Equating the terms containing x_n on both sides of (6.29) gives

$$\hat{D}_{nn}(x_n^2 + 2x_n \sum_{i=1}^{n-1} \hat{U}_{in}x_i)$$

$$= (\tilde{D}_{nn} - \alpha^{-1}v_n^2)x_n^2 - \frac{2}{\alpha} x_n v_n \sum_{j=1}^{n} x_j v_j,$$

which will hold for all x_n only if the coefficients of x_n^2 are equal, and the coefficients of x_n are equal i.e.,

$$\hat{D}_{nn} = \tilde{D}_{nn} - \alpha^{-1}v_n^2, \text{ and}$$

$$\hat{D}_{nn} \sum_{i=1}^{n-1} \hat{U}_{in} x_i = -\alpha^{-1}v_n \sum_{j=1}^{n-1} x_j v_j. \tag{6.30}$$

Since the second equality is to hold for all $(x_1, \dots, x_{n-1})$, we equate the coefficients of each x_i to get,

$$\hat{U}_{jn} = \frac{\alpha^{-1}v_j v_n}{\hat{D}_{nn}} \quad \text{for } j = 1, \dots, n-1. \tag{6.31}$$

Equations (6.30) and (6.31) specify the nth columns of $\hat{D}$ and $\hat{U}$.

We now obtain the $(n-1)$st columns of $\hat{D}$ and $\hat{U}$. Since we have already ensured the equality of the terms containing x_n on both sides of (6.29), we cancel these terms to get

$$\sum_{j=1}^{n-1} \hat{D}_{jj}(x_j + \sum_{i=1}^{j-1} \hat{U}_{ij} x_i)^2 + \hat{D}_{nn}(\sum_{i=1}^{n-1} \hat{U}_{in} x_i)^2$$

$$= \sum_{j=1}^{n-1} \tilde{D}_{jj} x_j^2 - \alpha^{-1}(\sum_{j=1}^{n-1} x_j v_j)^2.$$

Moving the term containing $\hat{D}_{nn}$ from the left to the right gives

$$\sum_{j=1}^{n-1} \hat{D}_{jj}(x_j + \sum_{i=1}^{j-1} \hat{U}_{ij} x_i)^2$$

$$= \sum_{j=1}^{n-1} \tilde{D}_{jj} x_j^2 - \alpha^{-1}(\sum_{j=1}^{n-1} x_j v_j)^2 - \hat{D}_{nn}(\sum_{i=1}^{n-1} \hat{U}_{in} x_i)^2.$$

Substituting for $\hat{U}_{in}$ from (6.31) gives

$$\sum_{j=1}^{n-1} \hat{D}_{jj}(x_j + \sum_{i=1}^{j-1} \hat{U}_{ij} x_i)^2$$

$$= \sum_{j=1}^{n-1} \tilde{D}_{jj} x_j^2 - [\alpha^{-1} + \frac{\alpha^{-2}v_n^2}{\hat{D}_{nn}}](\sum_{j=1}^{n-1} v_j x_j)^2. \tag{6.32}$$

We want to choose $\hat{U}_{ij}$ and $\hat{D}_{jj}$ such that equality holds above for all $(x_1, \dots, x_{n-1})$. However, (6.32) resembles (6.29) except that the α^{-1} of (6.29) is replaced by $[\alpha^{-1} + \frac{\alpha^{-2}v_n^2}{\hat{D}_{nn}}]$. Thus we can use the same procedure as above to choose the $(n-1)$st columns of $\hat{D}$ and $\hat{U}$. This gives us the following recursive scheme :

Initialization: Set $\alpha_n := \alpha$.

Repeat for $j = n, \ldots, 1$:

$$\hat{D}_{jj} := \widetilde{D}_{jj} - \alpha_j^{-1} v_j^2, \tag{6.33}$$

$$\hat{U}_{ij} := \frac{-\alpha_j^{-1} v_i v_j}{\hat{D}_{jj}} \quad \text{for } i = 1, \ldots, j - 1, \tag{6.34}$$

$$\alpha_{j-1}^{-1} := \alpha_j^{-1}[1 + \frac{\alpha_j^{-1} v_j^2}{\hat{D}_{jj}}]. \tag{6.35}$$

This recursive procedure can be improved. Define

$$v^{(j)} := (v_1, \ldots, v_j, 0, \ldots, 0)^T.$$

Then (6.34) means that

$$j\text{th column of } \hat{U} = \lambda_j v^{(j-1)} + e_j,$$

where e_j is the jth column of the identity matrix, and

$$\lambda_j := \frac{-v_j}{\alpha_j \hat{D}_{jj}} .$$

Hence

$$\hat{U} := I + [0, \lambda_2 v^{(1)}, \lambda_3 v^{(2)}, \ldots, \lambda_n v^{(n-1)}].$$

Thus from (6.24) we have

$$\overline{U} := \widetilde{U} \hat{U} = \widetilde{U} + [0, \lambda_2 \widetilde{U} v^{(1)}, \ldots, \lambda_n \widetilde{U} v^{(n-1)}].$$

Since $v^{(j)}$ merely consists of adding an element to the jth position of the vector $v^{(j-1)}$, we can simplify the computation by starting with $\widetilde{U} v^{(1)}$, then computing $\widetilde{U} v^{(2)}$, then $\widetilde{U} v^{(3)}$, and so on. For this purpose, define

$$K_j := \widetilde{U} v^{(j-1)}; \quad K_1 := 0.$$

Then

$$K_{j+1} = \widetilde{U}[v^{(j-1)} + v_j e_j] = K_j + v_j \widetilde{U}_j,$$

where $\widetilde{U}_j :=$ jth column of $\widetilde{U}$. Note also that the gain L is given from (6.2) and (6.25)-(6.27) by

$$L := \frac{\widetilde{U} \widetilde{D} \widetilde{U}^T C^T}{C \widetilde{U} \widetilde{D} \widetilde{U}^T C^T + r} = \frac{\widetilde{U} v}{\alpha} = \frac{\widetilde{U} v^{(n)}}{\alpha_n} = \frac{K_{n+1}}{\alpha_n} .$$

However, to do all this we need to start our recursive computation at $j = 1$ and then proceed to $j = 2, 3, \ldots, n$, in that order. But our recursive procedure (6.33)-(6.35) proceeds in the direction of decreasing j. So we need to reverse the direction of the algorithm, which we do as follows.

We need initial conditions at $j = 1$, rather than at $j = n$, so we have to compute α_1. By using (6.33) and then the expression $\widetilde{D}_{nn} = \dfrac{v_n}{f_n}$

from (6.27), we can simplify (6.35) to

$$\alpha_{j-1} := \alpha_j - v_j f_j. \tag{6.36}$$

Since

$$\alpha_n := \alpha = r + \sum_{i=1}^{n} f_i v_i \quad \text{from (6.25)-(6.27)},$$

it follows from (6.36) that

$$\alpha_1 := r + f_1 v_1.$$

This gives the appropriate initial condition to start the recursion at $j = 1$ rather than at $j = n$.

We can make one further numerical improvement. Equation (6.33) computes $\hat{D}_{jj}$ as the difference of two positive numbers, which is susceptible to a loss of accuracy. However, this computation can also be rewritten as

$$\hat{D}_{jj} = \frac{\widetilde{D}_{jj}}{\alpha_j} \alpha_{j-1}.$$

We have now the final form of the measurement update equation and the gain calculation.

Data: $C := C_{k+1}$, $r := \text{cov}(H_{k+1} v_{k+1})$, $\widetilde{U}$, $\widetilde{D}$, with $\Sigma_{k+1|k} = \widetilde{U}\widetilde{D}\widetilde{U}^T$.
Define $F := (f_1, \ldots, f_n)^T := \widetilde{U}^T C^T$, $v := (v_1, \ldots, v_n)^T$.

Initialization: Set

$$\alpha_1 := r + v_1 f_1,$$

$$\hat{D}_{11} := \frac{\widetilde{D}_{11} r}{\alpha_1},$$

$$K_2 := (v_1, 0, \ldots, 0)^T.$$

Recursive Step: For $j = 2, \ldots, n$ do

$$\alpha_j := \alpha_{j-1} + v_j f_j,$$

$$\hat{D}_{jj} := \frac{\widetilde{D}_{jj} \alpha_{j-1}}{\alpha_j},$$

$$\lambda_j := \frac{-f_j}{\alpha_{j-1}},$$

$$\overline{U}_j := \widetilde{U}_j + \lambda_j f_j,$$

$$K_{j+1} := K_j + v_j \widetilde{U}_j.$$

End Result: $\bar{U} := [\bar{U}_1, \ldots, \bar{U}_n]$, $L := \dfrac{K_{n+1}}{\alpha_n}$, with $\Sigma_{k+1 \mid k+1} = \bar{U}\hat{D}\bar{U}^T$ and $L_{k+1} = L$.

7. Linear non-Gaussian systems

Consider the linear stochastic system,

$$x_{k+1} = A_k x_k + B_k u_k + G_k w_k,$$

$$y_k = C_k x_k + H_k v_k.$$

Suppose the independent basic random variables $\{x_0, w_0, \ldots, v_0, \ldots\}$ are not Gaussian, but their means and covariances are as before,

$$Ex_0 = 0, \ \Sigma_0 := cov(x_0); \ \ Ew_k = 0, Q := cov(w_k); \quad \text{and}$$

$$Ev_k = 0, R := cov(v_k).$$

In Section 3.2 we saw that the second order properties (i.e., mean and covariance function) of the non-Gaussian processes $\{x_k\}$ and $\{y_k\}$ are the same as if the basic random variables were Gaussian. We see next that this property extends (in a weaker way) to the Kalman filter.

Let x and y be random variables with values in R^n and R^p, respectively. Let $\bar{x} := Ex$, $\bar{y} := Ey$. A *linear* estimate of x given y is any random variable of the form

$$\hat{x}(\alpha, L) := \alpha + L(y - \bar{y}),$$

where $\alpha \in R^n$, and $L \in R^{p \times n}$ are constants. The error corresponding to this estimate is

$$\tilde{x}(\alpha, L) := x - \hat{x}(\alpha, L),$$

and the covariance of this error is denoted $\Sigma_{\tilde{x}(\alpha, L)}$.

Exercise (7.1)

Show that the estimate

$$\hat{x} := \bar{x} + \Sigma_{xy} \Sigma_y^{-1}(y - \bar{y})$$

is the best linear estimate in the sense that

$$\Sigma_{\tilde{x}} = \Sigma_x - \Sigma_{xy} \Sigma_y^{-1} \Sigma_{xy}^T \le \Sigma_{\tilde{x}(\alpha, L)},$$

for all α, L. Here $\tilde{x} := x - \hat{x}$. The best linear estimate need not be the conditional expectation. Find an example such that $\hat{x} \ne E\{x \mid y\}$. [Hint: See the proof of Lemma (2.5).]

Thus the best linear estimate of x given y depends only on the second order properties of x and y, and, moreover, this estimate and the corresponding error covariance are the same as if x and y were jointly Gaussian.

As before let $g = \{g_0, g_1, ..\}$ be any feedback policy, and let

$$z^{g,k} := (y^{g,k}, u^{g,k-1}),$$

be the observations up to time k. Let $x^g_{k \mid k}$ and $x^g_{k+1 \mid k}$ respectively denote the best linear estimate of x^g_k and x^g_{k+1} given $z^{g,k}$; let $\Sigma_{k \mid k}$ and $\Sigma_{k+1 \mid k}$ denote the corresponding error covariances.

Exercise (7.2)

Show that $x^g_{k \mid k}$, $\Sigma_{k \mid k}$, and $\Sigma_{k+1 \mid k}$ are given by the Kalman filter.
[Hint: Use Exercise (7.1) and then observe that the proof of Theorem (3.11) applies without change.]

Three remarks are worth making. First, in Exercise (7.2), it is enough to assume that the basic random variables are uncorrelated rather than independent. Second, $x^g_{k \mid k}$ is only the best linear estimate and not the conditional mean. [The conditional mean may be regarded as the best nonlinear estimate by Exercise (2.29).] Third, from Lemma (5.1) we note that $Ex^T Sx$ depends only on the mean and covariance of x. This observation and the proof of Theorem (5.11) lead to the next result.

Exercise (7.3)

Suppose the basic random variables are not Gaussian and the cost is quadratic as in (5.4). Show that the optimum *linear* feedback law is given by $u^g_k = K_k x^g_{k \mid k}$, where $x^g_{k \mid k}$ is the best linear estimate and the gain matrix K_k is given by (5.14).

8. Minimum variance control of ARMAX models

It is sometimes convenient to work directly with the ARMAX models introduced in Chapter 5 rather than with state space models. Here we examine the problem of minimizing the variance of the output of an ARMAX model. To introduce the ideas we begin with the ARX model,

$$y_{k+1} = \sum_{i=0}^{p} [a_i y_{k-i} + b_i u_{k-i}] + w_{k+1}. \tag{8.1}$$

Assume that $b_0 \neq 0$, so the delay between the application of an input and its effect at the output is one time unit. $\{w_k\}$ is iid with $Ew_k = 0$ and $Ew_k^2 = \sigma^2$. Consider the problem of minimizing the variance of the output over $\{1, .., N\}$,

$$\sum_{k=1}^{N} Ey_k^2. \tag{8.2}$$

Let $g = \{g_0, g_1, ..\}$ be any feedback law, let $z^{g,k} := (y^{g,k}, u^{g,k-1})$. From (8.1), and the independence of w_{k+1} and $z^{g,k}$, we get

$$E\{y_{k+1}^2 \mid z^{g,k}\} = [\sum_{i=0}^{p} a_i y_{k-i} + b_i u_{k-i}]^2 + Ew_{k+1}^2.$$

Taking expectations,

$$Ey_{k+1}^2 = E[\sum_{i=0}^{p} a_i y_{k-i} + b_i u_{k-i}]^2 + Ew_{k+1}^2,$$

so the left-hand side is minimized if $\sum_{i=0}^{p}[a_i y_{k-i} + b_i u_{k-i}] = 0$, i.e., u_k is chosen as

$$u_k = -\frac{1}{b_0}[\sum_{i=0}^{p} a_i y_{k-i} + \sum_{i=1}^{p} b_i u_{k-i}]. \tag{8.3}$$

This is the optimal control law, and the minimum cost is $N\sigma^2$. Note that the control law (8.3) is independent of the time horizon N.

For some linear systems however, this control law may not be suitable, as the following example illustrates.

Example (8.4)
For the system

$$y_{k+1} = a_0 y_k + b_0 u_k + b_1 u_{k-1} + w_{k+1}, \tag{8.5}$$

the law (8.3) is

$$u_k = -\frac{1}{b_0}[a_0 y_k + b_1 u_{k-1}]. \tag{8.6}$$

Substituting into (8.5) gives $y_k = w_k$, and so

$$u_k = -\frac{1}{b_0}[a_0 w_k + b_1 u_{k-1}]. \tag{8.7}$$

Clearly if $|\frac{b_1}{b_0}| > 1$, then (8.7) is an unstable difference equation, so that the closed loop system (8.5),(8.6) is unstable. (The closed loop system is then said to be internally unstable.)

To analyze the stability of the minimum variance control law for the general ARMAX case we recall some facts about linear systems. Let z be the shift operator such that for any sequence $\{x_k\}$, $(zx)_k := x_{k-1}$. ($z = q^{-1}$ is just the inverse of the shift operator q of Section 5.2.) When the control law $u_k = \frac{H(z)}{K(z)} y_k$ is applied to the system

$$A(z)y_{k+1} = B(z)u_k + w_{k+1},$$

where $A(z), B(z), H(z)$, and $K(z)$ are polynomials, the resulting closed loop system is

$$[A(z)K(z) - zB(z)H(z)]y_{k+1} = K(z)w_{k+1}.$$

This closed loop system is said to be **internally stable** if

$[A(z)K(z) - zB(z)H(z)]$ has all roots strictly outside the unit circle.

For the minimum variance control law (8.3), we have

$$A(z) = 1 - \sum_{i=0}^{p} a_i z^{i+1}, \quad B(z) = \sum_{i=0}^{p} b_i z^i,$$

$$H(z) = \sum_{i=0}^{p} -a_i z^i, \quad K(z) = B(z),$$

so $[A(z)K(z) - zB(z)H(z)] = B(z)$. Hence the minimum variance control law (8.3) applied to the system (8.1) gives rise to an internally stable system if and only if

$$B(z) \text{ has all roots strictly outside the unit circle.} \tag{8.8}$$

Definition (8.9)

A system satisfying (8.8) is said to be (strictly) **minimum phase**.

Now consider the ARMAX model,

$$y_{k+1} = \sum_{i=0}^{p} [a_i y_{k-i} + b_i u_{k-i} + c_i w_{k-i}] + w_{k+1}, \tag{8.10}$$

and again suppose $b_0 \neq 0$.

Note that since $w_{k-p}, \ldots w_k$ are not observed, we only have partial observations of the state of the system. One can therefore rewrite the input-output description of the system (8.10) in a state space form and, as in Section (7.4), obtain the information state through the Kalman filter. The optimal control law can then be obtained in the separated form of Theorem (4.11). It will turn out to satisfy the certainty equivalence principle of Theorem (5.11) since the cost function (8.2) is a special case of the more general quadratic cost function (5.4). However, a more direct approach is appropriate.

Initially suppose w_k is observed at time k; then the control law

$$u_k = -\frac{1}{b_0} \left[\sum_{i=0}^{p} (a_i y_{k-i} + c_i w_{k-i}) + \sum_{i=1}^{p} b_i u_{k-i} \right] \tag{8.11}$$

minimizes the cost (8.2). Note that it does not depend on N and, under this law, $y_k = w_k$.

However, w_k cannot be observed. The certainty equivalence principle suggests using an estimate w_k in place of w_k. It is also reasonable to expect that this estimate will converge to the true value, as $k \to \infty$.

Consider therefore the modification of the control law (8.11) that results when w_k is replaced by y_k, which is what it would be if we could implement (8.11). This gives the implementable control law

$$u_k = -\frac{1}{b_0} \left[\sum_{i=0}^{p}(a_i + c_i)y_{k-i} + \sum_{i=1}^{p} b_i u_{k-i}\right]. \tag{8.12}$$

Of course this is not optimal. However, we can show that it is asymptotically optimal; i.e., it converges to the optimal control law (8.11).

To prove this, we examine the closed loop system (8.10) and (8.12) for internal stability. The closed loop system is

$$C(z)B(z)y_{k+1} = C(z)B(z)w_{k+1}, \tag{8.13}$$

where $C(z) := 1 + \sum_{i=0}^{p} c_i z^{i+1}$. This is internally stable if and only if

$B(z)$ and $C(z)$ have all roots strictly outside the unit

circle. $\tag{8.14}$

The requirement on $B(z)$ is just the minimum phase assumption. By the spectral factorization Theorem (5.5.12) we may assume without loss of generality that the roots of $C(z)$ are on or outside the unit circle, so that (8.14) only excludes roots on the unit circle.

From (8.13),

$$B(z)C(z)[y_{k+1} - w_{k+1}] = 0.$$

The solution to this homogeneous difference equation is determined by the $2p + 2$ initial conditions $(y_0-w_0), (y_1-w_1), \ldots, (y_{2p+1}-w_{2p+1})$. Moreover, under (8.14) the solution of the homogeneous difference equation converges to 0 (at an exponential rate). Hence

$$\lim_{k \to \infty} (y_k - w_k) = 0,$$

and so the control law (8.12) converges to the control law (8.11). Therefore the control law (8.12) is asymptotically optimal in the sense that if it is implemented from time 0 on, it minimizes

$$\sum_{k=K}^{N+K} E y_k^2 \tag{8.15}$$

arbitrarily closely for large K. Alternatively, we can see that this control law minimizes the average cost

$$\lim_{N \to \infty} \frac{1}{N} \sum_{k=1}^{N} E y_k^2. \tag{8.16}$$

The average cost criterion is studied in Chapter 8.

Of course, (8.12) does not minimize (8.15) for a finite K. To obtain the true optimal control law one could put the ARMAX model into state space form and then use Theorem (5.11). The resulting control law will be time varying.

Finally, consider the ARMAX model with delay $d \geq 1$ between the application of control and its effect on the output ($b_0 \neq 0$),

$$y_{k+1} = \sum_{i=0}^{p} [a_i y_{k-i} + b_i u_{k+1-i-d} + c_i w_{k-i}] + w_{k+1}. \tag{8.17}$$

Rewrite the model as

$$A(z)y_k = z^d B(z)u_k + C(z)w_k, \tag{8.18}$$

where

$$A(z) := 1 - \sum_{i=0}^{p} a_i z^{i+1}, \quad B(z) := \sum_{i=0}^{p} b_i z^i, \quad \text{and}$$

$$C(z) := 1 + \sum_{i=0}^{p} c_i z^{i+1}.$$

To obtain an asymptotically optimal law, begin by dividing the polynomial $C(z)$ by the polynomial $A(z)$, and conduct d steps of the long division process to obtain a quotient $F(z) = \sum_{i=0}^{d-1} f_i z^i$ and a remainder $z^d G(z)$; i.e.,

$$\frac{C(z)}{A(z)} = F(z) + z^d \frac{G(z)}{A(z)},$$

or equivalently,

$$C(z) = A(z)F(z) + z^d G(z). \tag{8.19}$$

The system (8.17) can be written as

$$y_k = z^d \frac{B(z)}{A(z)} u_k + \frac{C(z)}{A(z)} w_k,$$

and substituting for $\dfrac{C(z)}{A(z)}$ gives

$$y_k = z^d [\frac{B(z)}{A(z)} u_k + \frac{G(z)}{A(z)} w_k] + F(z)w_k$$

$$= [\frac{B(z)}{A(z)} u_{k-d} + \frac{G(z)}{A(z)} w_{k-d}] + F(z)w_k.$$

The term $\dfrac{B(z)}{A(z)} u_{k-d}$ depends only on $\{u_l \mid l \leq k-d\}$, and $\dfrac{G(z)}{A(z)} w_{k-d}$ depends only on $\{w_l \mid l \leq k-d\}$. Hence they are both independent of $F(z)w_k = \sum_{i=0}^{d-1} f_i w_{k-i}$. So

$$E\{y_k^2 \mid u_l, w_l, l \le k - d\} = [\frac{B(z)}{A(z)} u_{k-d} + \frac{G(z)}{A(z)} w_{k-d}]^2$$

$$+ (E\{F(z)w_k \mid u_l, w_l, l \le k - d\})^2$$

is minimized when

$$[\frac{B(z)}{A(z)} u_{k-d} + \frac{G(z)}{A(z)} w_{k-d}] = 0;$$

i.e., u_k is chosen to satisfy

$$u_k = -\frac{G(z)}{B(z)} w_k. \tag{8.20}$$

The notation (8.20) means that u_k is selected according to the difference equation

$$B(z)u_k = -G(z)w_k. \tag{8.21}$$

The resulting output satisfies

$$y_k = F(z)w_k. \tag{8.22}$$

The control law (8.20) is not implementable because w_k is not observed. As before we consider the implementable control law

$$u_k = -\frac{G(z)}{B(z)F(z)} y_k, \tag{8.23}$$

obtained by substituting the expression for w_k from (8.22) into (8.20); i.e., u_k is selected according to the difference equation

$$B(z)F(z)u_k + G(z)y_k = 0. \tag{8.24}$$

We now show that this control law is optimal for (8.16) (and so nearly optimal for (8.15) for large K). We begin with an analysis of the closed loop system (8.24) and (8.18). Multiply (8.18) by $F(z)$,

$$A(z)F(z)y_k = z^d B(z)F(z)u_k + C(z)F(z)w_k,$$

and substitute for $B(z)F(z)u_k$ from (8.24) to get

$$[A(z)F(z) - z^d G(z)]y_k = C(z)F(z)w_k.$$

Using (8.19) this reduces to

$$C(z)y_k = C(z)F(z)w_k.$$

Let $\epsilon_k := y_k - F(z)w_k$, so

$$C(z)\epsilon_k = 0. \tag{8.25}$$

The solution of this homogeneous difference equation is determined by the $p+1$ initial conditions $\epsilon_0, \ldots, \epsilon_p$. Assume as before that $C(z)$ has all its roots strictly outside the unit circle. Then the solution of (8.25) converges to zero at an exponential rate; i.e., there are constants $M < \infty$ and $0 < \gamma < 1$, so that

$$|\epsilon_k| \leq M\gamma^k, \quad k \geq 0. \tag{8.26}$$

In particular,

$$\lim_{k \to \infty} \{y_k - F(z)w_k\} = 0,$$

and so

$$\lim_{k \to \infty} \{B(z)u_k + G(z)w_k\}$$

$$= \lim_{k \to \infty} \{z^{-d}A(z)y_k - z^{-d}C(z)w_k + G(z)w_k\}$$

$$= \lim_{k \to \infty} \{z^{-d}A(z)y_k - z^{-d}[C(z) - z^d G(z)]w_k\}$$

$$= \lim_{k \to \infty} \{z^{-d}A(z)y_k - z^{-d}A(z)F(z)w_k\}$$

$$= \lim_{k \to \infty} \{z^{-d}[A(z)y_k - F(z)w_k]\} = 0.$$

Hence the control law (8.24) asymptotically satisfies the same relation as the control law (8.21).

To study internal stability, write the closed loop system as

$$[A(z)F(z)B(z) - z^d B(z)G(z)]y_k = C(z)w_k,$$

or

$$C(z)B(z)y_k = C(z)w_k, \tag{8.27}$$

so the closed loop system is internally stable if $C(z)$ and $B(z)$ have all roots strictly outside the unit circle.

Theorem (8.28)
Consider the ARMAX model (8.17) and suppose that the roots of $C(z)$ and $B(z)$ are strictly outside the unit circle. Then the time-invariant control law (8.24), (8.19) has the following properties:
(1) It converges asymptotically to the minimum variance control law (8.21) for the cost (8.15).
(2) The resulting closed loop system is internally stable.
(3) The output y_k converges asymptotically to a moving average of white noise, $F(z)w_k$.
(4) It minimizes the average cost (8.16).
Proof
It remains only to prove (4). With $F(z) = \sum_{i=0}^{d-1} f_i z^i$, we will show that

$\sigma^2 \sum\limits_{i=0}^{d-1} f_i^2$ is a lower bound for (8.16), and that the control law (8.24) attains this lower bound. Define γ_k and η_k by

$$A(z)\gamma_k := w_k, \quad A(z)\eta_k := u_k.$$

Assume that the initial conditions satisfy

$$y_{p+1} = \sum_{i=0}^{p} [b_i \eta_{p+1-i-d}, + c_i \gamma_{p+1-i}] + \gamma_{p+1}$$

which, with a slight abuse of notation, we rewrite as

$$y_{p+1} = z^d B(z)\eta_{p+1} + C(z)\gamma_{p+1}.$$

We now show by induction that a similar relation holds for all k, i.e., when $p + 1$ is replaced by any $k \geq 1$. Suppose it holds for k:

$$y_k = z^d B(z)\eta_k + C(z)\gamma_k. \tag{8.29}$$

Then from (8.17),

$$\begin{aligned}
y_{k+1} &= z^{-1}[1 - A(z)]y_k + z^{d-1}B(z)u_k + z^{-1}C(z)w_k \\
&= z^{d-1}[1 - A(z)]B(z)\eta_k + z^{-1}[1 - A(z)]C(z)\gamma_k \\
&\quad + z^{d-1}B(z)u_k + z^{-1}C(z)w_k \\
&= z^{d-1}[1 - A(z)]B(z)\eta_k + z^{-1}[1 - A(z)]C(z)\gamma_k \\
&\quad + z^{d-1}B(z)A(z)\eta_k + z^{-1}C(z)A(z)\gamma_k \\
&= z^{d-1}B(z)\eta_k + z^{-1}C(z)\gamma_k \\
&= z^d B(z)\eta_{k+1} + C(z)\gamma_{k+1},
\end{aligned}$$

and the induction is complete. Using (8.19), rewrite (8.29) as

$$\begin{aligned}
y_k &= z^d [B(z)\eta_k + G(z)\gamma_k] + F(z)A(z)\gamma_k \\
&= z^d [B(z)\eta_k + G(z)\gamma_k] + F(z)w_k.
\end{aligned}$$

The first term on the right depends only on the past of the system up to $k - d$, whereas the term $F(z)w_k$ depends on $w_{k-d+1}, \ldots, w_k$. Hence

$$Ey_k^2 \geq \sigma^2 \sum_{i=0}^{d-1} f_i^2.$$

Thus

$$\lim_{N \to \infty} \frac{1}{N} \sum_{k=1}^{N} Ey_k^2 \geq \sigma^2 \sum_{i=0}^{d-1} f_i^2$$

for any control law. On the other hand when the control law (8.24) is used, $y_k = \epsilon_k + F(z)w_k$, where ϵ_k satisfies (8.26). Hence using (8.26), and with deterministic initial conditions for simplicity, we have

$$\lim_{N \to \infty} \frac{1}{N} \sum_{k=1}^{N} E y_k^2$$

$$= \lim_{N \to \infty} \frac{1}{N} \sum_{k=1}^{N} E(\epsilon_k + F(z)w_k)^2$$

$$= \lim_{N \to \infty} \frac{1}{N} \sum_{k=1}^{N} \{\epsilon_k^2 + 2\epsilon_k E(F(z)w_k) + E(F(z)w_k)^2\}$$

$$= \lim_{N \to \infty} \frac{1}{N} \sum_{k=1}^{N} E(F(z)w_k)^2 = \sigma^2 \sum_{i=0}^{d-1} f_i^2,$$

concluding the proof.   □

Exercise   (8.30)

Consider the ARMA system

$$y_{k+1} = \sum_{i=0}^{p} [a_i y_{k-i} + c_i w_{k-i}] + w_{k+1}.$$

Suppose that $C(z)$ and $A(z)$ have all their roots strictly outside the unit circle. Let $\{\hat{y}_k\}$ be generated by

$$C(z)\hat{y}_k = z^d G(z)y_k,$$

where $C(z)$ and $G(z)$ are as above. Show that $\hat{y}_k$ converges to $y_{k \mid k-d}$— the minimum mean square error d-step ahead predictor of y_k given $\{y_l \mid l \le k - d\}$. Also show that the variance of the prediction error $(y_k - \hat{y}_k)$ converges to $\sigma^2 \sum_{i=0}^{d-1} f_i^2.$

9. Signal processing applications

In some applications, especially in signal processing, it is important to estimate the next value of a process as a linear combination of a fixed number of preceding observations.

Let $\{y_k\}$ be a scalar, stationary, Gaussian process with

$$Ey_k := 0, \tag{9.1}$$

$$cov(y_k, y_{k-i}) := \sigma_i, \ i \ge 0. \tag{9.2}$$

Note that $\sigma_i = \sigma_{-i}$. Suppose we want to predict y_k given y_l, $l \le k - 1$. From Chapter 6 we know that if $\{y_k\}$ has a rational spectral density, then it can be regarded as the output of a stable linear system of the form (2.1) and (2.2), and we can then obtain $y_{k \mid k-1}$ by the Kalman filter. This procedure is straightforward but computationally burdensome because we must first find the model (2.1) and (2.2) from the covariance function $\{\sigma_i\}$, and the Kalman filter calculations might be too complicated. We pursue here an alternative approach, which allows us to reduce the computational complexity at the expense of increased prediction error.

The basic idea is to fix an integer n and form the estimate

$$y_{k \mid k-1,\dots,k-n} := E\{y_k \mid y_{k-1}, \dots, y_{k-n}\}, \tag{9.3}$$

which is based only on the preceding n observations, rather than on the entire past. By Lemma (2.5), the estimate (9.3) is a linear combination of $y_{k-1}, \dots, y_{k-n}$, i.e., there exist constants $a_1, \dots, a_n$, such that

$$y_{k \mid k-1,\dots,k-n} = \sum_{i=1}^{n} a_i y_{k-i}, \tag{9.4}$$

which is easy to implement. By Exercise (2.29), the coefficients a_i minimize the error variance,

$$E(y_k - \sum_{i=1}^{n} a_i y_{k-i})^2. \tag{9.5}$$

Exercise (9.6)
Show that the coefficients a_i which minimize (9.5) are given as the solution of the following linear system of equations:

$$
\begin{bmatrix}
\sigma_0 & \sigma_1 & \cdot & \cdot & \sigma_{n-1} \\
\sigma_1 & \sigma_0 & \cdot & \cdot & \sigma_{n-2} \\
\cdot & & \cdot & \cdot & \cdot \\
\cdot & & & \cdot & \cdot \\
\sigma_{n-1} & \sigma_{n-2} & \cdot & \cdot & \sigma_0
\end{bmatrix}
\begin{bmatrix}
a_1 \\
a_2 \\
\cdot \\
\cdot \\
a_n
\end{bmatrix}
=
\begin{bmatrix}
\sigma_1 \\
\sigma_2 \\
\cdot \\
\cdot \\
\sigma_n
\end{bmatrix}. \tag{9.7}
$$

[Hint: Set to zero the partial derivatives of (9.5) with respect to each a_i.]

Equations (9.7) are called the **Yule-Walker** equations. Let

$$\epsilon_k := y_k - y_{k \mid k-1,\dots,k-n}$$

be the error. Its variance is

$$
\begin{aligned}
\sigma := E \epsilon_k^2 &= E(y_k - \sum_{i=1}^{n} a_i y_{k-i})^2 \\
&= E y_k^2 - 2E \sum_{i=1}^{n} a_i y_k y_{k-i} + E \sum_{i=1}^{n} \sum_{j=1}^{n} a_i a_j y_{k-i} y_{k-j} \\
&= \sigma_0 - 2 \sum_{i=1}^{n} a_i \sigma_i + \sum_{i=1}^{n} \sum_{j=1}^{n} a_i \sigma_{i-j} a_j
\end{aligned}
$$

$$= \sigma_0 - \sum_{i=1}^{n} a_i \sigma_i, \tag{9.8}$$

where the last equality follows from the use of (9.7).

The relation between the error and observation processes can be expressed in transfer function form as

$$\epsilon_k = H(q^{-1}) y_k,$$

where $H(q^{-1}) := 1 - \sum_{i=1}^{n} a_i q^{-i}$, and q^{-1} is the shift operator; see Section 5.2.

Exercise (9.9)
Show that $H(q^{-1})$ is minimum phase, i.e., all its roots are inside the unit circle.
[Hint: From (6.5.2) and (6.5.9),

$$E \epsilon_k^2 = \int_{-\pi}^{\pi} H(e^{-i\omega}) \Phi^y(\omega) H(e^{i\omega}) d\omega = \int_{-\pi}^{\pi} \Phi^y(\omega) |H(e^{i\omega})|^2 d\omega,$$

where $\Phi^y(\omega)$ is the spectral density function of $\{y_k\}$. Hence the problem of choosing the coefficients a_i to minimize the error variance is equivalent to choosing a polynomial $H(q^{-1}) := 1 - \sum_{i=1}^{\infty} a_i q^{-i}$, with real coefficients, and with 1 as its leading coefficient, so as to minimize the integral above. Let

$$H(q^{-1}) = (1 - \frac{q_1}{q}) \dots (1 - \frac{q_n}{q}).$$

If q_j is a real root of $H(q^{-1})$, with $|q_j| > 1$, then consider the alternative transfer function $\hat{H}(q^{-1})$ obtained by replacing the root q_j by $\bar{q}_j^{-1}$, where $\bar{q}_j$ is the complex conjugate of q_j. (This reflects q_j with respect to the unit circle.) Then

$$\hat{H}(q^{-1}) := H(q^{-1})(1 - \frac{1}{\bar{q}_j q})(1 - \frac{q_j}{q})^{-1},$$

and so

$$|\hat{H}(e^{i\omega})| = \frac{1}{|q_j|} |H(e^{i\omega})| < |H(e^{i\omega})|.$$

Thus the transfer function $\hat{H}(q^{-1})$ yields a lower variance, which is impossible. If q_j is complex, then there is also a conjugate root $\bar{q}_j$ of $H(q^{-1})$, which must also be reflected at the same time in order to ensure that $\hat{H}(q^{-1})$ has real coefficients.]

We now present the Levinson algorithm, which is a very efficient way of solving (9.7). It is convenient to combine equations (9.7) and (9.8) together as:

$$
\begin{bmatrix}
\sigma_0 & \sigma_1 & . & . & \sigma_n \\
\sigma_1 & \sigma_0 & . & . & \sigma_{n-1} \\
. & . & . & . & . \\
. & . & . & . & . \\
\sigma_n & \sigma_{n-1} & . & . & \sigma_0
\end{bmatrix}
\begin{bmatrix}
1 \\
-a_1 \\
. \\
. \\
-a_n
\end{bmatrix}
=
\begin{bmatrix}
\sigma \\
0 \\
. \\
. \\
0
\end{bmatrix}.
\tag{9.10}
$$

The complexity of the filter (9.4) is related to its order n. If the resulting error variance σ is too large, then one may wish to increase the order of the filter to $n + 1$, and thereby try to reduce the error variance. Thus n may be regarded as a parameter to be chosen by the designer. To emphasize the dependence of σ and a_i on n, we denote them by $\sigma_{.\,|\,n}$ and a_i^n.

When the filter order is increased from n to $n + 1$, one must calculate $\sigma_{.\,|\,n+1}$ and $a_1^{n+1}, \ldots, a_{n+1}^{n+1}$. It would be desirable to obtain these from $\sigma_{.\,|\,n}$ and $a_1^n, \ldots, a_n^n$. We can get such a recursive procedure by exploiting the Toeplitz structure of the matrix in (9.10). (A matrix A is said to be a **Toeplitz matrix** if it is constant along all the subdiagonals, i.e., $A_{ij} = A_{i+k,j+k}$.)

The vector $(1, -a_1^n, \ldots, -a_n^n, 0)$, obtained by augmenting the vector in (9.10) by adding a 0 as the last component, is not necessarily equal to $(1, -a_1^{n+1}, \ldots, -a_n^{n+1}, -a_{n+1}^{n+1})$. However,

$$
\begin{bmatrix}
\sigma_0 & \sigma_1 & . & . & \sigma_{n+1} \\
\sigma_1 & \sigma_0 & . & . & \sigma_n \\
. & . & . & . & . \\
. & . & . & . & . \\
\sigma_{n+1} & \sigma_n & . & . & \sigma_0
\end{bmatrix}
\begin{bmatrix}
1 \\
-a_1^n \\
. \\
-a_n^n \\
0
\end{bmatrix}
=
\begin{bmatrix}
\sigma_{.\,|\,n} \\
0 \\
. \\
0 \\
\beta_n
\end{bmatrix},
\tag{9.11}
$$

where

$$
\beta_n := \sigma_{n+1} - \sum_{i=1}^{n} a_i^n \sigma_{n-i+1}.
$$

If $\beta_n = 0$, then a comparison with (9.10) shows that the augmented vector would indeed be what we want. In the following we make $\beta_n = 0$ by a clever manipulation.

If A is a symmetric Toeplitz matrix and

$$
A
\begin{bmatrix}
x_1 \\
. \\
. \\
x_n
\end{bmatrix}
=
\begin{bmatrix}
y_1 \\
. \\
. \\
y_n
\end{bmatrix},
$$

then

$$A \begin{bmatrix} x_n \\ \vdots \\ x_1 \end{bmatrix} = \begin{bmatrix} y_n \\ \vdots \\ y_1 \end{bmatrix};$$

i.e., reversing the order of the elements in x leads to a reversal of the order of the elements in y. Hence from (9.11),

$$\begin{bmatrix} \sigma_0 & \sigma_1 & \cdot & \cdot & \sigma_{n+1} \\ \sigma_1 & \sigma_0 & \cdot & \cdot & \sigma_n \\ \cdot & \cdot & \cdot & \cdot & \cdot \\ \cdot & \cdot & \cdot & \cdot & \cdot \\ \sigma_{n+1} & \sigma_n & \cdot & \cdot & \sigma_0 \end{bmatrix} \begin{bmatrix} 0 \\ -a_n^n \\ \cdot \\ -a_1^n \\ 1 \end{bmatrix} = \begin{bmatrix} \beta_n \\ 0 \\ \cdot \\ 0 \\ \sigma_{.\,|\,n} \end{bmatrix}. \tag{9.12}$$

Let γ_n be some scalar. Adding γ_n times (9.12) to (9.11) gives

$$\begin{bmatrix} \sigma_0 & \sigma_1 & \cdot & \cdot & \sigma_{n+1} \\ \sigma_1 & \sigma_0 & \cdot & \cdot & \sigma_n \\ \cdot & \cdot & \cdot & \cdot & \cdot \\ \cdot & \cdot & \cdot & \cdot & \cdot \\ \sigma_{n+1} & \sigma_n & \cdot & \cdot & \sigma_0 \end{bmatrix} \begin{bmatrix} 1 \\ -a_1^n - \gamma_n a_n^n \\ \cdot \\ -a_n^n - \gamma_n a_1^n \\ \gamma_n \end{bmatrix} = \begin{bmatrix} \sigma_{.\,|\,n} + \gamma_n \beta_n \\ 0 \\ \cdot \\ 0 \\ \beta_n + \gamma_n \sigma_{.\,|\,n} \end{bmatrix}.$$

If we can choose γ_n so that the last component $(\beta_n + \gamma_n \sigma_{.\,|\,n})$ of the vector on the right is 0, we would obtain a solution satisfying (9.10). Hence we choose,

$$\gamma_n := -\frac{\beta_n}{\sigma_{.\,|\,n}}.$$

Moreover, $\sigma_{.\,|\,n+1}$ can then also be obtained as the first element of the vector on the right hand side above, [compare with (9.10)]. This is the **Levinson algorithm.**

Initialization: Set $a_1^1 := \dfrac{\sigma_1}{\sigma_0}$ and $\sigma_{.\,|\,1} := \sigma_0 - a_1^1 \sigma_1$.

Recursive Step: For $n = 1, 2, \ldots$, define

$$\beta_n := \sigma_{n+1} - \sum_{i=1}^n a_i^n \sigma_{n-i+1}, \quad \text{and} \quad \gamma_n := -\frac{\beta_n}{\sigma_{.\,|\,n}}.$$

Then,

$$\begin{bmatrix} \sigma_{.\,|\,n+1} \\ a_1^{n+1} \\ \cdot \\ \cdot \\ a_{n+1}^{n+1} \end{bmatrix} = \begin{bmatrix} \sigma_{.\,|\,n} \\ a_1^n \\ \cdot \\ a_n^n \\ 0 \end{bmatrix} + \gamma_n \begin{bmatrix} \beta_n \\ a_n^n \\ \cdot \\ a_1^n \\ -1 \end{bmatrix}. \tag{9.13}$$

This algorithm recursively provides a sequence of filters of growing order, as well as their error variances. Thus one can run this algorithm until the variance $\sigma_{.|n}$ is acceptable and then choose the corresponding filter. Another important advantage of the Levinson algorithm is that it is computationally very efficient, since it takes advantage of the Toeplitz structure of the problem.

The next exercise examines the solution when the stochastic process $\{y_k\}$ is vector-valued.

Exercise   (9.14)

Let $\{y_k\}$ be a vector-valued stationary, Gaussian process with $Ey_k \equiv 0$ and $cov(y_k, y_{k-i}) := E\, y_k y_{k-i}^T := \Sigma_i$. Let

$$E\{y_k \mid y_{k-1}, \ldots, y_{k-n}\} =: \sum_{i=1}^{n} A_i^n y_{k-i},$$

and denote the covariance of the error of the estimate by $\Sigma_{.|n}$. Obtain a recursive solution for the coefficients A_i^n, $1 \le i \le n$ and $\Sigma_{.|n}$.
[Hint: Show first that

$$[I, -A_1^n, \ldots, -A_n^n] = \begin{bmatrix} \Sigma_0 & \Sigma_1 & . & . & \Sigma_n \\ \Sigma_{-1} & \Sigma_0 & . & . & \Sigma_{n-1} \\ . & & . & . & . \\ . & & . & . & . \\ . & & . & . & . \\ \Sigma_{-n} & \Sigma_{-n+1} & . & . & \Sigma_0 \end{bmatrix} [\Sigma_{.|n}, 0, \ldots, 0].$$

The chief difference between the scalar and vector cases is that Σ_i is not equal to Σ_{-i} but to its transpose. Hence the matrix above is not symmetric. To overcome this difficulty introduce a reversed set of matrices as follows. Suppose, by induction, that there are matrices $\bar{A}_i^n$, $1 \le i \le n$, and $\bar{\Sigma}_{.|n}$ such that

$$[-\bar{A}_n^n, \ldots, -\bar{A}_1^n, I] \begin{bmatrix} \Sigma_0 & \Sigma_1 & . & . & \Sigma_n \\ \Sigma_{-1} & \Sigma_0 & . & . & \Sigma_{n-1} \\ . & & . & . & . \\ . & & . & . & . \\ . & & . & . & . \\ \Sigma_{-n} & \Sigma_{-n+1} & . & . & \Sigma_0 \end{bmatrix} = [0, \ldots, 0, \bar{\Sigma}_{.|n}].$$

Now take appropriate linear combinations of the two equations above to get recursions for $\{A_i^{n+1}\}$, $\{\bar{A}_i^{n+1}\}$, $\Sigma_{.|n+1}$ and $\bar{\Sigma}_{.|n+1}$. Finally, determine the appropriate initial conditions for the recursions.]

Above we assumed that $\{y_k\}$ is Gaussian. If it is not Gaussian, then by Exercise (7.1) the best linear estimate of y_k given y_{k-i}, $1 \le i \le n$, is still given by the Yule-Walker equations which can be solved by the Levinson algorithm.

The best estimate of y_k given y_{k-i}, $i \geq 1$, i.e., given the infinite past, is obtained by setting $n = \infty$ in (9.7). The coefficients a_i, $i \geq 1$, satisfy the infinite set of linear equations,

$$\sigma_i = \sum_{m=1}^{\infty} a_m \sigma_{i-m}, i \geq 1,$$

(noting that $\sigma_i = \sigma_{-i}$). These are the discrete time versions of the **Wiener-Hopf** equations, and the corresponding filter,

$$y_{k \mid k-1} = \sum_{i=1}^{\infty} a_i y_{k-i},$$

is the **Wiener filter**. The transfer function of the error resulting from the (infinite order) Wiener filter is also minimum phase.

The Wiener filter is an alternative representation of the asymptotic limit (2.40) of the Kalman filter, since it corresponds to the situation where an infinite number of past observations are available, and $\{y_k\}$ is stationary. To obtain the Kalman filter one needs a state space description of $\{y_k\}$, whereas to obtain the Wiener filter one needs the covariance sequence $\{\sigma_i\}$ or the spectral density function. These are equivalent since one can obtain a state space description from the spectral density and vice versa, as seen in Chapter 5.

Next we obtain the Wiener filter by means of spectral factorization.

It is convenient to define the following process, (see also Exercise (3.4.2)).

$$\nu_0 := y_0,$$

$$\nu_k := y_k - E\{y_k \mid y_0, y_1, \ldots, y_{k-1}\}$$

$$= y_k - \sum_{i=1}^{k} a_i^k y_{k-i}, \ k \geq 1. \tag{9.15}$$

Exercise (9.16)

(1) Show that $\nu^k := (\nu_0, \ldots, \nu_k)$ and $y^k := (y_0, \ldots, y_k)$ are linearly related, i.e., there exists a nonsingular matrix G_k such that

$$\nu^k = G_k y^k \ \text{and} \ y^k = G_k^{-1} \nu^k.$$

(2) Show that $\{\nu_k\}$ is an independent, Gaussian process.
[Hint: From (9.15) it follows that there is a matrix G_k such that $\nu^k = G_k y^k$. Since G_k is a lower triangular matrix with 1s on the main diagonal, it is invertible. By Lemma (2.5) ν_k is uncorrelated with y^{k-1}, from which follows the independence.]

We can interpret (9.15) as saying that v_k consists of that part of y_k which cannot be predicted from the past y^{k-1}. Hence $\{v_k\}$ is called the **innovations process** of $\{y_k\}$. Moreover part (1) shows that the innovations process $\{v_k\}$ and the process $\{y_k\}$ contain the same information.

The innovations process can be obtained by the Gram-Schmidt orthogonalization method.

Exercise (9.17)
Define the inner product of random variables x and y by $<x, y> := E(xy)$, and say that x and y are orthogonal if $<x, y> = 0$. (This defines the Hilbert Space of all random variables with finite variances.) Show that $\{v_0, \ldots, v_k\}$ is an orthogonal basis for the subspace generated by $\{y_0 \ldots, y_k\}$.

This orthogonalization can be identified with the problem of generating an orthogonal sequence of polynomials.

Exercise (9.18)
Let $\Phi^y(\omega)$ be the spectral density of $\{y_k\}$. Introduce the inner product of two polynomials $H(q)$ and $G(q)$, defined on the unit circle in the complex plane, by

$$<H, G> := \int_{-\pi}^{\pi} H(e^{-i\omega})\Phi^y(\omega)G(e^{i\omega})d\omega. \qquad (9.19)$$

Show that if $\{H_0, \ldots, H_k\}$ are orthogonal, monic polynomials,

$$H_n(q) := q^n - \sum_{i=1}^{n} a_i^n q^{n-i}, \qquad (9.20)$$

then the innovations process is given by

$$v_k = H_k(q)y_0.$$

As usual, q is the shift operator, $(q^j y)_0 = y_j$.

The polynomials $\{H_k(q)\}$ are characterized by the following three conditions:
(1) $\{H_k(q)\}$ are orthogonal with respect to the inner product (9.19).
(2) For each n, $\{H_k(q), k \leq n\}$ spans the subspace of polynomials of degree n or less.
(3) Each $H_k(q)$ is normalized to be a monic polynomial. (Monic polynomial means that the leading coefficient is 1.)
They can therefore be obtained by the Gram-Schmidt orthogonalization process, see Exercise (6.15). We can also use the Levinson algorithm (9.13) to develop a recursion for the polynomials. Define the reversed polynomial

$$\bar{H}_n(q) := 1 - \sum_{i=1}^{n} a_i^n q^i,$$

where the coefficients of increasing powers of q are written in the reverse of the order in which they appear in $H_n(q)$. Then the recursion (9.13) for the coefficients $\{a_i^n\}$ gives us the following recursion for the polynomials:

$$\begin{bmatrix} H_{n+1}(q) \\ \overline{H}_{n+1}(q) \end{bmatrix} = \begin{bmatrix} q & \gamma_n \\ \gamma_n q & 1 \end{bmatrix} \begin{bmatrix} H_n(q) \\ \overline{H}_n(q) \end{bmatrix}, \quad \begin{bmatrix} H_0(q) \\ \overline{H}_0(q) \end{bmatrix} = \begin{bmatrix} 1 \\ 1 \end{bmatrix}, \tag{9.21}$$

with $\gamma_0 := -\dfrac{\sigma_1}{\sigma_0}$. The polynomials $H_n(q)$ are called **Szego's orthogonal polynomials**.

The next exercise shows that the transfer function for the Wiener filter can also be obtained by a factorization of the spectral density function $\Phi^y(\omega)$ of $\{y_k\}$.

Exercise (9.22)

Supose $\Phi^y(\omega)$ is factored as in (6.5.10),

$$\Phi^y(\omega) = H(e^{-i\omega})H^*(e^{i\omega}),$$

where $H(q^{-1})$ is a rational transfer function with both numerator and denominator polynomials having all their roots strictly inside the unit circle. In power series form let $H(q^{-1}) =: \sum\limits_{i=0}^{\infty} h_i q^{-i}$. Show that the Wiener filter is given by

$$y_{k\,|\,k-1} = [\frac{H(q^{-1}) - h_0}{H(q^{-1})}]y_k.$$

[Hint: Show that $w_k := \dfrac{1}{H(q^{-1})} y_k$ is an independent, Gausssian process such that $\{w_s, s \le k\}$ can be obtained from $\{y_s, s \le k\}$ by linear operations, and vice versa. Hence $y_{k\,|\,k-1}$ can also be written as

$$E\{y_k \mid y_s, s \le k-1\} = E\{\sum_{i=0}^{\infty} h_i w_{k-i} \mid y_s, s \le k-1\}$$

$$= E\{\sum_{i=0}^{\infty} h_i w_{k-i} \mid w_s, s \le k-1\}$$

$$= \sum_{i=1}^{\infty} h_i w_{k-i}.]$$

Note that $\{w_k\}$ is also stationary. Hence it is a Gaussian white noise process. The filter $w_k = \dfrac{1}{H(q^{-1})} y_k$ is therefore called a **whitening filter**.

In the remainder of this section we present an implementation of the estimate (9.4) which is particularly suited for digital signal processing applications.

The estimate

$$y_{k \mid k-1,\ldots,k-n} = \sum_{i=1}^{n} a_i^n y_{k-i},$$

can be implemented using a sequence of delay elements as in the figure below. Such an implementation is called a **tapped delay line** or **transversal** filter.

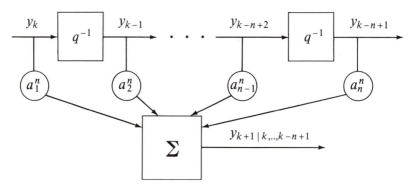

One disadvantage of this filter is that if its order n is changed, then the weights a_i^n of the taps also have to be altered, and so the entire filter is changed. It is desirable to have an implementation where an increase (or decrease) in the order of the predictor can be accomplished in a simple way. The difficulty lies in the fact that, in general, $a_i^{n+1} \neq a_i^n$. One special case where equality holds is given next.

Exercise   (9.24)

Suppose $\{z_k\}$ is an independent Gaussian process, and

$$E\{y_k \mid z_{k-1}, \ldots, z_{k-n}\} = \sum_{i=1}^{n} a_i^n z_{k-i}.$$

Show that $a_i^{n+1} = a_i^n$ for $1 \leq i \leq n$.

Usually, however, $\{y_k\}$ is not independent; but its innovations process is independent, and it has the same information as $\{y_k\}$. Hence it is preferable to use the innovations process to form the estimate. Motivated by this, we define the following process:

$$v_0^b(k) := y_k,$$

$$v_n^b(k) := y_{k-n} - E\{y_{k-n} \mid y_{k-n+1}, \ldots, y_k\}, \quad n \geq 1. \quad (9.25)$$

This is a backward version of the innovations process (9.15), i.e., the role of time is reversed, and the origin is moved from 0 to k. Accordingly, we call $\{v_n^b(k)\}$ the **backward** innovations process. It gives us the part of y_{k-n-1} which is not predictable from $\{y_{k-n}, \ldots, y_k\}$.

The next exercise shows that

$$E\{y_{k-n} \mid y_{k-n+1}, \ldots, y_k\} = \sum_{i=1}^{n} a_i^n y_{k-n+i}. \tag{9.26}$$

Exercise (9.27)

(1) Show that $\{y_k, \ldots, y_{k-n}\}$ and its reverse $\{y_{k-n}, \ldots, y_k\}$ have the same pd.

(2) Show that $E\{y_{k-n} \mid y_{k-n+1}, \ldots, y_k\}$ is given by (9.26), where the coefficients a_i^n are given by the Levinson algorithm (9.13).

[Hint: The two vectors have zero mean. Use the stationarity of $\{y_k\}$ to show that they also have the same covariance. Since they are both Gaussian, they have the same pd. The rest follows readily from this.]

A comparison of (9.26) with (9.20) shows that

$$v_n^b(k) = H_n(q^{-1})y_k.$$

We can use (9.21) to compute $\{v_n^b(k)\}$ recursively in n, and so we define the process

$$v_n^f(k) := \overline{H}_n(q^{-1})y_k.$$

We call $\{v_n^f(k)\}$ the **forward** innovations process, for reasons that are evident from the next exercise.

Exercise (9.28)

(1) Show that $v_n^f(k) = y_k - E\{y_k \mid y_{k-1}, \ldots, y_{k-n}\}$. (2) Show that the process $\{v_n^f(k), v_{n+1}^f(k+1), \ldots, v_{n+i}^f(k+i), \ldots\}$ is just the innovations process (9.15) with the time origin shifted from 0 to k.

Through (9.21) we now have the following recursions (in n) for $v_n^b(k)$ and $v_n^f(k)$:

$$v_{n+1}^b(k) = v_n^b(k-1) + \gamma_n v_n^f(k), \quad v_0^b(k) := y_k,$$

$$v_{n+1}^f(k) = \gamma_n v_n^b(k-1) + v_n^f(k), \quad v_0^f(k) := y_k.$$

The arrangement shown in the diagram on p.138 computes $v_{n+1}^b(k)$ and $v_{n+1}^f(k)$ from $v_n^b(k)$ and $v_n^f(k)$. It is called a **lattice section**. Clearly, by forming a cascade of such lattice sections, we can determine $\{v_n^b(k), v_n^f(k), n \geq 1\}$.

From the properties of the innovations process given in (9.16), which also hold for the backward innovations process, it follows that $E\{y_{k+1} \mid y_k, \ldots, y_{k-n}\}$, which is a linear combination of $y_k, \ldots, y_{k-n}$,

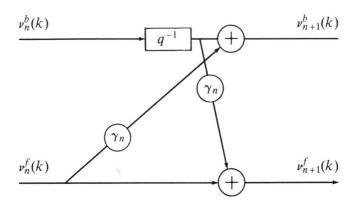

can also be written as a linear combination of $v_0^b(k), \ldots, v_n^b(k)$. Hence there exist constants $\lambda_0, \ldots, \lambda_n$, such that,

$$E\{y_{k+1} \mid y_k, \ldots, y_{k-n}\} = \sum_{i=0}^{n} \lambda_i v_i^b(k).$$

The following exercise exploits the independence of $v_0^b(k), \ldots, v_n^b(k)$ in calculating $\{\lambda_i\}$.

Exercise (9.29)

Since $v_0^b(k), \ldots, v_n^b(k)$ are independent, show that

$$E\{y_{k+1} \mid y_k, \ldots, y_{k-n}\} = \sum_{i=0}^{n} E\{y_{k+1} \mid v_i^b(k)\}$$

$$= \sum_{i=0}^{n} \frac{E[y_{k+1} v_i^b(k)]}{E[(v_i^b(k))^2]} v_i^b(k). \qquad (9.30)$$

[Hint: For zero mean, Gaussian vectors, w and z,

$$E\{z \mid w\} = [Ezw^T][Eww^T]^{-1}w.$$

In our context, with z and w appropriately chosen, Eww^T is diagonal.]

It only remains to compute $E[y_{k+1} v_i^b(k)]$ and $E[(v_i^b(k))^2]$.

Exercise (9.31)

(1) Show that $E[y_{k+1} v_i^b(k)] = -\beta_i$, where β_i is as defined in the Levinson algorithm.

(2) Show that $E[(v_i^b(k))^2] = \sigma_{.|i}$.

[Hint: For (1), substitute (9.25) and (9.26) for $v_i^b(k)$. For (2), use (1) of Exercise (9.27), as suggested in the proof of that exercise.]

From (9.30) it follows that,

$$E\{y_{k+1} \mid y_k, \ldots, y_{k-n}\} = \sum_{i=0}^{n} \gamma_i v_i^b(k),$$

where γ_i is as given in the Levinson algorithm. Hence the estimate can be obtained as a weighted combination of the backward innovations. Since the backward innovations process can be obtained as a cascade of lattice sections, we have the the **lattice filter** implementation shown in the accompanying diagram, where LS denotes a lattice section.

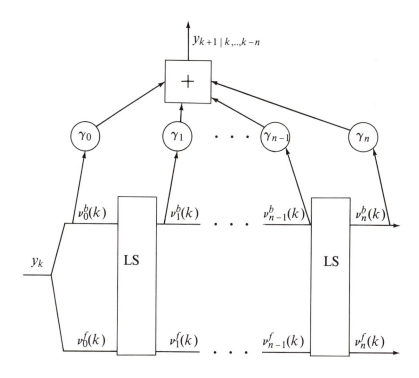

The prediction is simply the weighted combination of the outputs of the various lattice sections. The important advantage of this arrangement is that the order n of the predictor is increased simply by the addition of another lattice section, and adding the weighted output of the section to the prediction. Thus the lattice arrangement yields predictions of various orders simultaneously.

The lattice filter is completely determined by the specification of the sequence $\{\gamma_i\}$. In analogy with circuits of this form, the $\{\gamma_i\}$ are called **reflection coefficients**, or, in statistics, **partial correlation coefficients**. The next exercise shows that, in conformity with the usual usage of the term, the reflection coefficients are smaller than one in magnitude.

Exercise   (9.32)

Show that $|\gamma_n| \leq 1$ for all n.

[Hint: Show that the recursion for the error variance in the Levinson algorithm can also be written as $\sigma_{.|n+1} = (1 - \gamma_n^2)\sigma_{.|n}.$]

10. Notes

1. In Theorem (2.38), the conditions of reachability and observability can be replaced by the weaker conditions of stabilizability and detectability; see Anderson and Moore (1979). More detailed behavior of the Riccati equation is studied in Goodwin and Sin (1984).

2. The minimum variance control law is studied in Astrom (1970) which also treats Exercise (8.30). If the system is not minimum phase, then the control law derived in Section 8 does not give an internally stable system. One can then ask for the control law which minimizes the variance of the output within the class of control laws which give rise to an internally stable system. This constrained minimization problem is solved in Peterka (1972); its multivariable version is examined in Shaked and Kumar (1986).

3. The U-D factorization of Section 6 is from Bierman (1975, 1977) and Thornton and Bierman (1977).

4. The Levinson algorithm is originally due to Levinson (1946). The lattice filter is generally credited to Itakura and Saito (1971). For further developments in this area, see Dewilde and Dym (1981). More on the topic of fast algorithms can be found in Morf (1974), Morf and Kailath (1975), Kailath, Morf and Sidhu (1974), Lindquist (1974), and Kailath, Vieira and Morf (1979). For a survey of these algorithms see Favier (1982). The role of the innovations process in estimation theory is explained in Kailath (1970).

5. The lattice filter for the process $\{y_k\}$ is completely characterized, or parameterized, by the reflection coefficients. In fact the spectral density function of any stochastic process can be so parameterized by a sequence of reflection coefficients $\{k_i\}$. This is useful, since whereas the covariance function, $\{\sigma_i\}$, is constrained by a complicated relation, namely, that the resulting spectrum generated by (6.5.1) is positive for all ω, the reflection coefficients are constrained simply to be less than 1 in absolute magnitude, as in Exercise (9.32). This parametrization is useful in spectral estimation. For example, suppose one has a finite sequence $\{\sigma_i, i \leq n\}$ of covariances, and one wants to extend this to an infinite sequence. This can be done by simply calculating the first $n-1$ reflection coefficients (which the Levinson algorithm shows to be dependent only on the first few correlation coefficients), and then choosing the remaining

reflection coefficients arbitrarily to lie in $[-1, 1]$. Of course there are several possible extensions, and one may wish to choose between them by imposing some selection criterion. For example, in the *maximum entropy method* of spectral estimation, one extends the spectrum so that it has maximum entropy. It can be shown that the resulting process is an autoregressive process. An excellent tutorial survey of the connections between the Levinson algorithm, Wiener filtering, spectral estimation, Wold decomposition, and the maximum entropy method is given in Papoulis (1985).

6. An illuminating historical survey of the field of estimation is given in Kailath (1974). For an account of the implementation of estimators by very large scale integrated circuits (VLSI), and of their connections with signal processing, see Kung, Whitehouse and Kailath (1985). That reference also discusses **systolic array** implementations, which use synchronized or rhythmic parallel computation for high sampling rates.

11. References

[1] B.D.O. Anderson and J.B. Moore (1979), *Optimal filtering*, Prentice Hall, Englewood Cliffs, NJ.

[2] K. Astrom (1970), *Introduction to stochastic control theory*, Academic Press, New York.

[3] G.J. Bierman (1975), "Measurement updating using the U-D factorization," *Proceedings of the IEEE Conference on Decision and Control*, Houston, 337-346.

[4] G.J. Bierman (1977), *Factorization methods for discrete sequential estimation*, Academic Press, New York.

[5] P. Dewilde and H. Dym (1981), "Schur recursions, error formulas, and convergence of rational estimators for stationary stochastic sequences," *IEEE Transactions on Information Theory*, vol IT-27(4), 446-461.

[6] G. Favier (1982), *Filtrage, modelisation et identification de systemes lineaires stochastiques a temps discret*, Editions du CNRS, Paris.

[7] G. Goodwin and K. Sin (1984), *Adaptive filtering, prediction and control*, Prentice Hall, Englewood Cliffs, NJ.

[8] F. Itakura and S. Saito (1971), "Digital filtering techniques for speech analysis and synthesis," *Proceedings of the Seventh International Congress on Acoustics*, vol 25-C-1, 261-274, Budapest.

[9] T. Kailath (1970), "The innovations approach to detection and estimation theory," *Proceedings of the IEEE*, vol 58, 680-695.

[10] T. Kailath (1974), "A view of three decades of linear filtering

theory," *IEEE Transactions on Information Theory,* vol IT-20(2), 146-181.

[11] T. Kailath, M. Morf and G. Sidhu (1974), "Some new algorithms for recursive estimation in constant discrete-time linear systems," *IEEE Transactions on Automatic Control,* Vol AC-19(4), 315-323.

[12] T. Kailath, A. Vieira and M. Morf (1979), "Orthogonal transformation (square-root) implementations of the generalized Chandrasekhar and generalized Levinson algorithms," in A. Bensoussan and J.L. Lions (eds), *International Symposium on Systems Optimization and Analysis,* Lecture Notes in Control and Information Sciences, Vol 14, 81-91, Springer Verlag, Berlin.

[13] S. Y. Kung, H. J. Whitehouse and T. Kailath, (Editors) (1985), *VLSI and modern signal processing,* Prentice Hall, Englewood Cliffs, NJ.

[14] N. Levinson (1946), "The Wiener RMS (root mean square) error criterion in filter design and prediction," *Journal of Mathematics and Physics,* vol XXV, 261-278.

[15] A. Lindquist (1974), "A new algorithm for optimal filtering of discrete-time stationary processes," *SIAM J. Control,* Vol 12(4), 736-746.

[16] M. Morf (1974), "Fast algorithms for multivariable systems," *Ph. D. Thesis,* Department of Electrical Engineering, Stanford University.

[17] M. Morf and T. Kailath (1975), "Square root algorithms for least squares estimation," *IEEE Transactions on Automatic Control,* Vol AC-20(4), 487-497.

[18] A. Papoulis (1985), "Levinson's algorithm, Wold's decomposition, and spectral estimation," *SIAM Review,* vol 27(3), 405-441.

[19] V. Peterka (1972), "On steady state minimum variance control strategy," *Kybernetica,* vol. 8, 219-232.

[20] U. Shaked and P. R. Kumar (1986), "Minimum variance control of multivariable ARMAX systems," *SIAM J. Control and Optimization,* to appear.

[21] C.L. Thornton and G.J. Bierman (1977), "Gram-Schmidt algorithms for covariance propogation," *International Journal of Control,* Vol 25(2), 243-260.

CHAPTER 8
INFINITE HORIZON DYNAMIC PROGRAMMING

We consider feedback control over the infinite horizon with the criteria of discounted cost and the average cost per unit time. Optimal control laws are stationary, and the value function satisfies certain functional equations. Algorithms for the value function and the optimal law are derived. In some problems, these algorithms allow us to deduce qualitative features of the optimal law without performing computations.

1. Types of cost criteria

Consider a stochastic system modeled as a controlled Markov chain whose state x_k takes on finitely many values $i = 1, 2, . . , I$. The control u_k is selected from the finite control constraint set U. The system is described by the state transition probability matrix $P(u) := \{P_{ij}(u), 1 \leq i, j \leq I\}$,

$$P_{ij}(u) := Prob\ \{x_{k+1} = j \mid x_k = i, u_k = u\}.$$

The state is observed, so a control law is any sequence $g = \{g_0, g_1, . . \}$ with

$$u_k = g_k(x^k) \in U$$

for all $x^k = (x_0, . . , x_k)$; G denotes the set of all control laws.

In Chapter 6 we studied the cost

$$E^g \sum_{k=0}^{N-1} c_k(x_k, u_k) + c_N(x_N), \tag{1.1}$$

which considers the system behavior up to time N. If, however, we are interested in the system behavior for all time, it seems natural to consider the cost

$$E^g \sum_{k=0}^{\infty} c_k(x_k, u_k).$$

To simplify the problem suppose that the cost functions c_k do not depend on k, and so the (infinite horizon) total cost is

$$E^g \sum_{k=0}^{\infty} c(x_k, u_k). \tag{1.2}$$

Note that the sum in (1.2) may be infinite or undefined.

Exercise   (1.3)
Construct a system with two states and one control value for which the total cost is infinite. Construct another such system where the total cost is undefined.

[Hint: If $a_k = (-1)^k$, then $\sum_{k=0}^{\infty} a_k$ is undefined.]

Exercise   (1.4)
If $E^g \sum_{k=0}^{\infty} c(x_k, u_k)$ is well defined and finite, show that $c(x, u) = 0$ for some (x, u).

[Hint: If $\sum_{k=0}^{\infty} a_k$ is finite, then $\lim_{k \to \infty} a_k = 0$.]

There is another way to extend the finite horizon cost (1.1). The cost

$$E^g \frac{1}{N} \sum_{k=0}^{N-1} c(x_k, u_k)$$

is equivalent to (1.1). Its infinite horizon analogs are

$$E^g \lim_{N \to \infty} \frac{1}{N} \sum_{k=0}^{N-1} c(x_k, u_k) \qquad (1.5)$$

or

$$\lim_{N \to \infty} \frac{1}{N} \sum_{k=0}^{N-1} E^g c(x_k, u_k); \qquad (1.6)$$

these coincide if the limit and expectation can be interchanged. Since $\lim_{N \to \infty} \frac{1}{N} \sum_{k=0}^{N} c(x_k, u_k)$ is the time average of the cost incurred over the infinite horizon, (1.5) or (1.6) is called the **average cost** (per unit time). Only one of the two costs above is meaningful as the next exercise shows.

Exercise   (1.7)
Show that if the total cost (1.2) is finite, then the average cost (1.5) is zero.

Another useful criterion is the **discounted cost**,

$$E^g \sum_{k=0}^{\infty} \beta^k c(x_k, u_k), \qquad (1.8)$$

where $\beta \in (0, 1)$ is the discount factor. Costs incurred at time k are weighted by the factor β^k, so present costs are more important than future costs.

2. Functional equation for the discounted cost problem

Consider first the problem of minimizing the finite horizon cost

$$E^g \sum_{k=0}^{N-1} \beta^k \, c(x_k, u_k), \tag{2.1}$$

with $0 < \beta < 1$. By Theorem (6.3.2), the (optimal) value function $V_k(i)$, $k = 0, \ldots, N$, $i = 1, \ldots, I$, satisfies the dynamic programming equations,

$$V_N(i) = 0, \tag{2.2}$$

$$V_k(i) = \inf_{u \in U} \{\beta^k \, c(i, u) + \sum_{j=1}^{I} V_{k+1}(j) \, P_{ij}(u)\}. \tag{2.3}$$

$V_k(i)$ is the minimum value of $E^g \sum_{n=k}^{N-1} \beta^n \, c(x_n, u_n)$ starting in state i at time k. It is more convenient to work not with $V_k(i)$, but the related quantity,

$$W_n(i) := \beta^{n-N} \, V_{N-n}(i), \quad 0 \leq n \leq N, 1 \leq i \leq I. \tag{2.4}$$

The index n in $W_n(i)$ is the time to go.

Exercise (2.5)

Show that for $1 \leq i \leq I$ and $0 \leq n \leq N$,

$$W_0(i) := 0, \tag{2.6}$$

$$W_n(i) := \inf_{u \in U} \{c(i, u) + \beta \sum_{j=1}^{I} W_{n-1}(j) \, P_{ij}(u)\}. \tag{2.7}$$

Evidently $W_N(i) = V_0(i)$ is the minimum value of $E^g \sum_{k=0}^{N-1} \beta^k \, c(x_k, u_k)$ starting in state $x_0 = i$. For $n < N$, $W_n(i) = \beta^{n-N} V_{N-n}(i)$ is β^{n-N} times the minimum value of $E^g \sum_{k=N-n}^{N-1} \beta^k \, c(x_k, u_k)$ starting in state $x_{N-n} = i$, i.e., it is the minimum value of $E^g \sum_{k=N-n}^{N-1} \beta^{k-(N-n)} \, c(x_k, u_k)$. But since the transition probabilities are not dependent on time, we can translate the interval $\{N-n, \ldots, N-1\}$ to the interval $\{0, \ldots, n-1\}$ and get the following result.

Lemma (2.8)

Let $W_n(i) := \min_{g} E^g \{\sum_{k=0}^{n-1} \beta^k \, c(x_k, u_k) \mid x_0 = i\}$. Then $W_n(i)$ satisfies

(2.6) and (2.7).

For the discounted cost $E^g \sum_{k=0}^{\infty} c(x_k, u_k)$ it is natural to expect that its minimum value starting in state $x_0 = i$, denoted by $W_\infty(i)$, will satisfy (2.7) with ∞ replacing n. So it is plausible that W_∞ will satisfy

$$W_\infty(i) = \inf_{u \in U} \{c(i,u) + \beta \sum_{j=1}^{I} W_\infty(j) P_{ij}(u)\}, \text{ all } i. \qquad (2.9)$$

This is the dynamic programming equation for the discounted cost criterion.

Equation (2.9) is a system of I equations in the I unknowns $W_\infty(1), \ldots, W_\infty(I)$. Next we examine existence and uniqueness of solutions to these equations, and show that the solution indeed gives the value function.

The key to these questions rests on the fact that the discount factor β permits application of the next result.

Theorem (contraction mapping)   (2.10)
Let F be a complete normed space with norm denoted by $||\cdot||$. Let $T: F \to F$ satisfy the following inequality for some $0 < \beta < 1$,

$$||Ty - Tz|| \le \beta||y - z|| \text{ for all } y, z \in F. \qquad (2.11)$$

Then:
(1) There is a unique w satisfying $Tw = w$.
(2) For any $z \in F$, let $\{z, Tz, T^2 z, ..\}$ be the sequence of successive iterations—i.e., $T^2 z = T(Tz)$, etc;—then $\lim_{n \to \infty} ||T^n z - w|| = 0$ for all $z \in F$.
[Here, $Ty := T(y)$. A mapping T satisfying (2.11) is called a **contraction mapping**. If $Tw = w$, w is called a **fixed point** of T.]

Proof
Fix $z \in F$. Let $\alpha := ||z - Tz||$; then

$$||Tz - T^2 z|| \le \beta ||z - Tz|| = \beta\alpha,$$

$$||T^2 z - T^3 z|| \le \beta ||Tz - T^2 z|| \le \beta^2\alpha,$$

and iterating,

$$||T^n z - T^{n+1} z|| \le \beta ||T^{n-1} z - T^n z||$$

$$\le .. \le \beta^{n-1} ||z - Tz|| = \beta^n \alpha.$$

Next we show that $\{z, Tz, T^2 z, ..\}$ is a Cauchy sequence—i.e., given $\epsilon > 0$, we can find N such that whenever $n \ge N$, then $||T^n z - T^{n+i} z|| < \epsilon$

for all $i \geq 1$. Let N be so large that $\frac{\beta^N \alpha}{1 - \beta} < \epsilon$, which is possible since $0 < \beta < 1$. Then, for $n \geq N, i \geq 1$,

$$\| T^n z - T^{n+i} z \|$$
$$= \| T^n z - T^{n+1} z + T^{n+1} z - T^{n+2} z + T^{n+2} z$$
$$- .. + T^{n+i-1} z - T^{n+i} z \|$$
$$\leq \| T^n z - T^{n+1} z \| + \| T^{n+1} z - T^{n+2} z \|$$
$$+ .. + \| T^{n+i-1} z - T^{n+i} z \|$$
$$\leq \beta^n \alpha + \beta^{n+1} \alpha + .. + \beta^{n+i-1} \alpha < \frac{\beta^n \alpha}{1 - \beta} < \epsilon.$$

This shows that $\{z, Tz, .. \}$ is Cauchy. Since F is complete, the sequence converges. Let $w := \lim_{n \to \infty} T^n z$. Also T is continuous since $\| Tx - Ty \| \leq \beta \epsilon$ whenever $\| x - y \| \leq \epsilon$. Thus $Tw = T(\lim_{n \to \infty} T^n z) = \lim_{n \to \infty} T(T^n z) = \lim_{n \to \infty} T^{n+1} z = w$. Suppose $\widetilde{w}$ is also a fixed point. Then $\| w - \widetilde{w} \| = \| Tw - T\widetilde{w} \| \leq \beta \| w - \widetilde{w} \|$ which implies $\| w - \widetilde{w} \| = 0$. This proves (1). To prove (2) take any $y \in F$. The previous argument, with y replacing z, shows that $\lim_{n \to \infty} T^n y$ exists and is a fixed point, but since there is only one fixed point w, it must be that $\lim_{n \to \infty} T^n y = w$.   □

(In the proof above it is enough for F to be a complete metric space.)

Exercise   (2.12)

Let $c \in R^n$ and let $P = \{P_{ij}\} \in R^{n \times n}$ with $\sum_{j=1}^{n} | P_{ij} | \leq \gamma$ for $i = 1, .. ,$ n. If $\beta > 0$ is such that $\beta \gamma < 1$, then there is a unique solution w to the vector equation $w = c + \beta Pw$.
[Hint: For $v \in R^n$, define its norm as $\| v \| := \max_{1 \leq i \leq n} | v_i |$. See Exercise (4.5.3).]

Exercise   (2.13)
Let F be a complete normed space and let $T : F \to F$ be continuous. T is an N-stage contraction if there exists $\beta \in (0, 1)$ such that for $y, z \in F$, $\| T^N y - T^N z \| \leq \beta \| y - z \|$. Show that the conclusions of Theorem (2.10) are valid if T is an N-stage contraction.
[Hint: Show first that there is a unique w such that $T^N w = w$. Then $Tw = T(T^N w) = T^N(Tw)$, showing Tw is also fixed point of T^N.]

Theorem (2.10) suggests a procedure for obtaining w as a limit of $T^n z$ starting with any z. We will take advantage of this in studying (2.9).

Lemma   (2.14)

Let F be the class of all functions $z: \{1, \ldots, I\} \to R$. Define $||z|| :=$ $\max\limits_{1 \leq i \leq I} |z(i)|$ and $T: F \to F$ by

$$(Tz)(i) := \inf_{u \in U} \{c(i, u) + \beta \sum_{j=1}^{I} z(j) P_{ij}(u)\}; \qquad (2.15)$$

then T is a contraction.

Proof

This is just a matter of checking the definition. For $y, z \in F$,

$$(Tz)(i) - (Ty)(i) = \inf_{u \in U} \{c(i, u) + \beta \sum_{j=1}^{I} z(j) P_{ij}(u)\}$$

$$- \inf_{u \in U} \{c(i, u) + \beta \sum_{j=1}^{I} y(j) P_{ij}(u)\}.$$

Let $v = Arg \min\limits_{u \in U} \{c(i, u) + \beta \sum_{j=1}^{I} y(j) P_{ij}(u)\}$; then

$$(Tz)(i) - (Ty)(i)$$

$$= \inf_{u \in U} \{c(i, u) + \beta \sum_{j=1}^{I} z(j) P_{ij}(u)\}$$

$$- [c(i, v) + \beta \sum_{j=1}^{I} y(j) P_{ij}(v)]$$

$$\leq [c(i, v) + \beta \sum_{j=1}^{I} z(j) P_{ij}(v)]$$

$$- [c(i, v) + \beta \sum_{j=1}^{I} y(j) P_{ij}(v)]$$

$$= \beta \sum_{j=1}^{I} [z(j) - y(j)] P_{ij}(v)$$

$$\leq \beta \max_{1 \leq j \leq I} |z(j) - y(j)| = \beta ||z - y||.$$

By reversing the roles of z and y, we get

$$(Ty)(i) - (Tz)(i) \leq \beta ||y - z||.$$

Hence $|(Tz)(i) - (Ty)(i)| \leq \beta ||z - y||$ for $1 \leq i \leq I$. So,

$$||Tz - Ty|| = \max_{1 \leq i \leq I} |(Tz)(i) - (Ty)(i)| \leq \beta ||z - y||,$$

and T is a contraction. $\square$

With T as in (2.15), we can rewrite (2.9) as

$$W_\infty = TW_\infty. \tag{2.16}$$

Since T is a contraction, there is a unique $W_\infty := (W_\infty(1), \ldots, W_\infty(I))$ that satisfies (2.16) or, equivalently, (2.9).

Moreover, one can obtain W_∞ as $\lim_{n\to\infty} T^n h_0 = W_\infty$ starting with any h_0 in F. More explicitly, let h_0 be an initial guess for the function W_∞. Define h_n recursively by

$$h_{n+1}(i) := \inf_{u \in U} \{c(i,u) + \beta \sum_{j=1}^{I} P_{ij}(u) h_n(j)\}. \tag{2.17}$$

Then

$$\lim_{n\to\infty} \max_{1\le i \le I} |h_n(i) - W_\infty(i)| = \lim_{n\to\infty} ||h_n - W_\infty|| = 0,$$

which shows that $\lim_{n\to\infty} h_n(i) = W_\infty(i)$ for every $i = 1, 2, \ldots, I$.

The scheme (2.17) for obtaining W_∞ is called fixed point iteration, or successive approximation. In dynamic programming it is called the **value iteration** method for reasons we will see presently.

Exercise (2.18)
Let $h_0(i) := 0$ for $1 \le i \le I$, and let $h_n = T^n h_0$ where T is as in (2.15), or equivalently, let h_n be defined recursively as in (2.17). Show that $h_n(i) = W_n(i)$, where $W_n(i) := \min_g E^g \{ \sum_{k=0}^{n-1} \beta^k c(x_k, u_k) \mid x_0 = i \}$.

Exercise (2.19)
Suppose $h_0 \ne 0$. What is the interpretation of $(T^n h_0)(i)$?

The Nth iteration of (2.17) gives us the optimal value function W_n for the N-stage finite horizon optimal control problem with cost (2.1), and

$$W_\infty(i) = \lim_{N\to\infty} W_N(i), \quad 1 \le i \le I, \tag{2.20}$$

is the unique solution of (2.9).

We show finally that $W_\infty(i)$ is the minimum value of $E \sum_{k=0}^{\infty} \beta^k c(x_k, u_k)$ starting in state $x_0 = i$. Assume for the time being that $0 \le c(i,u) \le M$ for all i, u.

Clearly, with $x_0 = i$, and for all N,

$$\underset{g \in G}{Inf} E^g \sum_{k=0}^{\infty} \beta^k c(x_k, u_k) \geq \underset{g \in G}{Inf} E^g \sum_{k=0}^{N-1} \beta^k c(x_k, u_k)$$

$$= W_N(i).$$

Hence

$$\underset{g \in G}{Inf} E^g \sum_{k=0}^{\infty} \beta^k c(x_k, u_k) \geq \lim_{N \to \infty} W_n(i) = W_{\infty}(i). \tag{2.21}$$

Conversely, for all N,

$$\underset{g \in G}{Inf} E^g \sum_{k=0}^{\infty} \beta^k c(x_k, u_k) \leq \underset{g \in G}{Inf} E^g \sum_{k=0}^{N-1} \beta^k c(x_k, u_k) + \sum_{k=N}^{\infty} \beta^k M$$

$$= W_N(i) + \frac{\beta^N M}{1 - \beta},$$

and so

$$\underset{g \in G}{Inf} E^g \sum_{k=0}^{\infty} \beta^k c(x_k, u_k) \leq \lim_{N \to \infty} [W_N(i) + \frac{\beta^N M}{1 - \beta}]$$

$$= W_{\infty}(i). \tag{2.22}$$

Hence

$$\underset{g \in G}{Inf} E^g \sum_{k=0}^{\infty} \beta^k c(x_k, u_k) = W_{\infty}(i). \tag{2.23}$$

Exercise (2.24)
Suppose c can take both positive and negative values. Show that (2.23) is still true.
[Hint: If $-\dfrac{M}{2} \leq c(i, u) \leq \dfrac{M}{2}$ for all i, u, define a new cost function
$d(i, u) := c(i, u) + \dfrac{M}{2}$.]

This proves the following theorem.

Theorem (2.25)
(1) There exists a unique solution $\{W_{\infty}(1), \ldots, W_{\infty}(I)\}$ to the set of equations

$$W_{\infty}(i) = \underset{u \in U}{inf} \{c(i, u) + \beta \sum_{j=1}^{I} P_{ij}(u) W_{\infty}(j)\}, \quad 1 \leq i \leq I.$$

(2)

$$W_{\infty}(i) = \underset{g \in G}{inf} E^g \{\sum_{k=0}^{\infty} \beta^k c(x_k, u_k) \mid x_0 = i\}.$$

(3) The mapping T defined by

$$(Tz)(i) = \inf_{u \in U} \{c(i,u) + \beta \sum_{j=1}^{I} P_{ij}(u) z(j)\}, \quad 1 \le i \le I,$$

is a contraction with respect to the norm $||z|| = \max_{1 \le i \le I} |z(i)|$.

(4) For any z, $\lim_{n \to \infty} ||T^n z - W_{\infty}|| = 0$ and so

$$\lim_{n \to \infty} (T^n z)(i) = W_{\infty}(i) \quad \text{for all } i.$$

(5) If $z(i) = 0$ for all i, then $(T^n z)(i) = W_n(i)$, where

$$W_n(i) = \inf_{g \in G} E^g \{\sum_{k=0}^{n-1} \beta^k c(x_k, u_k) \mid x_0 = i\}.$$

Exercise   (2.26)
Show that Theorem (2.25) continues to hold when (1) the state space is countable, (2) U is compact subset of R^p, (3) $c(i,u)$ is a bounded function, continuous in u for each i, and (4) $P_{ij}(u)$ is continuous in u uniformly in j.

3. Optimal policies for the discounted cost problem

In the previous section we studied the problem of obtaining the value function $W_{\infty}(\cdot)$. We now show that the optimal policy is stationary.

Definition   (3.1)
        A Markovian policy $\{g_0, g_1, . . \}$ is **stationary** or time invariant if $g_0 \equiv g_1 \equiv . . \equiv g$. We refer to g itself as the stationary policy.

Exercise   (3.2)
Let g be stationary and let $W^g_{\infty}(i)$ be the discounted cost incurred using g and starting in state $x_0 = i$. Show that $W^g_{\infty}(i)$ is the unique solution of

$$W^g_{\infty}(i) = c(i, g(i)) + \beta \sum_{j=1}^{I} P_{ij}(g(i)) W^g_{\infty}(i), \quad 1 \le i \le I.$$

[Hint: See Exercise (4.5.3).]

        From (2.9) we now construct a particular stationary policy g as follows. Let

$$g(i) := Arg \min_{u \in U} \{c(i,u) + \beta \sum_{j=1}^{I} P_{ij}(u) W_{\infty}(j)\}; \tag{3.3}$$

then

$$W_\infty(i) = c(i, g(i)) + \beta \sum_{j=1}^{I} P_{ij}(g(i)) \, W_\infty(j). \tag{3.4}$$

But by Exercise (3.2), the unique solution of (3.4) is $W_\infty^g(\cdot)$. Hence we have proved the following theorem.

Theorem (3.5)

For each $i \in \{1, \ldots, I\}$ let $g(i)$ satisfy (3.3); then the stationary policy g is optimal for the cost criterion (1.8).

Exercise (3.6)

Show that Theorem (3.5) is valid under the more general assumptions of Exercise (2.26).

Exercise (3.7)

In Section 1 a control law is a sequence $\{g_0, g_1, \ldots\}$ with $g_k : (x^k) \to U$. A **randomized** control law is a sequence with $g_k : (x^k) \to P(U)$, where $P(U)$ is the set of probability distributions over U. Such a law specifies at each stage k a probability distribution according to which u_k is picked. Call the class of randomized control laws $\bar{G}$. Of course $G \subset \bar{G}$. Show that the stationary policy g of (3.3) is optimal even within $\bar{G}$.
[Hint: If the optimal value function for the finite horizon cost criterion is still W_N where W_N satisfies (2.6) and (2.7), then the same proof applies.]

4. Algorithms for the optimal control law

Combining Theorem (2.25) with Theorem (3.5) we get an algorithm for obtaining the value function W_∞ as well as the optimal stationary policy g. This value iteration procedure proceeds as follows. Let h_0 be an initial guess for the optimal value function. Obtain successive refinements h_n by the recursion equation (2.17). Then $h_n \to W_\infty$. Once we have W_∞, then we can construct an optimal stationary policy as in (3.3).

In many problems one can use this procedure to deduce useful properties of the optimal value function and the optimal stationary policy without explicit calculations.

Let F and T be as in Lemma (2.14). Suppose there is a closed subset $H \subset F$ which is invariant under T, i.e.,

$$z \in H \implies Tz \in H. \tag{4.1}$$

Then, since $Tz \in H$, applying (4.1) repeatedly shows that $T^n z \in H$, $n \geq 1$. Moreover, $T^n z \to W_\infty$ and so, since H is closed, $W_\infty \in H$. Thus we have the following lemma.

Lemma (4.2)

Let F and T be as in Lemma (2.14). If $H \subset F$ is a closed subset of functions invariant under T, then $W_\infty \in H$.

Practical considerations can often lead us to guess invariant subsets of T and allow us to deduce useful properties of W_∞ without performing computations. The following is a simple example.

Exercise (4.3)

Let the state space for $\{1, 2, 3\}$ and the control set be $U = \{a, b\}$. The transition probabilities are shown in the diagram below.

$$P_{11}(a)= \qquad P_{21}(a)=1 \qquad P_{23}(b)=1 \qquad P_{33}(b)=$$
$$P_{11}(b)=1 \qquad c(2,a)=1 \qquad c(2,b)=0 \qquad P_{33}(b)=1$$

$$1 \qquad\qquad 2 \qquad\qquad 3$$

$$c(1,a)=c(1,b) \qquad\qquad c(3,a)=c(3,b)=0$$

Show that the stationary policy $g(i) = a$ for $i = 1, 2, 3$ is optimal if $c(1,a) < c(3,a)$. The discount factor $\beta \in (0,1)$.
[Hint: The only relevant part of the control law is $g(2)$. Now $g(2) = a$ will be optimal if $W_\infty(1) \le W_\infty(3)$. Let $H := \{z : z(1) \le z(3)\}$. Show that H is a closed invariant set of T.]

In general, the value iteration procedure converges only asymptotically, and most often will not converge in a finite number of iterations. This should be clear from (2.19) and (2.20), and the likelihood that the W_n will usually be all different.

This suggests that if we could generate approximations within a finite set, then we could obtain algorithms that converge within a finite number of iterations. The set of stationary policies is such a finite set since both I and U are finite. We now present the **policy iteration**, or policy improvement, algorithm which at each iteration produces a better stationary policy.

First, we introduce some useful notation. As before, F is the set of functions mapping $\{1, .., I\}$ into R. We can think of $z \in F$ either as a function or a vector $z = (z(1), .., z(I))^T$. The second viewpoint is more convenient and so we identify F with R^I, and $z \in F$ with a vector in R^I. Write $z \le y$ if $z(i) \le y(i)$ for $i = 1, .., I$; and $z < y$ if $z \le y$ but $z \ne y$.

For a stationary g, let $c^g := (c(1,g(1)),\ c(2,g(2)),\ \ldots ,c(I,g(I)))^T$ and let $P^g := \{P_{ij}(g(i))\}$. Rewrite (2.15) as

$$Tz := \min_{g} \{c^g + \beta\, P^g z\},$$

where the vector minimization is done component by component, i.e., $g(1)$ minimizes the first component, $g(2)$ minimizes the second component, and so on. For each stationary g, define the mapping T_g by

$$T_g z := c^g + \beta\, P^g z.$$

Exercise (2.12) shows that T_g is a contraction and Exercise (3.2) shows that W_∞^g is the unique solution of

$$T_g W_\infty^g = W_\infty^g.$$

In this notation we can also define T as

$$Tz = \min_{g} T_g z.$$

This notation is used repeatedly in what follows.

Exercise (4.4)
Show that if $z \le y$, then $T_g z \le T_g y$ and $Tz \le Ty$, (T_g and T are said to be monotone operators.)

Lemma (4.5)
If $T_h W_\infty^g \le W_\infty^g$, then $W_\infty^h \le W_\infty^g$.

Proof
By Exercise (4.4), $T_h(T_h W_\infty^g) \le T_h(W_\infty^g)$, i.e. $T_h^2 W_\infty^g \le T_h W_\infty^g$. Iterating gives $T_h^n W_\infty^g \le T_h^{n-1} W_\infty^g$. Hence $T_h^n W_\infty^g \le W_\infty^g$. However, T_h is a contraction and its unique fixed point is W_∞^h from Exercise (3.2). Hence $W_\infty^h = \lim_{n \to \infty} T_h^n W_\infty^g \le W_\infty^g$.   □

Exercise (4.6)
If $T_h W_\infty^g < W_\infty^g$, then $W_\infty^h < W_\infty^g$ and so h is a better policy than g.

Hence to improve a policy g, one has only to find a policy h such that $T_h W_\infty^g < W_\infty^g$. Now we are ready for the policy iteration algorithm.

Begin with any stationary policy g_0. Recursively generate new policies $g_0, g_1, \ldots , g^*$ as follows.
Step 1
Solve the set of I simultaneous linear equations $T_{g_n} W_\infty^{g_n} = W_\infty^{g_n}$ to obtain $W_\infty^{g_n}$.

Step 2

If $T W_\infty^{g_n} = W_\infty^{g_n}$, then stop and set $g^* = g_n$.

Step 3

If $T W_\infty^{g_n} < W_\infty^{g_n}$, then let g_{n+1} be such that $T_{g_{n+1}} W_\infty^{g_n} = T W_\infty^{g_n}$, and return to Step 1 with $n + 1$ replacing n.

Theorem   (4.7)

The algorithm converges to an optimal stationary policy g^* in a finite number of iterations. Moreover, at each stage n, g_{n+1} is a better policy than g_n, i.e., $W_\infty^{g_{n+1}} < W_\infty^{g_n}$.

Proof

At stage n, if $T W_\infty^{g_n} = W_\infty^{g_n}$, then $W_\infty^{g_n} = W_\infty$ and so $g_n = g^*$ is optimal. If not, then $T_{g_{n+1}} W_\infty^{g_n} = T W_\infty^{g_n} < W_\infty^{g_n}$ and so, by Exercise (4.6), $W_\infty^{g_{n+1}} < W_\infty^{g_n}$, i.e., g_{n+1} is a strictly better policy than g_n. The algorithm has to terminate in a finite number of iterations, since the number of stationary policies is finite and each g_{n+1} is strictly better than all g_i, $i \leq n$.   □

Exercise   (4.8)

In Exercise (4.3) let $g_0(i) = b$ for $i = 1, 2, 3$. Use the policy iteration algorithm to find the optimal policy. After how many iterations does it terminate?

Finally, observe that W_∞ can be obtained as the solution of a linear programming problem.

Exercise   (4.9)

Show that $W_\infty(i)$, $i = 1, \ldots, I$ is the solution of the linear program:

$$\text{Maximize } \sum_{i=1}^{I} z(i)$$

subject to

$$z(i) \leq c(i, u) + \beta \sum_{j=1}^{I} P_{ij}(u) z(j) \text{ for all } u, i.$$

5. The average cost criterion

We now consider the average cost per unit time

$$J(g) := \text{lim sup}_{N} \sum_{k=0}^{N-1} E^g c(x_k, u_k),$$ (5.1)

where the lim in (1.6) is replaced by lim sup, since for some g the former might not exist.

We begin by reexamining the finite horizon cost

$$E^g \sum_{k=0}^{N-1} c(x_k, u_k),$$

for which the value function $W_N(\cdot)$ satisfies

$$W_N(i) = \min_{u \in U} \{c(i,u) + \sum_{j=1}^{I} P_{ij}(u) \, W_{N-1}(j)\}, \quad \text{all } i. \tag{5.2}$$

[The subscript N in W_N is the time to go; see (2.4).] Since the average cost criterion is the limit as $N \to \infty$ of

$$\frac{1}{N} E^g \sum_{k=0}^{N-1} c(x_k, u_k),$$

it is more convenient to rewrite (5.2) as

$$\frac{W_N(i)}{N} = \min_{u \in U} \{c(i,u)$$

$$+ \sum_{j=1}^{I} P_{ij}(u) \, [W_{N-1}(j) - W_N(i) + \frac{W_N(i)}{N}]\}. \tag{5.3}$$

Assume now that for some J^*, $w(1), \ldots, w(I)$, the following limit exists:

$$\lim_N [W_N(i) - NJ^*] := w(i), \quad i = 1, \ldots, I. \tag{5.4}$$

This implies in particular that

$$\lim_N \frac{W_N(i)}{N} = J^*, \quad i = 1, \ldots, I,$$

and the limit does not depend on the initial state i which is to be expected since the average cost measures only the asymptotic behavior and not the initial behavior. Rewrite (5.3) as

$$\frac{W_N(i)}{N} = \min_{u \in U} \{c(i,u) + \sum_{j=1}^{I} P_{ij}(u) \, [(W_{N-1}(j) - (N-1)J^*)$$

$$- (W_N(i) - NJ^*) + (\frac{W_N(i)}{N} - J^*)]\},$$

and take the limit as $N \to \infty$ to get

$$J^* + w(i) = \min_{u \in U} \{c(i,u) + \sum_{j=1}^{I} P_{ij}(u) \, w(j)\}, \quad \text{all } i. \tag{5.5}$$

Thus we expect that for the average cost problem there will exist J^* and $w(1), \ldots, w(I)$ satisfying (5.5). Moreover J^* is the minimum average cost that does not depend on the initial state.

We also have an interpretation for $w(i)$. By (5.4)

$$W_N(i) = NJ^* + w(i) + o(1).$$

(The notation means that $o(1) \to 0$ as $N \to \infty$.) Thus $w(i)$ is the deviation of $W_N(i)$ from NJ^* for large N. Since (5.5) holds if each $w(i)$ is replaced by $w(i) + \alpha$ for any constant α, we should more properly interpret $w(k) - w(i)$ as the relative advantage of starting in state $x_0 = i$ instead of in state $x_0 = k$.

Equation (5.5) is the dynamic programming equation for the average cost problem. Later we will establish that it holds under conditions more general than (5.4).

We begin by showing that any J^* satisfying (5.5) is the minimal average cost and further that if g^* is a stationary policy such that $g^*(i)$ achieves the minimum on the right-hand side of (5.5) for every i, then g^* is optimal. Suppose that

$$J^* + w(i) = c(i, g^*(i)) + \sum_{j=1}^{I} P_{ij}(g^*(i)) w(j), \quad \text{all } i. \tag{5.6}$$

Let g be any policy, possibly nonstationary. Rearrange (5.6) as

$$c(i, u) \geq J^* + w(i) - \sum_{j=1}^{I} P_{ij}(u) w(j),$$

and so

$$E^g c(x_k, u_k) \geq J^* + E^g w(x_k) - E^g \sum_{j=1}^{I} P_{x_k, j}(u_k) w(j).$$

But since $E^g \sum_{j=1}^{I} P_{x_k, j}(u_k) w(j) = E^g w(x_{k+1})$, we get

$$E^g c(x_k, u_k) \geq J^* + E^g w(x_k) - E^g w(x_{k+1}).$$

Summing and dividing by N gives

$$\frac{1}{N} \sum_{k=0}^{N-1} E^g c(x_k, u_k) \geq J^* + \frac{1}{N} [E^g w(x_0) - E^g w(x_N)]. \tag{5.7}$$

The last term in (5.7) vanishes as $N \to \infty$, and so

$$\lim_{N \to \infty} \text{Inf} \frac{1}{N} \sum_{k=0}^{N-1} E^g c(x_k, u_k) \geq J^*,$$

However when g^* replaces g equality holds throughout and so g^* is optimal. This proves the following lemma.

Lemma (5.8)

Suppose there exist J^* and $w(1), \ldots, w(I)$ satisfying

$$J^* + w(i) = \min_{u \in U} \{c(i,u) + \sum_{j=1}^{I} P_{ij}(u) \, w(j)\}, \quad \text{all } i. \tag{5.9}$$

Let g^* be a stationary policy such that $g^*(i)$ attains the minimum in the right hand side of (5.9) for every i; then g^* is an optimal policy and J^* is the minimum average cost.

The following exercise examines the infinite horizon average cost case of the LQG problem of Section 7.5. Though the state space is not finite, the arguments leading to Lemma (5.8) yield the optimal policy.

Exercise (5.10)

Consider the time invariant linear stochastic system,

$$x_{k+1} = A \, x_k + B \, u_k + G \, w_k.$$

The basic random variables $x_0, w_0, w_1, \ldots$, are independent and Gaussian with

$$x_0 \sim N(\bar{x}_0, \Sigma_0), \quad w_k \sim N(0, Q).$$

We wish to find a policy g which minimizes the average cost

$$\limsup_{N} \frac{1}{N} \sum_{k=0}^{N-1} E^g[x_k^T P x_k + u_k^T T u_k],$$

where $P \geq 0$ and $T > 0$. Assume that A, B is reachable and A, Δ is observable, where $\Delta^T \Delta = P$.

(1) Show that there exists a unique nonnegative definite solution S to the matrix equation

$$S = P + A^T[S - SB(T + B^T SB)^{-1}B^T S]A$$

and that the matrix $[A - B(T + B^T SB)^{-1}B^T SA]$ is stable.

(2) Show that $J^* := Tr(SGQG^T)$ and $w(x) := x^T Sx$ satisfy the equation

$$J^* + w(x)$$
$$= \min_{u} \{x^T P x + u^T T u + E\{w(x_{k+1}) \mid x_k = x, u_k = u\}\},$$

and that

$$u = g^*(x) := -[T + B^T SB]^{-1}B^T SA \, x$$

achieves the minimum.

(3) Show that the stationary policy g^* is optimal compared to all policies g which satisfy $\lim_{N \to \infty} \frac{1}{N} E^g |x_N|^2 = 0$.

[Hint: For (1) make the appropriate identifications and compare with

Theorem (7.2.38) and Exercise (7.2.43). To show (3) follow the argument used in (5.7) leading to Lemma (5.8).]

We now show that there exist J^* and $w(1), w(2), \ldots, w(I)$ satisfying (5.5). We have already done so under the assumption that (5.4) holds. However this is a very restrictive assumption, as the following exercise shows.

Exercise   (5.11)
Construct a simple example where (5.4) is' violated, even though (5.5) holds.
[Hint: Take $I = 2$, $U = \{u\}$, $P_{ii}(u) = 0$ for $i = 1, 2$. Now choose $c(i, u)$.]

Next we examine more general conditions under which (5.5) can be established. Consider first the average cost criterion for a fixed stationary policy g. Let $P^g := \{P_{ij}(g(i))\}$, and $c^g := (c(1, g(1)), \ldots, c(I, g(I)))^T$. By (4.5.4), the following limit exists,

$$J(g) = \lim_N \frac{1}{N} \sum_{k=0}^{N-1} E^g c(x_k, g(x_k)),$$

but it may depend on the initial state x_0. However, if P^g is irreducible (see Section 4.5), then, by Lemma (4.5.7),

$$J(g) = \pi^g c^g, \tag{5.12}$$

where $\pi^g = (\pi^g(1), \ldots, \pi^g(I))$ is the unique steady state pd given by

$$\pi^g P^g = \pi^g, \quad \pi^g e = 1,$$

and $e := (1, \ldots, 1)^T$; in particular $J(g)$ does not depend on the initial state. Moreover,

$$\pi^g(i) > 0, \quad \text{for all } i. \tag{5.13}$$

So for the remainder of this section we make the following assumption.

Assumption   (5.14)
For every stationary g the transition probability matrix P^g is irreducible.

We study (5.12) in more detail. Rewrite it as

$$\pi^g [J(g)e - c^g] = 0.$$

This shows that the vector $J(g)e - c^g$ is orthogonal to π^g. Since π^g generates the null space of $[P^g - I]^T$ (Lemma (4.5.7)), it follows that $J(g)e - c^g$ belongs to the range space of $[P^g - I]$. Hence there exists a vector w_g such that

$$J(g)e - c^g = [P^g - I]w_g, \quad \text{or}$$

$$J(g) + w_g(i) = c(i, g(i)) + \sum_{j=1}^{I} P_{ij}(g(i)) w_g(j), \quad 1 \le i \le I. \tag{5.15}$$

This equation resembles (5.5). From Lemma (5.8) we surmise that if g^* is an optimal stationary policy, then $J(g^*)$ and $w_{g^*}(i)$ satisfy (5.5). Accordingly let $G_S \subset G$ be the set of all stationary policies (G is the set of all policies), and let g^* be the best stationary policy, i.e.,

$$J(g^*) \leq J(g), \quad \text{for all } g \in G_S. \tag{5.16}$$

Lemma (5.17)
If g^* satisfies (5.15), then for all i

$$J(g^*) + w(i) = \min_{u \in U} \{c(i,u) + \sum_{j=1}^{I} P_{ij}(u) \, w_{g^*}(j)\}. \tag{5.18}$$

Proof
From (5.15), with $g = g^*$,

$$J(g^*) + w_{g^*}(i) \geq \min_{u \in U} \{c(i,u) + \sum_{j=1}^{I} P_{ij}(u) \, w_{g^*}(j)\}. \tag{5.19}$$

Suppose that $g(i)$ attains the minimum in the right-hand side of (5.19). Then g is stationary and

$$J(g^*) + w_{g^*}(i) \geq c(i,g(i)) + \sum_{j=1}^{I} P_{ij}(g(i)) \, w_{g^*}(j). \tag{5.20}$$

Suppose that strict inequality holds in (5.19), or equivalently in (5.20), for some i. Then multiplying (5.20) by $\pi^g(i)$ and summing over i, and noting the strict positivity in (5.13), we get

$$J(g^*) + \sum_{i=1}^{I} \pi^g(i) \, w_{g^*}(i) > \sum_{i=1}^{I} \pi^g(i) \, c(i,g(i))$$
$$+ \sum_{i=1}^{I} \pi^g(i) \sum_{j=1}^{I} P_{ij}(g(i)) w_{g^*}(j).$$

Using (5.12) and (5.13) in the preceding inequality gives $J(g^*) > J(g)$, contradicting (5.16). Hence equality holds in (5.20) for every i, and the lemma is proved.   □

Combining Lemmas (5.8) and (5.17), gives the following Theorem.

Theorem (5.21)
Suppose assumption (5.14) holds. Then:
(1) There exist $(w(1),\ldots,w(I))$ and J^* such that

$$J^* + w(i) = \min_{u \in U} \{c(i,u) + \sum_{j=1}^{I} P_{ij}(u) \, w(j)\}, \quad \text{all } i. \tag{5.22}$$

(2) If $g^*(i)$ attains the minimum in (5.22) for every i, then g^* is an optimal policy.
(3) The minimum average cost is J^*.

Equation (5.22) is the dynamic programming equation for the average cost criterion.

Exercise   (5.23)
Show that a stationary policy g^* is optimal if and only if it satisfies (2) of Theorem (5.21).

Exercise   (5.24)
Show that if w satisfies (5.22) so does $w + \alpha e$ for all real α. Conversely, if w and $\hat{w}$ satisfy (5.22), then $w - \hat{w} = \alpha e$ for some α. (However, J^* satisfying (5.22) is unique.)

Exercise   (5.25)
Show that g^* of Theorem (5.21) is optimal even within the class of randomized control laws. (See Exercise (3.7) for the definition of a randomized law.)

It can be shown that the optimal stationary policy g^* is optimal in a strong sample path sense. This requires use of the martingale stability theorem.

Theorem (martingale stability)   (5.26)
Suppose $\{x_k, F_k\}$ is a **martingale difference** sequence, (i.e., $F_k \subset F_{k+1}$ is an increasing sequence of σ-algebras, x_k is F_k-measurable, and $E\{x_{k+1} \mid F_k\} = 0$ a.s.) Also suppose

$$\sum_{k=1}^{\infty} \frac{E\{ \mid x_k \mid^p \mid F_{k-1}\}}{k^p} < \infty \ a.s. \quad \text{for some } 0 < p \le 2;$$

then

$$\lim_{N \to \infty} \frac{1}{N} \sum_{k=1}^{N} x_k = 0 \ a.s.$$

We will frequently use this theorem as follows.

Exercise   (5.27)
Suppose y_k is F_k-measurable and either $\mid y_k \mid \le M$ a.s. for some constant M or, more generally, there is a constant $p > 1$, and a random variable $M < \infty$ a.s., such that $E\{\mid y_{k+1} \mid^p \mid F_k\} < M$ a.s. Show that $\{y_k - E\{y_k \mid F_{k-1}\}, F_k\}$ is a martingale difference sequence and

$$\lim_{N \to \infty} \frac{1}{N} \sum_{k=1}^{N} [y_k - E\{y_k \mid F_{k-1}\}] = 0 \ a.s.$$

Exercise (5.28)

Show that for any policy $g \in G$,

$$\liminf \frac{1}{N} \sum_{k=0}^{N-1} c(x_k, u_k) \geq J^* \quad a.s.$$

with equality holding a.s. for stationary policies g^* of Theorem (5.21). Thus g^* is optimal in the much stronger sample path sense than in the expected value sense of (1.6).

[Hint: Let $y_{k+1} := c(x_k, u_k) - J^* + w(x_{k+1}) - w(x_k) - g(x_k, u_k)$, where

$$g(i, u) := c(i, u) + \sum_{j=1}^{I} P_{ij}(u) w(j) - J^* - w(i).$$

Then $E\{y_{k+1} \mid x^k, u^k\} = 0$, and so $\{y_k\}$ is a martingale difference sequence, which is also uniformly bounded. Hence by Theorem (5.27),

$$\lim \frac{1}{N} \sum_{k=0}^{N-1} y_k = 0 \text{ a.s. Moreover, } g(i, u) \geq 0 \text{ for all } i, u.]$$

By Theorem (5.21) the search for an optimum policy can be limited to stationary policies. We conclude this section with the policy iteration algorithm.

Step 1

Start with any stationary policy g_0.

Step 2

At the nth iteration solve the set of simultaneous linear equations

$$J(g_n) e + w_{g_n} = c^{g_n} + P^{g_n} w_{g_n},$$

for $J(g_n)$ and w_{g_n}. [This is a set of I equations in $I+1$ unknowns. By Exercise (5.24) there is one degree of freedom and so one can fix, for example, $w_{g_n}(1)$, at any arbitrary value.]

Step 3

If

$$J(g_n) e + w_{g_n} = \min_{g \in G_S} \{c^g + P_g w_{g_n}\},$$

then stop and set $g^* = g_n$; g^* is an optimal stationary policy and $J_{g_n} = J^*$ is the minimum cost. Otherwise, let g_{n+1} attain the minimum in

$$\min_{g \in G_S} \{c^g + P_g w_{g_n}\},$$

and return to Step 2 with n replaced by $n + 1$.

Exercise (5.29)

Show that this policy iteration algorithm converges in a finite number of iterations to an optimal stationary policy g^*. Also show that $J(g_0) >$

$J(g_1) > .. > J(g^*)$.

[Hint: Suppose $J(g_n)e + w_{g_n} > c^{g_{n+1}} + P^{g_{n+1}} w_{g_n}$; then multiply both sides by $\pi^{g_{n+1}}$.]

Exercise   (5.30)

Show that the minimum cost is the solution of the linear program:

$$\text{Maximize } J^*$$

subject to

$$J^* + w(i) \leq c(i,u) + \sum_{j=1}^{I} P_{ij}(u) w(j), \ 1 \leq i \leq I, \ u \in U.$$

6. The average cost criterion as a limit

In this section we obtain (5.22) as the limit of discounted cost problems as $\beta \to 1$. Denote by W^β the value function for the problem with discounted cost

$$E^g \sum_{k=0}^{\infty} \beta^k \, c(x_k, u_k),$$

so that

$$W^\beta(i) = \min_{u \in U} \{c(i,u) + \beta \sum_{j=1}^{I} P_{ij}(u) \, W^\beta(j)\}.$$

This can be written as

$$(1-\beta) \, W^\beta(i) = \min_{u \in U} \{c(i,u)$$
$$+\beta \sum_{j=1}^{I} P_{ij}(u) \, [W^\beta(j) - W^\beta(i)]\}. \tag{6.1}$$

Assume that

$$| W^\beta(j) - W^\beta(i)| \leq M < \infty, 0 < \beta < 1, 1 \leq i, j \leq I. \tag{6.2}$$

As a consequence $(1 - \beta)W^\beta$ is bounded, since

$$|(1 - \beta)W^\beta(i)| \leq (1 - \beta)E^g \sum_{k=0}^{\infty} \beta^k [\max_{j,u} |c(j,u)|]$$

$$= \max_{j,u} |c(j,u)|.$$

Hence we can find a sequence $\{\beta_l\}$, with $\lim_{l \to \infty} \beta_l = 1$, so that

$$\lim_{l\to\infty} (1 - \beta_l)\, W^{\beta_l}(1) =: J^*, \tag{6.3}$$

and

$$\lim_{l\to\infty} [W^{\beta_l}(j) - W^{\beta_l}(1)] =: w(j), \quad 1 \le j \le I. \tag{6.4}$$

It also follows from (6.2) and (6.3) that for all i

$$\lim_{l\to\infty} (1 - \beta_l)\, W^{\beta_l}(i)$$

$$= \lim_{l\to\infty} [(1 - \beta_l)W^{\beta_l}(1) + (1 - \beta_l)(W^{\beta_l}(i) - W^{\beta_l}(1))] = J^*.$$

Taking limits in (6.1) along the sequence $\{\beta_l\}$, we get (5.22), which we summarize as the next theorem.

Theorem (6.5)
Let W^β be the optimal value function for the discounted cost problem with discount factor $\beta \in (0, 1)$. If (6.2) holds, there is a sequence $\{\beta_l\}$, $0 < \beta_l < 1$, $\beta_l \to 1$, for which the limits in (6.3) and (6.4) exist. Moreover, J^* and w satisfy the dynamic programming equation (5.22).

Exercise (6.6)
Under the assumptions of Exercise (2.26) [see also Exercise (3.6)] show that Theorem (6.5) is valid under (6.2).

Exercise (6.7)
Suppose the same stationary policy g^* is optimal for all discounted cost criteria with discount $\beta \in (1 - \epsilon, 1)$ for some $\epsilon > 0$. Show then that g^* is also optimal for the average cost criterion, under the assumptions of Theorem (6.5).

Exercise (6.8)
Let τ be the random time when the system first enters a prespecified state, for example 1—i.e.,

$$\tau = \min\{k \ge 1 \mid x_k = 1\}.$$

Suppose that

$$E^g\{\tau \mid x_0 = i\} \le T < \infty \quad \text{for all } g \in G_s \text{ and all } i.$$

Show that (6.2) holds.
[Hint: It is enough to show that $|W^\beta(i) - W^\beta(1)| \le M < \infty$ for all i, β. Let g be a stationary policy which is optimal when the discount factor is β. We can assume without loss of generality that $0 \le c(i, u) \le M$ for all i, u and some constant M. Now

$$W^\beta(i) = E^g\{\sum_{k=0}^{\tau-1} \beta^k c(x_k, u_k) + \beta^\tau W^\beta(1) \mid x_0 = i\}$$

$$\leq M \, E^g \{ \tau \mid x_0 = i \} + W^\beta(1).$$

Also $W^\beta(i) \geq E^g \{ \beta^\tau \mid x_0 = i \} W^\beta(1) \geq \beta^T W^\beta(1)$ by Jensen's inequality, which states that for any convex function f, random variable x, and σ-field F, $E \{ f(x) \mid F \} \geq f(E \{ x \mid F \})$.]

7. Notes

1. Classic references for the dynamic programming problem with the discounted and average cost criteria are Blackwell (1965a and 1962). Books covering this material in detail are Howard (1960), Ross (1970), Derman (1970), Hordijk (1974), Bertsekas (1976), Kushner (1971), Dubins and Savage (1976), and Dynkin and Yushkevich (1975).

2. The total cost criterion (1.2) is considered in Blackwell (1965b), Strauch (1966), and Ornstein (1969). The cases where $c(i,u) \geq 0$ for all i,u or $c(i,u) \leq 0$ for all i,u, exhibit different properties.

3. In some applications, for example in the optimal control of queuing systems, the generalization suggested in Exercise (2.13) to N-stage contractions is useful; see Lippman (1973) and Denardo (1967).

4. The technique of identifying invariant sets of T and thus deducing qualitative properties of the optimal control law, as suggested in Lemma (4.2) and Exercise (4.3), is often very useful. See Hajek (1984) and Lin and Kumar (1984) for some applications to the control of queuing systems.

5. For more on Exercise (5.27) as well as a central limit theorem, see Mandl (1974).

6. A recent paper on the linear programming approach is Hordijk and Kallenberg (1979).

7. For a weaker assumption than (5.14), which still yields a result such as Theorem (5.21), see Bather (1973).

8. For more on Exercises (6.6) and (6.8), counterexamples in the countable state case, and a treatment of the continuous time semi-Markov case, see Ross (1970). Recurrence conditions such as in Exercise (6.8) are explored in depth in Federgruen, Hordijk and Tijms (1978).

9. The determination of policies optimal in the sense of Exercise (6.7) is treated in Blackwell (1962), Miller and Veinott (1969), and Veinott (1969).

10. For an operator-theoretic approach to the average cost problem see Kumar (1983) and Wijngaard (1977).

11. If the state space is uncountable, then measurability problems arise. These are treated in Bertsekas and Shreve (1978).

12. For a treatment of dynamic programming problems under continuity and compactness assumptions such as in Exercises (2.26), (3.6) and (6.6), see Schal (1973 and 1975).

13. The martingale stability theorem (5.26) can be found in Stout (1974) as Theorem 3.3.1.

8. References

[1] J. Bather (1973), "Optimal decision procedures for finite Markov chains," *Advances in Applied Probability,* (3 parts), vol. 5, 328-339, 521-540, 541-553.

[2] D. Bertsekas (1976), *Dynamic programming and stochastic control,* Academic Press, New York.

[3] D. Bertsekas and S. Shreve (1978), *Stochastic optimal control: the discrete time case,* Academic Press, New York.

[4] D. Blackwell (1962), "Discrete dynamic programming," *Annals of Mathematical Statistics,* vol 33, 719-726.

[5] D. Blackwell (1965a), "Discounted dynamic programming," *Annals of Mathematical Statistics,* vol 36, 226-235.

[6] D. Blackwell (1965b), "Positive dynamic programming," *Proceedings of the Fifth Berkeley Symposium on Mathematical Statistics and Probability,* University of California Press, Berkeley, 415-418.

[7] E. Denardo (1967), "Contraction mappings in the theory underlying dynamic programming," *SIAM Review,* vol 9, 165-177.

[8] C. Derman (1970), *Finite state Markovian decision processes,* Academic Press, New York.

[9] L. Dubins and L. Savage (1976), *Inequalities for stochastic processes (How to gamble if you must)* Dover, New York.

[10] E. Dynkin and A. Yushkevich (1975), *Controlled Markov processes and their applications,* Springer-Verlag, New York.

[11] A. Federgruen, A. Hordijk and H. Tijms (1978), "A note on simultaneous recurrence conditions on a set of denumerable stochastic processes," *Journal of Applied Probability,* vol 15, 356-373.

[12] B. Hajek (1984), "Optimal control of two interacting service stations," *IEEE Transactions on Automatic Control,* vol. AC-29, 491-498.

[13] A. Hordijk (1974), *Dynamic programming and Markov potential theory,* Mathematical Center Tracts, vol. 51, Amsterdam.

[14] A. Hordijk and L. C. M. Kallenberg (1979), "Linear programming and Markov decision chains," *Management Science,* vol 25, 352-362.

[15] R. Howard (1960), *Dynamic programming and Markov processes,* MIT Press, Cambridge.

[16] P.R. Kumar (1983), "Simultaneous identification and adaptive control of unknown systems over finite parameter sets," *IEEE Transactions on Automatic Control,* vol AC-28(1), 68-76.

[17] H. Kushner (1971), *Introduction to stochastic control,* Holt, New York.

[18] W. Lin and P. R. Kumar (1984), "Optimal control of a queueing system with two heterogeneous servers," *IEEE Transactions on Automatic Control,* vol AC-29, 696-703.

[19] S. Lippman (1973), "Semi-Markov processes with unbounded rewards," *Management Science,* vol 19, 717-731.

[20] P. Mandl (1974), "Estimation and control in Markov chains," *Advances in Applied Probability,* vol 6, 40-60.

[21] B. Miller and A. Veinott (1969), "Discrete dynamic programming with a small interest rate," *Annals of Mathematical Statistics,* vol 40, 366-370.

[22] D. Ornstein (1969), "On the existence of stationary optimal strategies," *Proceedings of the American Mathematical Society,* vol 20, 563-569.

[23] S. Ross (1970), *Applied probability models with optimization applications,* Holden-Day, San Francisco.

[24] M. Schal (1973), "Dynamic programming under continuity and compactness assumptions," *Advances in Applied Probability,* vol 5, 24-25.

[25] M. Schal (1975), "Conditions for optimality in dynamic programming and for the limit of N-stage optimal policies to be optimal," *Z. Wahrscheinlichkietstheorie vew. Gebiete,* vol 32, 179-196.

[26] W. F. Stout (1974), *Almost sure convergence,* Academic Press, New York.

[27] R. Strauch (1966), "Negative dynamic programming," *Annals of Mathematical Statistics,* vol 3, 871-890.

[28] A. Veinott (1969), "Discrete dynamic programming with sensitive optimality criteria," *The Annals of Mathematical Statistics,* vol 40, 1635-1660.

[29] J. Wijngaard (1977), "Stationary Markovian decision problems and perturbation theory of quasi-compact linear operators," *Mathematics of Operations Research,* vol 2, 91-102.

CHAPTER 9
INTRODUCTION TO SYSTEM IDENTIFICATION

We start with the identification of memoryless systems. After introducing the idea of parametrization, we review some of the major concepts from statistics. Then we turn to dynamic systems. We formulate the system identification problem and discuss it in the context of the examples given in Chapter 1. Finally, we consider parametrization of linear stochastic systems.

1. Identification of memoryless systems

We begin by considering a static, or memoryless, stochastic system. Such a system accepts an input u (for example, $u \in R$) and produces a random output y (for example, $y \in R$). Thus there is a conditional probability distribution (cd)

$$F(z \mid u) := Prob(y \leq z \mid u)$$

of y given u. Hence a static, or memoryless, system is characterized by the cd $F(\cdot \mid \cdot)$. F is also called a *model* of the system.

Example (1.1)

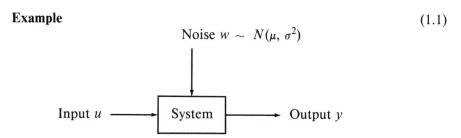

Noise $w \sim N(\mu, \sigma^2)$

Input $u \longrightarrow$ System $\longrightarrow$ Output y

Suppose $y = u + w$, where $w \sim N(\mu, \sigma^2)$, and w is independent of u. Then

$$F(z \mid u) = \frac{1}{\sqrt{2\pi\sigma^2}} \int_{-\infty}^{z} \exp\left[\frac{-1}{2\sigma^2}(y - u - \mu)^2\right] dy.$$

Thus to understand the behavior of the system, one should determine the cd $F(\cdot \mid \cdot)$. This is the problem of identification.

2. Parametrization

The set W of all conditional probability distributions is infinite-dimensional, and frequently we assumes or knows that the particular cd for the system at hand belongs to a smaller subset $\hat{W} \subset W$.

Example (2.1)

Suppose the system is as in Example (1.1) except that μ and σ^2 are unknown. Then we can confine attention to the set

$$\hat{W} := \{F(\cdot \mid \cdot) \mid \text{for some real numbers } \mu \text{ and } \sigma,$$

$$F(z \mid u) = \frac{1}{\sqrt{2\pi\sigma^2}} \int_{-\infty}^{z} \exp\left[\frac{-1}{2\sigma^2} (y - u - \mu)^2\right] dy \}.$$

In the example above, to every $F \in \hat{W}$ there corresponds a pair of numbers (μ, σ) which characterizes it. Frequently, there is a set Θ (typically a subset of R^p) with the property that every cd in $\hat{W}$ can be associated with an element of Θ. More precisely there is a function $G: \Theta \to \hat{W}$ from Θ onto G. (Θ, G) is said to be a **parametrization** of $\hat{W}$.

Example (2.2)

In Example (2.1) let $\Theta = R^2$ and for each $\theta = (\theta_1, \theta_2) \in \Theta$, let G^θ be the cd

$$G^\theta(z \mid u) = \frac{1}{\sqrt{2\pi\theta_2^2}} \int_{-\infty}^{z} \exp\left[\frac{-1}{2\theta_2^2} (y - u - \theta_1)^2\right] dy. \qquad (2.3)$$

We call Θ the parameter set and our goal is to find $\theta \in \Theta$ such that $G^\theta(\cdot \mid \cdot)$ is the cd for the system at hand. If $G: \Theta \to \hat{W}$ is not one-to-one, then there may be two (or more) values $\theta' \neq \theta''$ such that $G^{\theta'}(\cdot \mid \cdot)$ and $G^{\theta''}(\cdot \mid \cdot)$ both correspond to the cd for the system at hand. This motivates the next definition.

Definition (2.4)

(Θ, G) is a **canonical** parametrization of $\hat{W}$ if $G: \Theta \to \hat{W}$ is one-to-one and onto.

Example (2.5)

The parametrization in Example (2.2) is not canonical since both $\theta' = (\theta_1, \theta_2)$ and $\theta'' = (\theta_1, -\theta_2)$ give the same cd, i.e., $G^{\theta'}(\cdot \mid \cdot) = G^{\theta''}(\cdot \mid \cdot)$. However, if $\Theta := R \times (0, \infty)$ and G is as in (2.3), then (Θ, G) is canonical.

If the actual system model is in $\hat{W}$, then there will be $\theta° \in \Theta$ such that $G^{\theta°}$ is the cd of the output given the input. Such a value $\theta°$ is called the *true* parameter.

We can now state the parameter identification problem more precisely. We perform some experiment on the system, that is we apply an input u and observe the resulting output y. Thus our data is (y, u).

Given this data, we make an estimate $\hat{\theta} = \tau(y, u)$. The function τ is called the **estimator** and $\hat{\theta}$ is the **estimate**. The identification problem is to find an estimator τ such that $\hat{\theta}$ is close to the true parameter $\theta°$. [Sometimes there is no input, i.e., y is freely evolving. In such situations we replace the cd $G^\theta(y \mid u)$ by the unconditional distribution $G^\theta(y)$, and the same discussion applies.]

Let us note one difficulty in the statement of the identification problem. The estimator $\tau(y) \equiv \theta°$ is obviously the best estimator. It is not of much use since $\theta°$ is not known. So in finding good estimators we should avoid relying on knowledge that we don't have. There are two ways in which this can be done: the Bayesian approach, and the classical approach of statistical inference.

3. The Bayesian approach

It will be notationally convenient to assume that the cd $G^\theta(y \mid u)$ admits a density denoted $p(y \mid u, \theta)$. It will also be convenient to assume that the input u has been chosen and is fixed.

In the Bayesian approach one is assumed to possess some prior knowledge of $\theta°$ expressed as a prior pd, say $p(\theta)$, on the parameter set Θ— $p(\theta)$ is our belief (before making any observation) of the likelihood that $\theta = \theta°$. From the prior density $p(\theta)$ we can now obtain the posterior density $p(\theta \mid y, u)$ by Bayes' rule,

$$p(\theta \mid y, u) = \frac{p(y \mid u, \theta) \, p(\theta)}{p(y \mid u)}$$
$$= \frac{p(y \mid u, \theta) \, p(\theta)}{\int p(y \mid u, \theta) \, p(\theta) \, d\theta}.$$

Next suppose given a loss or cost function $L(\theta', \theta)$ representing the loss incurred by estimating the parameter to be θ' when its true value is θ. Let τ be any estimator. Then the expected loss incurred by τ when (y, u) is the data available is

$$J(\tau \mid y, u) := \int_\Theta L(\tau(y, u), \theta) \, p(\theta \mid y, u) \, d\theta.$$

The (optimum) **Bayes estimator** τ^* is the one which minimizes the expected loss,

$$J(\tau^* \mid y, u) \le J(\tau \mid y, u), \text{ for all } \tau.$$

The calculation of the Bayes estimator is thus reduced to a minimization problem, namely,

$$\tau^*(y, u) = Arg \min_{\theta' \in \Theta} \int_\Theta L(\theta', \theta) \, p(\theta \mid y, u) \, d\theta.$$

The optimum Bayes estimator depends on the choice of the prior density $p(\theta)$ and the loss function L. In many situations the choice of $p(\theta)$ is not self-evident, and then the classical approach may be more attractive. In some problems the Bayesian approach is quite meaningful, as in the following example.

Exercise (3.1)

Consider a communication channel in which the unknown transmitted signal θ is either 0 or 1, so $\theta \in \Theta = \{0,1\}$. The received signal is $y = \theta + w$ and the channel noise is $w \sim N(0, \sigma^2)$. The receiver observes y and has to decide which of the two possible signals θ was transmitted. So the receiver's decision is represented by a function $\tau: y \rightarrow \{0,1\}$. Suppose the loss function is $L(\theta', \theta) = |\theta' - \theta|^2$, and the prior probability of the transmitted signal is $p_0 = Prob\ \{\theta=0\}$, $p_1 = Prob\ \{\theta = 1\} = 1 - p_0$. Calculate the optimum τ in terms of p_0, p_1, σ.

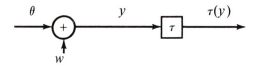

Note that no input u is needed in this model.

4. The classical approach

If there is no prior distribution over the set of model parameters Θ, one cannot conceive of optimum estimators as in the Bayesian approach. Instead we look for estimators with desirable properties such as unbiasedness, minimum variance, etc. We suppose as before that each parameter specifies a probability density $p(y \mid u, \theta)$ of the observation y when θ is the true parameter.

Let $\tau: (y, u) \rightarrow \tau(y, u) \in R^p$ be any estimator. (We are assuming that $\theta \in R^p$.) Then

$$E(\tau \mid u, \theta) := \int \tau(y, u)\, p(y \mid u, \theta)\, dy$$

is the expected value of the estimate when θ is the true parameter and the input u is applied.

Definition (4.1)

τ is said to be an **unbiased** estimator for the input u if $E(\tau \mid u, \theta) = \theta$ for every $\theta \in \Theta$.

Definition (4.2)

The (error) covariance corresponding to an estimator τ and parameter θ, when the input u is applied, is

$$\Sigma(\tau \mid u, \theta) := \int [\tau(y, u) - E(\tau \mid u, \theta)] \times$$
$$[\tau(y, u) - E(\tau \mid u, \theta)]^T \, p(y \mid u, \theta) dy.$$

Given two unbiased estimators τ^1, τ^2, it is reasonable to say that τ^1 is better than τ^2 if $\Sigma(\tau^2 \mid u, \theta) \geq \Sigma(\tau^1 \mid u, \theta)$ [i.e., if $\Sigma(\tau^2 \mid u, \theta) - \Sigma(\tau^1 \mid u, \theta)$ is a positive semidefinite matrix] for every $\theta \in \Theta$. An unbiased estimator τ is a "minimum variance unbiased estimator" (MVUE) (for u) if $\Sigma(\tau \mid u, \theta) \leq \Sigma(\tau^1 \mid u, \theta)$ for every $\theta \in \Theta$ and unbiased estimator τ^1. Unfortunately, a MVUE does not always exist. More typically, given two good unbiased estimators τ^i, $i = 1, 2$, one finds $\Sigma(\tau^1 \mid u, \theta) \leq \Sigma(\tau^2 \mid u, \theta)$ for one set of parameter values and the inequality is reversed for another set of values. In this case a MVUE will not exist. However, there is a well-known lower bound on the achievable variance.

Definition (4.3)
The **Fisher information matrix** is the $p \times p$ matrix $I(u, \theta)$ with elements

$$I_{ij}(u, \theta) :=$$
$$\int [\frac{\partial \ln p(y \mid u, \theta)}{\partial \theta_i} \frac{\partial \ln p(y \mid u, \theta)}{\partial \theta_j}] \, p(y \mid u, \theta) dy. \qquad (4.4)$$

Theorem (Cramer-Rao inequality) (4.5)
Let τ be any unbiased estimator. Then $[I(u, \theta)]^{-1} \leq \Sigma(\tau \mid u, \theta)$, for all θ in Θ.

Definition (4.6)
An unbiased estimator is **efficient** if $[I(u, \theta)]^{-1} = \Sigma(\tau \mid u, \theta)$ for all $\theta \in \Theta$.

Remark (4.7)
The Cramer-Rao inequality suggests that the larger the information matrix, the smaller the covariance. The information matrix becomes large (more positive definite) as the sensitivity of the observation y to the parameter θ increases, which means that small changes in the parameter induce appreciable changes in the output pd. Note that $p(y \mid u, \theta)$ depends on the input u. In some situations we may have freedom in selecting the input. To improve identification u should be chosen so as to make $I(u, \theta)$ as large as possible. One measure of its size is its trace, $Tr \, I(u, \theta)$, a less convenient measure is its determinant, $\det I(u, \theta)$. Suppose we are free to select u from some specified set U. Then a good choice would be

$$u = Arg \max_{u' \in U} [\min_{\theta \in \Theta} Tr \, I(u', \theta)].$$

In many situations, we can perform repeated experiments on the system—i.e., we can apply an input u_1, then observe y_1—and then we can apply a second input u_2 and observe y_2, and so on. Thus we obtain data (y^k, u^k), where $y^k = (y_1, \ldots, y_k)$ and $u^k = (u_1, \ldots, u_k)$. For such situations, for each k, let τ_k be an estimator, i.e., $\tau_k(y^k, u^k) \in \Theta$. Then, as the number of experiments or observations grows to infinity, we may be able to find estimates that are asymptotically good.

Definition (4.8)

$\{\tau_k\}$ is **asymptotically unbiased** if $\lim_{k \to \infty} E(\tau_k \mid u^k, \theta) = \theta$ for all $\theta \in \Theta$; it is **consistent** if $\tau_k(y^k, u^k) \to \theta$ in probability for all θ; it is **strongly consistent** if $\tau_k(y^k, u^k) \to \theta$ wp 1 for all θ.

We introduce the important notion of a sufficient statistic. It is notationally convenient to assume that the system is freely evolving, i.e., one observes y^k without having to apply u^k. A **statistic** is any function of the observations, y^k. Thus an estimator is a particular kind of statistic. Let $g(y^k)$ be a statistic. We say that it is **sufficient** for Θ if the conditional density of y^k given $g(y^k)$ does not depend on θ, i.e., if $p(y^k \mid g(y^k), \theta)$ does not depend on $\theta \in \Theta$.

As an example, suppose for each θ the y_i are independent and $p(y_i) \sim N(\theta, 1)$; then

$$p(y^k \mid \theta) = [2\pi]^{-\frac{1}{2} k} \exp \{-\frac{1}{2} \sum (y_i - \theta)^2\}.$$

Since

$$\exp \{-\frac{1}{2} \sum (y_i - \theta)^2\} = \exp \{-\frac{1}{2} \sum y_i^2\}$$
$$\times \exp \{-\frac{1}{2} (-2\theta \sum y_i + k\theta^2)\},$$

the next result shows that $\sum y_i$ is a sufficient statistic for $\theta \in R$.

Theorem (factorization criterion) (4.9)

$g(y^k)$ is a sufficient statistic for Θ if and only if there are functions h, q such that

$$p(y^k \mid \theta) = h(g(y^k), \theta) q(y^k), \quad \theta \in \Theta.$$

[In words: g is sufficient for θ if and only if $p(y^k \mid \theta)$ depends on θ only through $g(y^k)$.]

It is intuitively evident that in finding estimators it is enough to search over those based on sufficient statistics. This is made precise next.

Theorem (Rao-Blackwell) (4.10)

Suppose $y^k = (y_1, \ldots, y_k)$ and the y_i are iid with marginal pd $p(y_i \mid \theta)$. Let $\tau(y^k)$ be any unbiased estimator of $\gamma(\theta)$ where γ is any function [including, in particular, the function $\gamma(\theta) = \theta$]. Let $g(y^k)$ be sufficient for θ. Let

$$\bar{\tau} = \bar{\tau}(g(y^k)) := \int \tau(y^k) p(y^k \mid g(y^k), \theta) \, dy^k.$$

Then $\bar{\tau}$ is an unbiased estimate of $\gamma(\theta)$, and $\Sigma(\bar{\tau} \mid \theta) \le \Sigma(\tau \mid \theta)$. [Note that $\bar{\tau}$ is based on g.]

5. The maximum likelihood estimator

This is a well-known estimator with remarkable asymptotic properties. The density $p(y \mid u, \theta)$ considered as a function of θ in Θ for the given value of the input u and the observation y is called the **likelihood** function. Let $\tau(y, u) \in \Theta$ maximize the likelihood function, i.e.,

$$\tau(y, u) = Arg\max \{p(y \mid u, \theta) \mid \theta \in \Theta\}. \tag{5.1}$$

τ is called the **maximum likelihood estimator** (MLE): for any observed value y, $\tau(y)$ is that value of θ which maximizes the likelihood that y is observed given that u is applied.

Suppose $y^k = (y, \ldots, y_k)$ is observed. (For notational convenience, suppose no input.) Suppose $p(y^k \mid \theta) = \prod p(y_i \mid \theta)$, i.e., the y_i are independent. It is then more convenient to maximize the log-likelihood function $\ln p(y^k \mid \theta) = \sum \ln p(y_i \mid \theta)$ since it is separable in the y_i. In the case of independent observations, more generally when y_i and y_{i+j} are almost independent for large j, it can be shown that the MLE is strongly consistent and asymptotically efficient. Hence the popularity of the MLE. However, carrying out the maximization in (5.1) can be computationally expensive; also, for small sample size (k small), the MLE may not be a reasonable estimator.

6. Parametrization of system models

We now turn to dynamic systems and first motivate the problem of system identification. The analysis of optimal control starts with a given state space model

$$x_{k+1} = f_k(x_k, u_k, w_k),$$

$$y_k = h_k(x_k, v_k).$$

Saying that the model is given means that the functions f_k, h_k, and the pd of the basic random variables $\{x_0, w_0, \ldots, v_0, \ldots\}$ are all known. Equivalently, we know the cd of (x_{k+1}, y_k) given (x_k, u_k). This is therefore the prior or *off-line*, information, in contrast to the *on-line* information which at time k consists of the observations $z^k = (y^k, u^{k-1})$. Thus until now we have assumed that the off-line information specifies a unique system model.

However, the off-line information is often insufficient to characterize a model completely. That is, the system model is not known initially, and we are faced with the question of making a rational choice of the control values $u_0, u_1, \ldots$ when the model is unknown. In Chapter 6 we saw that this question can be reformulated as a problem with partial information. But the resulting computational burden is then so great as to make this formulation impractical except in the case when the unknown system model is known to belong to a finite set, as in Section 6.8. Here we consider the simpler, related problem of system identification without any regard to choice of control. In later chapters on adaptive control we study computationally attractive ways of combining system identification with control.

Suppose the system model is unknown. As we make more observations z^k, i.e., as k increases, we ought to be able to characterize better the system model. How can we describe and analyze this learning process? The abstract framework is quite simple. A particular model is specified by a triple $M := (f, h, P)$, where f is the function in the state equation, h is the function in the observation equation (for notational convenience we ignore the possibility of time varying systems), and P is the pd of the basic random variables. We will assume that the off-line information is such as to guarantee that the true system model belongs to the family of models $M^\theta = (f^\theta, h^\theta, P^\theta)$ parametrized by the finite-dimensional vector θ which is known to belong to the set Θ. Thus we know a priori that the true system model corresponds to some true parameter $\theta° \in \Theta$. Initially we do not know $\theta°$. At time k, the on-line information obtained is z^k. Based on this observation we make an estimate θ_k of the true parameter. If the estimation or identification procedure is a good one, then θ_k should approach $\theta°$ as k increases. The model M^θ need not be in state space form. In fact for identification purposes an input-output model is preferable.

The initial uncertainty about the system is reflected in the parametrization—i.e., the function $\theta \rightarrow (f^\theta, h^\theta, P^\theta)$,—and the size of the parameter set Θ, in particular, the dimension of θ. The dimension of θ may be large or small, and the parametrization may be more or less complex (linear vs. nonlinear, one-to-one vs. many-to-one). In practice, of course, the set of models $\{M^\theta\}$ can only approximately represent the true system. Better approximations will lead to more complex parametrizations and a larger model set Θ. But this also makes the identification procedures more complex. Thus the choice of parametrization must make a balance between the demand for accuracy and the need to limit the computational burden.

In the identification techniques discussed in Chapter 10 the choice of parametrization is supposed to have been made, and we study alternative ways of making a good estimate θ_k based on the on line information z^k where good means that θ_k is close to $\theta°$.

7. Examples

We consider some of the examples of Chapter 1. The dealer's stock of gasoline at the beginning of day $k + 1$ is

$$x_{k+1} = x_k + u_k - w_k =: f(x_k, u_k, w_k). \tag{7.1}$$

Physical considerations (conservation of mass and no leakage) lead us to believe in the accuracy of (7.1), so that f is known. However, we may be unable to specify the pd P of the random sequence of demands $\{w_k\}$. We want to represent parametrically our uncertainty about P. As a first step and for simplicity we may accept that the sequence $\{w_k\}$ is iid so that

$$p(w_0, w_1, \ldots) = p(w_0) p(w_1) \ldots,$$

and the question reduces to the parametrization of the marginal density $p(w_k)$. That is, we must select a finite dimensional family of densities that will approximate the true density p sufficiently well. (Note that the set of all densities is infinite dimensional.) For example, we may suppose that

$$w_k \sim N(\overline{w}, \sigma^2), \tag{7.2}$$

where the mean $\overline{w}$ and variance σ^2 are unknown. The prior uncertainty is now parametrized by the two dimensional vector $\theta = (\overline{w}, \sigma^2)$. The set of parameters would be $\Theta = \{(\overline{w}, \sigma^2) \mid \overline{w} \geq 0\}$. For each θ in Θ we get a system model

$$x_{k+1} = x_k + u_k - (\overline{w} + \sigma v_k) =: f^\theta(x_k, u_k, v_k). \tag{7.3}$$

where $\{v_k\}$ is white noise, i.e., an iid sequence with $v_k \sim N(0, 1)$.

To pursue the example further, the assumption that $\{w_k\}$ is an independent sequence may be untenable. For instance, if a lot of gasoline is sold on day k, w_k is large, then w_{k+1} is likely to be small. w_k and w_{k+1} are then negatively correlated, hence dependent. Suppose this dependence is believed not to extend beyond one day so that w_{k+j} and w_k are independent for $j \geq 2$. This could be specified by

$$w_k = \overline{w} + c_0 v_k + c_1 v_{k-1}, \tag{7.4}$$

where $\{v_k\}$ is white noise as before. This leads to the parametrization

$$x_{k+1} = f^\theta(x_k, u_k, v_k, v_{k-1}) = x_k + v_k - (\overline{w} + c_0 v_k + c_1 v_{k-1}), \tag{7.5}$$

where the unknown parameter $\theta = (\overline{w}, c_0, c_1)$ must satisfy $\overline{w} \geq 0$. The identification model is more complex since there are more parameters to be estimated.

Remark (7.6)

The model (7.5) is not directly suitable for control since the input noise $\{\overline{w} + c_0 v_k + c_1 v_{k-1}\}$ is not independent. It is quite easy to recover independence by augmenting the state to (x_k, z_{k-1}) where $z_k := v_{k-1}$ in terms of which we get the state space model

$$
\begin{bmatrix} x_{k+1} \\ z_{k+1} \end{bmatrix} = \begin{bmatrix} 1 & -c_1 \\ 0 & 0 \end{bmatrix} \begin{bmatrix} x_k \\ z_k \end{bmatrix} + \begin{bmatrix} 1 \\ 0 \end{bmatrix} u_k + \begin{bmatrix} -1 \\ 0 \end{bmatrix} \overline{w}
$$
$$
+ \begin{bmatrix} -c_0 \\ 1 \end{bmatrix} v_k, \tag{7.7}
$$

$$ y_k := x_k. $$

The input noise is now the independent sequence $\{v_k\}$. Note that since $\{w_k\}$ is not independent, its pd is more complex. In (7.7) this complexity reappears in the fact that one component of the state, namely z_k, is not observed.

In the second example of Chapter 1, the amount of crop grown in season $k + 1$ is x_{k+1} and depends on x_k, the input u_k, and the unknown effects of weather w_k according to the relation

$$ x_{k+1} = (1 + w_k) g(x_k, u_k) =: f(x_k, u_k, w_k). \tag{7.8} $$

This relation is more complicated than (7.1) since f is nonlinear, and it is likely that the form of g and the pd of $\{w_k\}$ are both unknown. We have to parametrize this uncertainty. Recall that in Chapter 1, w_k is defined in such a way that it has zero mean. So as a first step we may suppose that

$$ w_k = \sigma v_k, \tag{7.9} $$

where $\{v_k\}$ is white noise. This gives one unknown parameter σ. To obtain a parametrization of g we may argue as follows. Clearly we would expect $g(x_k, u_k) = 0$ if $x_k = 0$—i.e., nothing grows if nothing is sown—and g should increase in each of its arguments. This suggests a relationship of the form

$$ g(x, u) = \alpha \, x^\beta \, u^\gamma, \tag{7.10} $$

which involves three additional parameters. This yields the identification model

$$ x_{k+1} = f(x_k, u_k, v_k) := (1 + \sigma v_k)\alpha \, x_k^\beta \, u_k^\gamma, \tag{7.11} $$

with the unknown parameter $\theta = (\alpha, \beta, \gamma, \sigma)$. One unpleasant aspect of this model (as well as in the gasoline example) is that since $\{w_k\}$ is assumed Gaussian, the event $\{w_k < -1\}$ has positive probability. This permits $x_k < 0$ in (7.8) which is unrealistic. Thus the pd of v_k needs to be changed to distributions which do not permit negative values. Two common families are the chi-square and log-normal. Recall that w is chi-square (with 1 degree of freedom) if

$$w = v^2, v \sim N(0, 1),$$

and it is log-normal if

$$\ln w = v, v \sim N(0, 1).$$

Suppose we choose the latter family. Then in place of (7.11) we obtain

$$x_{k+1} = f(x_k, u_k, v_k) := (\exp \sigma v_k) \, \alpha \, x_k^\beta \, u_k^\gamma. \tag{7.12}$$

The multiplicative form of f^σ suggests that we can achieve a degree of linearity by taking logarithms:

$$\ln x_{k+1} = \ln \alpha + \beta \ln x_k + \gamma \ln u_k + \sigma v_k.$$

This is nice since by defining the state variable as $\xi = \ln x$ and the input as $\mu = \ln u$, and similarly by redefining the unknown parameters, we get a linear system model which is also linear in the parameters. The identification procedures are then much simpler. As a general remark we may note here that in these procedures it is not so much the nonlinearity in the observed input and output variables that makes matters difficult but rather the nonlinearity in the unknown parameters.

8. Dependence between input and noise

Recall the memoryless identification model $M^\theta = (f^\theta, P^\theta)$, $\theta \in \Theta$, where

$$y_i = f^\theta(u_i, w_i), \tag{8.1}$$

and P^θ is the pd of $\{w_i\}$. We have implicitly assumed till now that $\{u_i\}$ is deterministic. This implies, in particular, that the processes $\{u_i\}$ and $\{w_i\}$ are independent for all θ. Indeed, whenever $\{u_i\}$ is a random and observed sequence, so long as it is independent of $\{w_i\}$—i.e., P^θ $(w_1, w_2, \ldots \mid u_1, u_2, \ldots) = P^\theta$ $(w_1, w_2, \ldots)$—then the preceding discussion remains valid. The only change is that one interprets

$$p(y^k \mid \theta) := p(y^k \mid u^k, \theta).$$

To see this observe that

$$P\{y^k \in Y \mid u^k, \theta\} = P\{w^k \in W \mid u^k, \theta\},$$

$$W := \{w^k \mid (f^\theta(u_1, w_1), \ldots, f^\theta(u_k, w^k)) = y^k \in Y\}.$$

If $\{u_i\}$ and $\{w_i\}$ are independent for all θ, then $P\{w^k \in W \mid u^k, \theta\} = Prob\{w^k \in W \mid \theta\}$ and so

$$P\{y^k \in Y \mid u^k, \theta\} = P^\theta(W),$$

which is the same expression as if u^k were deterministic. (Of course W depends on the value of u^k but not on its distribution.)

We can see in a simple example what happens when the input and noise are not independent. Consider the linear model

$$y_k = \alpha u_k + w_k, \tag{8.2}$$

where $\theta := \alpha \in R$ is the unknown parameter and $\{w_i\}$ is white noise. Assuming input and noise are independent; then

$$p(y^k \mid \alpha, u^k) = K \exp -\tfrac{1}{2} \sum (y_i - \alpha u_i)^2,$$

where K is a positive constant. Hence the MLE $\bar{\alpha}$ is obtained by minimizing $\sum (y_i - \alpha u_i)^2$, which gives

$$\bar{\alpha} = \left(\sum y_i u_i\right)\left(\sum u_i^2\right)^{-1}. \tag{8.3}$$

Substitution of (8.2) into (8.3) gives

$$\bar{\alpha} = \alpha + \left(\sum u_i y_i\right)\left(\sum u_i^2\right)^{-1},$$

and so

$$E(\bar{\alpha} \mid \alpha, u^k) = \alpha + \left[\sum u_i E(w_i \mid \alpha, u^k)\right]\left(\sum u_i^2\right)^{-1}. \tag{8.4}$$

If $\{u_i\}$, $\{w_i\}$ are indeed independent for all α, then (8.4) implies $E(\bar{\alpha} \mid \alpha, u^k) = \alpha$, and so $\bar{\alpha}$ is an unbiased estimate.

If, however, the input and noise are related as

$$u_i = g w_i + v_i,$$

where v_i is also white and independent of w_i, then

$$E(w_i \mid \alpha, u^k) = g(1 + g^2)^{-1},$$

which shows that $\bar{\alpha}$ is a biased estimate if $g \neq 0$.

9. Parametrization of linear systems

Now we turn to the special situation of linear systems for which we examine identification methods in Chapter 10. A linear (time invariant) stochastic system with input $u_k \in R^m$ and output $y_k \in R^p$, $-\infty < k < \infty$, can be represented in three different ways. The state space representation takes the form

$$x_{k+1} = A\, x_k + B\, u_k + G\, w_k, \tag{9.1a}$$

$$y_k = C\, x_k + H\, v_k. \tag{9.1b}$$

An ARMAX representation has the form

$$y_{k+1} + \sum_{i=1}^{n} A_i\, y_{k+1-i} = \sum_{i=0}^{n} B_i\, u_{k-i} + w_{k+1} + \sum_{i=1}^{n} C_i\, w_{k+1-i}. \tag{9.2}$$

Lastly, the impulse response model has the form

$$y_{k+1} = \sum_{i=0}^{\infty} F_i\, u_{k-i} + w_{k+1} + \sum_{i=1}^{\infty} G_i\, w_{k+1-i}. \tag{9.3}$$

The processes $\{y_k\}$ and $\{u_k\}$ are observed and are the same for the three models—i.e., they have the same joint pd. This implies, in particular, that the conditional distribution

$$P\{y_{k+1} \mid y_s, u_s, s \leq k\}$$

is the same for all models and, in particular,

$$y_{k+1 \mid k} := E\{y_{k+1} \mid y_s, u_s, s \leq k\}$$

is also the same for the three models. We call $y_{k+1 \mid k}$ the (one-step) output prediction, and the prediction error is

$$\widetilde{y}_{k+1 \mid k} := y_{k+1} - y_{k+1 \mid k}.$$

We calculate the output prediction and error for each model. We use the notation of Section 5.2: for any sequence $\{\xi_k\}$, $q\xi_k := \xi_{k+1}$, and $q^{-1} \xi_k := \xi_{k-1}$.

Express the impulse response model (9.3) as

$$y_{k+1} = F(q^{-1})\, u_k + G(q^{-1})w_{k+1}, \tag{9.4}$$

with

$$F(q^{-1}) := \sum_{i \geq 0} q^{-i}\, F_i, \quad G(q^{-1}) := \sum_{i \geq 0} q^{-i}\, G_i, \; G_0 := I. \tag{9.5}$$

We assume that

$$\det G(z) = 0 \quad \text{implies} \quad |z| > 1, \tag{9.6}$$

so $[G(q^{-1})]^{-1}$ is stable, and

$$[G(q^{-1})]^{-1} = \sum_{i \geq 0} q^{-i}\, \Gamma_i,$$

where $\sum \| \Gamma_i \| < \infty$. Then (9.4) can be solved for w_{k+1} to get

$$y_{k+1} = [I - (G(q^{-1}))^{-1}]\, y_{k+1} + [G(q^{-1})]^{-1}\, F(q^{-1})\, u_k$$
$$+ w_{k+1}. \tag{9.7}$$

This is well defined because $G(0) = I$, so the first term on the right depends only on y_s, $s \leq k$, and not on y_{k+1}. Next we assume

$$E\{w_{k+1} \mid y_s, u_s, s \leq k\} = 0. \tag{9.8}$$

[(9.8) certainly holds if the noise sequence $\{w_k\}$ is independent and if past values of the input $\{u_s, s \leq k\}$ are independent of future values of the noise $\{w_s, s \geq k\}$.] From (9.7) and (9.8) the next result is immediate.

Lemma   (9.9)

For the impulse response model (9.4), and under assumptions (9.6) and (9.8),

$$y_{k+1 \mid k} = [I - (G(q^{-1}))^{-1}] y_{k+1}$$
$$+ [G(q^{-1})]^{-1} F(q^{-1}) u_k, \tag{9.10}$$

$$\widetilde{y}_{k+1 \mid k} = w_{k+1}. \tag{9.11}$$

Next express the ARMAX model (9.2) as

$$A(q^{-1}) y_{k+1} = B(q^{-1}) u_k + C(q^{-1}) w_{k+1}; \tag{9.12}$$

with

$$A(q^{-1}) := \sum_{i=0}^{n} q^{-i} A_i, \, A_0 = I,$$

$$B(q^{-1}) := \sum_{i=0}^{n} q^{-i} B_i, \quad C(q^{-1}) := \sum_{i=0}^{n} q^{-i} C_i, \, C_0 = I.$$

We assume

$$\det A(z) = 0 \text{ implies } \mid z \mid > 1, \tag{9.13}$$

so (9.12) can be put in impulse response form

$$y_{k+1} = [A(q^{-1})]^{-1} B(q^{-1}) u_k$$
$$+ [A(q^{-1})]^{-1} C(q^{-1}) w_{k+1}. \tag{9.14}$$

Note that $[A(0)]^{-1} C(0) = I$. Assume further that

$$\det C(z) = 0 \text{ implies } \mid z \mid > 1. \tag{9.15}$$

Then as a corollary to Lemma (9.9) we get the next result.

Lemma   (9.16)

For the ARMAX model (9.2), and under assumptions (9.8), (9.13), and (9.15),

$$y_{k+1 \mid k} = [I - (C(q^{-1}))^{-1} A(q^{-1})] y_{k+1}$$
$$+ [C(q^{-1})]^{-1} B(q^{-1}) u_k, \tag{9.17}$$

$$\widetilde{y}_{k+1 \mid k} = w_{k+1}.$$

It is useful to know that for the ARMAX model one can obtain the prediction equation without the stability condition (9.13). Observe from (9.12) and (9.15) that

$$w_{k+1} = [C(q^{-1})]^{-1} [A(q^{-1}) y_{k+1} - B(q^{-1}) u_k],$$

which shows that w_k is a function of (measurable with respect to)

$\{y_s, u_{s-1}, s \le k\}$. Hence taking conditional expectations in (9.2) and using (9.8), we get

$$y_{k+1 \mid k} + \sum_{i=1}^{n} A_i y_{k+1-i} = \sum_{i=0}^{n} B_i u_{k-i} + \sum_{i=1}^{n} C_i w_{k+1-i}, \tag{9.18a}$$

$$\widetilde{y}_{k+1 \mid k} = w_{k+1}. \tag{9.18b}$$

Substituting for w_{k+1-i} from (9.18b) into (9.18a) gives the next result.

Lemma (9.19)

For the ARMAX model (9.2) and under assumptions (9.8) and (9.15),

$$y_{k+1 \mid k} + \sum_{i=0}^{n} A_i y_{k+1-i} = \sum_{i=0}^{n} B_i u_{k-i}$$

$$+ \sum_{i=0}^{n} C_i [y_{k+1-i} - y_{k+1-i \mid k-i}], \tag{9.20}$$

$$\widetilde{y}_{k+1 \mid k} = w_{k+1}.$$

The equation (9.20) is called the innovations representation or the **pre-whitening filter** for the process $\{y_k\}$. [See Section 7.9.]

Finally consider the state space model (9.1) and assume that

$\{w_k\}, \{v_k\}$ are independent,

$E\{(w_k, v_k) \mid y_s, u_s, s \le k\} = (0, 0)$,

$E\{w_k w_k^T \mid y_s, u_s, s \le k\} = Q$, and

$$E\{v_k v_k^T \mid y_s, u_s, s \le k\} = R. \tag{9.21}$$

Further assume that the Kalman filter has a steady state so that the predictor equation is

$$x_{k+1 \mid k} = A x_{k \mid k-1} + B u_k + L \widetilde{y}_{k \mid k-1}, \tag{9.22a}$$

$$y_{k \mid k-1} = C x_{k \mid k-1}, \tag{9.22b}$$

$$\widetilde{y}_{k \mid k-1} = y_k - y_{k \mid k-1} =: e_k. \tag{9.22c}$$

Exercise (9.23)

Show that under these assumptions

$$x_{k+1 \mid k} = E\{x_{k+1} \mid y_s, u_s, s \le k\},$$

$$y_{k+1 \mid k} = E\{y_{k+1} \mid y_s, u_s, s \le k\}.$$

From (9.22), (9.23) we get the impulse response form

$$y_{k+1} = C[I - q^{-1}A]^{-1} B u_k$$

$$+ \{I + q^{-1} C[I - q^{-1}A]^{-1}L\} e_{k+1}. \tag{9.24}$$

Lemma (9.25)

If the models (9.1), (9.2), and (9.3) have the same input and output processes, then

$$F(z) = [A(z)]^{-1} B(z) = C[I - zA]^{-1}B,$$

$$G(z) = [A(z)]^{-1} C(z) = I + zC[I - zA]^{-1}L, \qquad (9.26)$$

where $F(z)$, and $G(z)$ are as in (9.5); $A(z)$, $B(z)$, and $C(z)$ are as in (9.12); A, B, C, and L are as in (9.22).

Proof

The models must give the same prediction. From (9.10) we see that this determines uniquely the coefficients of the impulse response model. The result follows by comparing (9.4), (9.14), and (9.24).   □

10. Identifiability

The impulse response, ARMAX, and state space identification models correspond, respectively, to the parametrizations

$$\theta \rightarrow (F^\theta(z), G^\theta(z), P_w^\theta), \qquad (10.1)$$

$$\theta \rightarrow (A^\theta(z), B^\theta(z), P_w^\theta), \qquad (10.2)$$

$$\theta \rightarrow (A^\theta, B^\theta, C^\theta, L^\theta, P^\theta), \qquad (10.3)$$

together with the restriction $\theta \in \Theta$. As before θ° denotes the true parameter value. When it is convenient to do so we write $F(z, \theta)$, $A(z, \theta)$, etc., and the true values by $F^\circ(z) = F(z, \theta^\circ)$, etc.

From the observations of the processes $\{u_k\}$, $\{y_k\}$ we can at most determine their joint pd and then we can infer the exact predictor equation. We say that the identification model is **identifiable** if knowledge of the predictor equation is sufficient to identify the true parameter. From the last section we know that the predictor equation specifies a unique impulse response model, i.e., the triple $(F(z), G(z), P_w)$ in (9.4). It follows that the noise process is always identifiable. Furthermore an impulse response model of the form (9.1) is always identifiable no matter how large Θ is.

From Lemma (9.25) we see that if the correct predictor equation is known, then the ARMAX model (10.2) must satisfy

$$[A(z, \theta)]^{-1} B(z, \theta) = F(z, \theta^\circ),$$

$$[A(z, \theta)]^{-1} C(z, \theta) = G(z, \theta^\circ), \qquad (10.4a)$$

$$\theta \in \Theta, \qquad (10.4b)$$

and so identifiability requires that (10.4) have a unique solution. It is clear that (10.4a) by itself cannot guarantee uniqueness, since the equations do not change if $A(z, \theta)$, $B(z, \theta)$, and $C(z, \theta)$ are all multiplied by the same polynomial in z. Thus Θ must be restricted to guarantee uniqueness. We say that the parametrization is canonical if

$$[A(z, \theta)]^{-1} B(z, \theta) = [A(z, \theta')]^{-1} B(z, \theta'),$$

$$[A(z, \theta)]^{-1} C(z, \theta) = [A(z, \theta')]^{-1} C(z, \theta'),$$

and θ, θ' in Θ imply that $\theta = \theta'$.

Similarly the parametrization for (9.3) is canonical if

$$C(\theta)[I - zA(\theta)]^{-1} B(\theta) = C(\theta')[I - zA(\theta)]^{-1} B(\theta'),$$

$$C(\theta)[I - zA(\theta)]^{-1} L(\theta) = C(\theta')[I - zA(\theta)]^{-1} L(\theta'),$$

and θ, θ' in Θ imply that $\theta = \theta'$.

Thus to obtain a canonical representation for ARMAX and state space models the set Θ must be restricted. Unfortunately, when such a restriction is made is set Θ no longer has a nice shape. To see this consider the single input single output nth order ARMAX model

$$y_{k+1} + \sum_{i=1}^{n} a_i y_{k+1-i} = \sum_{i=0}^{n} b_i u_{k-1} + w_{k+1} + \sum_{i=1}^{n} c_i w_{k+1-i},$$

where w_k is white noise and $\theta = (a_1, .., a_n, b_1, .., b_n, c_1, .., c_n)$. Let

$$a(z, \theta) = 1 + \sum a_i z^i, \, b(z, \theta) = \sum b_i z^i, \, \text{and}$$

$$c(z, \theta) = 1 + \sum c_i z^i.$$

The simplest set Θ would be the unconstrained case $\Theta = R^{3n}$, or where Θ is specified by linear equality and inequality constraints on θ. However, neither of these specifications of Θ is canonical. One canonical parametrization is the following

$$\Theta = \{\theta \mid a(z, \theta), b(z, \theta), c(z, \theta)$$

$$\text{have no common factors}\}.$$

But the condition no common factors cannot be expressed through any simple restriction on Θ.

We shall see later that a canonical representation is desirable since it guarantees convergence of identification procedures. However computational simplicity demands that the set Θ be specified in a simple way as well. These two objectives are in conflict when we use an ARMAX and state space identification model. In Chapter 10 we consider in detail the identification of the ARMAX system

$$y_{k+1} + \sum_{i=1}^{n} A_i y_{k+1-i} = \sum_{i=0}^{n} B_i u_{k-i} + w_{k+1} + \sum_{i=1}^{n} C_i w_{k+1-i}.$$

11. Notes

1. For more on the problem of parametrization of linear stochastic systems for use in identification procedures, see Glover and Willems (1974), Gertz, Gevers and Hannan (1982) and the references therein.

2. The problem of parameter estimation and the associated topics such as biasedness, consistency, maximum likelihood estimates, Cramer bounds etc., are well covered in books dealing with statistics; see Kendall and Stuart (1964), Cramer (1946), Rao (1973), and Rohatgi (1976).

3. Goodwin and Payne (1977) pay particular attention to the application of these concepts in the identification of dynamic systems. Issues relating to the design of good input signals and system parametrizations are also covered there. For design of good input signals discussed in Remark (4.7), also see Mehra (1976).

4. For more on the type of communication problem of Exercise (3.1), see Van Trees (1968).

5. The Bayesian approach is covered in Blackwell and Girshick (1954) and DeGroot (1970).

12. References

[1]   D. Blackwell and M.A. Girshick (1954). *Theory of games and statistical decisions,* John Wiley, New York.

[2]   H. Cramer (1946), *Mathematical methods of statistics,* Princeton University Press, Princeton.

[3]   M. H. DeGroot (1970), *Optimal statistical decision,* McGraw Hill, New York.

[4]   V. Gertz, M. Gevers and E. Hannan (1982), "The determination of optimum structures for the state space representation of multivariate stochastic processes," *IEEE Transactions on Automatic Control,* vol AC-27, 1200-1211.

[5]   G. Goodwin and R. Payne (1977), *Dynamic system identification: experiment design and data analysis,* Academic Press, New York.

[6]   K. Glover and J. C. Willems (1974), "Parametrizations of linear dynamical systems: canonical forms and identifiability," *IEEE Transactions on Automatic Control,* vol. AC-19.

[7]   M. G. Kendall and A. Stuart (1964), *The advanced theory of statistics,* Griffin, London.

[8]   R.K. Mehra, "Synthesis of optimal inputs for multiinput multioutput systems with process noise," in R.K. Mehra and D.G. Lainiotis (eds), *System identification: Advances and case studies,* Academic Press, New York, 211-250.

[9]   C. R. Rao (1973), *Linear statistical inference and its applications,* Wiley, New York.

[10] V. K. Rohatgi (1976), *An introduction to probability theory and mathematical statistics,* John Wiley, New York.

[11] H. Van Trees (1968), *Detection, estimation and modulation theory,* Wiley, New York.

CHAPTER 10
LINEAR SYSTEM IDENTIFICATION

The chapter covers the major techniques for identification of linear systems. The least squares method is suitable when the disturbance is white noise. If the noise is colored, more complex methods are needed to avoid bias and to identify the disturbance process. The maximum likelihood and prediction error estimates are studied closely. Several modifications that yield recursive algorithms are presented, including instrumental variables, pseudolinear regression, and the recursive prediction error estimate. Lastly, a promising approach for analyzing the asymptotic behavior of recursive algorithms is introduced.

1. The least squares method

Suppose we have the data $\{\phi_0, \ldots, \phi_{n-1}, y_1, \ldots, y_n\}$ and we believe that

$$y_{k+1} \approx \phi_k^T \theta, \ k = 0, 1, 2, \ldots, n-1, \tag{1.1}$$

i.e., y_{k+1} is approximately a linear function of ϕ_k. Here the ϕ_i are vectors and the y_i are scalars. The least squares method is to choose θ to minimize

$$V_n(\theta) := \sum_{k=0}^{n-1} (y_{k+1} - \phi_k^T \theta)^2. \tag{1.2}$$

Exercise (1.3)

Let $\theta \in R^p$. Setting $\dfrac{\partial V_n(\theta)}{\partial \theta} = 0$, show that $V_n(\theta)$ is minimized by $\hat{\theta}_n$ where

$$\hat{\theta}_n = \left(\sum_{k=0}^{n-1} \phi_k \phi_k^T \right)^{-1} \sum_{k=0}^{n-1} \phi_k \, y_{k+1}, \tag{1.4}$$

assuming the inverse exists.

The next exercise considers the case when the inverse does not exist.

Exercise (1.5)

Let $Y_n := (y_1, y_2, \ldots, y_n)^T$ and let $X_n^T := [\phi_0, \phi_1, \ldots, \phi_{n-1}]$. Show that there exists a solution $\hat{\theta}_n$ to

$$(X_n^T X_n)\, \hat{\theta}_n = X_n^T Y_n, \tag{1.6}$$

and every solution minimizes

$$||Y_n - X_n\, \theta||^2 = \sum_{k=0}^{n-1} (y_{k+1} - \phi_k^T\, \theta)^2.$$

The estimate $\hat{\theta}_n$ of (1.4) or (1.6) is the **least squares estimate** (LSE). Note that the LSE does not require any probability structure; it is just the result of fitting a linear model to data in a certain way.

Suppose one more datum $(\phi_n,\, y_{n+1})$ becomes available so that we now have $Y_{n+1} := (Y_n^T,\, y_{n+1})^T$ and $X_{n+1}^T := [X_n^T,\, \phi_n]$; then we can obtain the new LSE, $\hat{\theta}_{n+1} = (X_{n+1}^T X_{n+1})^{-1} X_{n+1}^T Y_{n+1}$, assuming the inverse exists. $\hat{\theta}_{n+1}$ can also be obtained from $\hat{\theta}_n$ as follows.

Exercise (1.7)
Verify that the LSE can be written as

$$\hat{\theta}_{n+1} = \hat{\theta}_n + R_n^{-1}\, \phi_n(y_{n+1} - \phi_n^T\, \hat{\theta}_n), \tag{1.8}$$

$$R_{n+1} = R_n + \phi_n \phi_n^T. \tag{1.9}$$

[Hint: Define $R_n := \sum_{k=0}^{n} \phi_k \phi_k^T$ and substitute $\sum_{k=0}^{n-1} \phi_k y_{k+1} = R_{n-1}\, \hat{\theta}_n$ and

$R_{n-1} = R_n - \phi_n \phi_n^T$ in $\hat{\theta}_{n+1} = R_n^{-1} \sum_{k=0}^{n} \phi_k y_{k+1}.$]

The recursion (1.8) requires inversion of the $p \times p$ matrix R_n. This can be avoided using the next result.

Exercise (matrix inversion lemma) (1.10)
Show that

$$(S + G\, Q\, H)^{-1} = S^{-1} - S^{-1} G\, (H\, S^{-1} G + Q^{-1})^{-1} H\, S^{-1},$$

where the matrices are of appropriate size and the required inverses exist. [Hint: Multiply by (S + GQH) from the right.]

Exercise (1.11)
Let $P_n := R_n^{-1}$. Show that

$$P_n = P_{n-1} - \frac{P_{n-1}\, \phi_n \phi_n^T\, P_{n-1}}{1 + \phi_n^T\, P_{n-1}\, \phi_n}, \tag{1.12}$$

$$\hat{\theta}_{n+1} = \hat{\theta}_n + \frac{P_{n-1}\, \phi_n}{1 + \phi_n^T\, P_{n-1}\, \phi_n}\, (y_{n+1} - \phi_n^T\, \hat{\theta}_n). \tag{1.13}$$

Note that the recursions (1.8) and (1.9) or (1.12) and (1.13) are valid only for those n for which $\sum_{k=0}^{n-1} \phi_k \phi_k^T$ is strictly positive definite. In practice this is achieved by starting the recursion with P_0 or R_0 chosen as some multiple of the identity matrix. We will see later that the asymptotic analysis is not affected by the choice of such initial conditions.

The recursions (1.8) and (1.9), or (1.12) and (1.13), are called the **recursive least squares** (RLS) estimates.

If more importance is attached to recent data than to data received in the remote past, then one chooses $\hat{\theta}_n$ to minimize

$$\sum_{k=0}^{n-1} \lambda^{n-k} (y_{k+1} - \phi_k^T \theta)^2 \tag{1.14}$$

over all θ; and the *forgetting factor* λ is chosen with $0 < \lambda < 1$.

Exercise (1.15)

If $\hat{\theta}_n$ minimizes (1.14), show that it can be recursively written as

$$\hat{\theta}_{n+1} = \hat{\theta}_n + R_{n,\lambda}^{-1} \, \phi_n \, (y_{n+1} - \phi_n^T \hat{\theta}_n),$$

$$R_{n,\lambda} = \lambda \, R_{n-1,\lambda} + \phi_n \, \phi_n^T.$$

Also,

$$R_{n,\lambda}^{-1} = \frac{1}{\lambda} \, R_{n-1,\lambda}^{-1} - \frac{1}{\lambda} \frac{R_{n-1,\lambda}^{-1} \, \phi_n \, \phi_n^T \, R_{n-1,\lambda}^{-1}}{\lambda + \phi_n^T \, R_{n-1,\lambda}^{-1} \, \phi_n}.$$

2. Prediction error, maximum likelihood, and conditional mean estimates

Above, we did not model the way $\{\phi_0, \ldots, \phi_{n-1}, y_1, \ldots, y_n\}$ are generated. We now consider models of the form

$$y_{k+1} = \phi_k^T \theta + w_{k+1}, \tag{2.1}$$

where y_{k+1} and ϕ_k are random variables linearly related through the parameter θ, except for the perturbation w_{k+1}. The joint pd of the three stochastic processes $\{y_k\}$, $\{\phi_k\}$, and $\{w_k\}$ depends on θ and we denote it by P^θ. Expectation with respect to P^θ is denoted by E^θ, and θ° denotes the true parameter. Given $\{y_s, \phi_s, s \leq n\}$ we want to form an estimate of the true parameter.

We begin with the prediction error estimate.

We can predict y_{k+1} for the model (2.1) given the past $\{y_s, \phi_s, s \leq k\}$ by $E^\theta\{y_{k+1} \mid y_s, \phi_s, s \leq k\}$. Since this prediction depends on the parameter value θ, we denote it by

$$\hat{y}_{k+1 \mid k}(\theta) := E^\theta\{y_{k+1} \mid y_s, \phi_s, s \leq k\}.$$

From the examples in Chapter 9 we see that the primary function of a model such as (2.1) is that it tells us how y_{k+1} is generated. Hence the model (2.1) with parameter θ is a good fit for the data $\{y_s, \phi_s, s \leq k\}$ if the prediction errors, $y_{k+1} - \hat{y}_{k+1 \mid k}(\theta)$, are small. The quantity

$$L_n(\theta) := \sum_{k=0}^{n-1} (y_{k+1} - \hat{y}_{k+1 \mid k}(\theta))^2 \tag{2.2}$$

is a measure of the prediction error and it seems reasonable to estimate the true parameter as that value $\hat{\theta}_n$ that minimizes $L_n(\theta)$.

Definition (2.3)
The estimate $\hat{\theta}_n$ that minimizes $L_n(\theta)$ is called the **prediction error estimate** (PEE).

So far we have not made any special assumptions regarding the disturbance or noise process $\{w_k\}$. Suppose that

$$E^\theta \{w_{k+1} \mid y_s, \phi_s, s \leq k\} = 0, \tag{2.4}$$

so

$$\hat{y}_{k+1 \mid k}(\theta) = \phi_k^T \theta, \tag{2.5}$$

and we have the following result.

Lemma (2.6)
If $\{w_k\}$ satisfies (2.4), then the PEE that minimizes (2.2) is just the LSE.

One important case where (2.4) holds for (2.1) is when (under P^θ) $\{w_k\}$ is a white noise process (i.e., a zero mean iid process,) and the past $\{y_s, \phi_s, s \leq k\}$ is independent of the future noise $\{w_s, s \geq k+1\}$.

It is important to note that if $\{w_k\}$ does not satisfy (2.4), then the PEE and the LSE will differ. A typical situation where (2.4) is not satisfied is when $\{w_k\}$ is not white noise.

Exercise (2.7)
Suppose that

$$y_{k+1} = \theta u_k + w_{k+1}, \quad k = 0, 1, \dots,$$

and $\{w_k\}$ satisfies

$$w_1 = v_1,$$

$$w_{k+1} = -cv_k + v_{k+1}, \quad k = 1, 2, \dots,$$

where (under P^θ) $\{v_k\}$ is iid with mean 0 and variance 1, and $\{v_k\}$ is independent of $\{u_k\}$. (Note that $\{w_k\}$ is not white since $E^\theta w_k w_{k+1} = -c$). Show that

$$E^\theta \{y_{k+1} \mid u_s, y_s, s \leq k\} = \theta u_k - c \sum_{j=0}^{k-1} c^{k-1-j}(y_{j+1} - \theta u_j).$$

Hence the PEE that minimizes

$$\sum_{k=0}^{n-1} (y_{k+1} - \hat{y}_{k+1 \mid k}(\theta))^2$$

$$= \sum_{k=0}^{n-1} [y_{k+1} - \theta u_k - c \sum_{j=0}^{k-1} c^{k-1-j} (y_{j+1} - \theta u_j)]^2$$

differs from the LSE which minimizes

$$\sum_{k=0}^{n-1} (y_{k+1} - \theta \, u_k)^2.$$

However, if $c = 0$—i.e., $\{w_k\}$ is white noise—then the PEE and the LSE coincide.

[Hint: Show that $E^\theta\{v_1 \mid u_s, y_s, s \leq 1\} = v_1 = y_1 - \theta u_0$ and recursively show that for $k \geq 2$

$$E^\theta\{v_k \mid u_s, y_s, s \leq k\} = v_k = \sum_{j=0}^{k-1} c^{k-1-j} (y_{j+1} - \theta u_j).$$

Then determine

$$E^\theta\{y_{k+1} \mid u_s, y_s, s \leq k\} = \theta u_k - c \, E^\theta\{v_k \mid u_s, y_s, s \leq k\}$$

$$+ E^\theta\{v_{k+1} \mid u_s, y_s, s \leq k\}$$

$$= \theta u_k - c \, E^\theta\{v_k \mid u_s, y_s, s \leq k\}.]$$

We turn now to the **maximum likelihood estimate** (MLE). By definition, the MLE $\hat{\theta}_n$ maximizes the likelihood function

$$p^\theta(\phi_s, y_s , s \leq n),$$

where p^θ denotes the conditional joint density. This is also equal to

$$p_{0 \mid -1}^\theta(\phi_0) q_{1 \mid 0}^\theta(y_1 \mid \phi_0) p_{1 \mid 0}^\theta(\phi_1 \mid \phi_0, y_1)$$

$$\times q_{2 \mid 1}^\theta(y_2 \mid \phi_0, y_1, \phi_1) . . p_{n \mid n-1}^\theta(\phi_n \mid \phi_0, y_1, . . , y_n)$$

where $q_{k \mid k-1}^\theta$ and $p_{k \mid k-1}^\theta$ are conditional probability densities. We strengthen our assumptions about (2.1) as follows.

Assumption   (2.8)
Under P^θ,
(1) $\{w_k\}$ is iid with known probability density $f(w)$;
(2) ϕ_k is independent of $\{w_s, s \geq k + 1\}$ for $k = 0, 1, . .$; and
(3) $p_{k \mid k-1}^\theta(\phi_k \mid \phi^0, y_1, . . , \phi_{k-1}, y_k)$ does not depend on θ.

We have seen assumptions (1) and (2) before. Assumption (3) will be satisfied if each ϕ_k is a function of the past, with the function not depending on θ. We will see that this condition is satisfied by models of linear systems.

Under (2.8), the MLE $\hat{\theta}_n$ maximizes

$$M_n(\theta) = q_{1\,|\,0}^{\theta}(y_1 \mid \phi_0)\, q_{2\,|\,1}^{\theta}(y_2 \mid \phi_0, y_1, \phi_1) \cdot\cdot$$

$$q_{n\,|\,n-1}^{\theta}(y_n \mid y_1, \ldots, y_{n-1}, \phi_s, s \leq n-1).$$

In view of (2.1) and (2.8),

$$M_n(\theta) = f(y_1 - \phi_0^T\,\theta)\,f(y_2 - \phi_1^T\,\theta) \cdot\cdot f(y_n - \phi_{n-1}^T\theta),$$

$$= \prod_{k=0}^{n-1} f(y_{k+1} - \phi_k^T\,\theta).$$

Incidentally, since the MLE maximizes M_n, it also maximizes

$$\ln M_n(\theta) = \sum_{k=0}^{n-1} \ln f(y_{k+1} - \phi_k^T\,\theta). \tag{2.9}$$

Note now that under (2.8), $\phi_k^T\,\theta = \hat{y}_{k+1\,|\,k}(\theta)$, and so the MLE is related to the PEE in the sense that the function $\ln f(\xi)$ in (2.9) is replaced by $-\xi^2$ to get the PEE.

When does the MLE coincide with the LSE? Clearly if we add the assumption that

$$f(w) = \frac{1}{\sqrt{2\,\pi\,\sigma^2}}\, \exp -\frac{w^2}{2\,\sigma^2}\,, \tag{2.10}$$

i.e., $\{w_i\}$ is Gaussian white noise, then (2.9) reduces to

$$\ln M_n(\theta) = -\frac{n}{2}\,\ln(2\pi\sigma^2) - \frac{1}{2\sigma^2}\sum_{k=0}^{n-1}(y_{k+1} - \phi_k^T\,\theta)^2,$$

and since the first term on the right does not depend on θ, we get the following result.

Lemma (2.11)

If for every θ, the processes $\{\phi_k\}$ and $\{w_k\}$ in (2.1) satisfy (2.8) and (2.10), then the MLE coincides with the LSE.

The coincidence of the MLE and the LSE rests crucially on the assumption that $\{w_k\}$ is a Gaussian white noise process as seen in the following exercise.

Exercise (2.12)

In Exercise (2.7) impose the additional assumption that $v_k \sim N(0, \sigma^2)$. Show that the MLE coincides with the PEE, and therefore also differs from the LSE.

We now turn to the Bayesian identification problem. Recall that in a Bayesian framework, there is a prior distribution for the true parameter. Suppose that $\theta^\circ \sim N(\hat{\theta}, \Sigma)$. Then the Bayesian conditional mean estimate θ_n is the parameter value that minimizes $E\{\|\theta^\circ - \theta\|^2 \mid \phi_0, \ldots, \phi_{n-1}, y_1, \ldots, y_n\}$.

Exercise   (2.13)
Assume that $\theta^\circ \sim N(\bar{\theta}, \Sigma)$, $\{y_k\}$ is determined by $y_{k+1} = \phi_k^T \theta^\circ + w_{k+1}$, $\{w_k\}$ is iid with $w_k \sim N(0, \sigma^2)$, and $\{y_s, \phi_s, s \le k\}$ is independent of $\{w_s, s \ge k+1\}$. Show that $\hat{\theta}_n := E\{\theta^\circ \mid \phi_0, \ldots, \phi_{n-1}, y_1, \ldots, y_n\}$ minimizes $E\{\|\theta^\circ - \theta\|^2 \mid \phi_0, \ldots, \phi_{n-1}, y_1, y_2, \ldots, y_n\}$. Show also that $\hat{\theta}_n$ and $P_n := E\{(\theta^\circ - \hat{\theta}_n)(\theta^\circ - \hat{\theta}_n)^T \mid \phi_0, \ldots, \phi_{n-1}, y_1, \ldots, y_n\}$ satisfy the recursions of Exercise (1.11) with $\hat{\theta}_0 := \bar{\theta}$ and $P_0 := \Sigma$.
[Hint: Write down the Kalman filter for the system

$$\theta_{n+1} = \theta_n, \qquad \theta_0 \sim N(\bar{\theta}, \Sigma),$$

$$y_{n+1} = \theta_n^T \phi_n + w_{n+1},$$

and regard $\{\theta_n\}$ as the state process and $\{y_n\}$ as the observation process.]

Thus if $\{w_k\}$ is Gaussian white noise, the recursions for the conditional mean of θ° under a Bayesian formulation coincide with the LSE. If the initial conditions $\hat{\theta}_0$ and P_0 also coincide, then the estimates also agree. The Bayesian viewpoint also provides an interpretation and basis for the selection of the initial conditions $\hat{\theta}_0$ and P_0 for the RLS scheme of Exercise (1.11). If the initial uncertainty about the true parameter is large, then choose P_0 large, otherwise choose P_0 small. Moreover start the recursion with $\hat{\theta}_0$ as the best guess for the true parameter.

3. LSE for linear systems: a simple ARX model

We study the LSE estimate for the parameters of the ARX model (see Chapter 5),

$$y_{k+1} = \left(\sum_{i=0}^{p} a_i y_{k-i} + b_i u_{k-i}\right) + w_{k+1},$$

where $\{w_k\}$ is white noise. In this section we examine the special case where

$$a_0 = \ldots = a_p = 0; \quad c_0 = \ldots = c_p = 0,$$

and in the next section we analyze the general ARX case.

Thus our model is

$$y_{k+1} = b_0 u_k + b_1 u_{k-1} + \ldots + b_p u_{k-p} + w_{k+1}. \tag{3.1}$$

Assume that $\theta^\circ := (b_0, b_1, \ldots, b_p) \in \Theta = R^{p+1}$ is unconstrained. Let

$$\phi_k := (u_k, \ldots, u_{k-p})^T, \tag{3.2}$$

and suppose that the data available at time n is $(\phi_0, \ldots, \phi_{n-1}, y_1, \ldots, y_n)$, so (3.1) can be written as

$$y_{k+1} = \phi_k^T \theta^\circ + w_{k+1}. \tag{3.3}$$

which is in the form (2.1).

Hence the LSE is

$$\hat{\theta}_n = (\sum_{k=0}^{n-1} \phi_k \phi_k^T)^{-1} \sum_{k=0}^{n-1} \phi_k y_{k+1}, \tag{3.4}$$

which we analyze next. For simplicity of notation we write E for E^{θ°.

Substituting for y_{k+1} from (3.3) into (3.4) gives

$$\hat{\theta}_n = \theta^\circ + P_{n-1} \sum_{k=0}^{n-1} \phi_k w_{k+1}, \tag{3.5}$$

where

$$P_{n-1} := [\sum_{k=0}^{n-1} \phi_k \phi_k^T]^{-1}, \tag{3.6}$$

and so the behavior of $\hat{\theta}_n$ depends crucially on the last term $P_{n-1} \sum_{k=0}^{n-1} \phi_k w_{k+1}$ in (3.5). We study the properties of $\hat{\theta}_n$ beginning with bias.

Recall that $\hat{\theta}_n$ is unbiased if $E\hat{\theta}_n = \theta^\circ$ (for all $\theta^\circ \in R^{p+1}$). From (3.5),

$$E\hat{\theta}_n = \theta^\circ + E P_{n-1} \sum_{k=0}^{n-1} \phi_k w_{k+1}. \tag{3.7}$$

Hence it might be conjectured that the LSE is unbiased if

$$E\{w_{k+1} \mid \phi_s, s \le k\} = 0, \quad k = 0, 1, \dots \tag{3.8}$$

However, this conjecture is false.

Exercise (3.9)

Let $y_{k+1} = b_0 u_k + w_{k+1}$, for $k = 0, 1$, and consider the LSE

$$\hat{\theta}_2 = b_0 + [\sum_{k=0}^{1} u_k^2]^{-1} \sum_{k=0}^{1} u_k w_{k+1}.$$

Suppose $u_k = w_k$ for $k = 0, 1$. Regarding the noise, suppose that $w_2 = 0$ while w_0 and w_1 are independent with $P(w_k = -1) = \dfrac{2}{3}$, $P(w_k = 2) = \dfrac{1}{3}$ for $k = 0, 1$. Then (3.8) holds. Show that $\hat{\theta}_2 = b_0 + (w_0^2 + w_1^2)^{-1} w_0 w_1$ is biased.

This shows that even if $\{w_k\}$ is an independent, zero mean process, i.e., white noise, the LSE can be biased if u_k depends on $\{w_s, s \leq k\}$, as will be the case when u_k is selected by feedback from $\{y_s, s \leq k\}$.

We therefore strengthen (3.8) to

$$E\{w_{k+1} \mid \phi_s, s < \infty\} = 0, \quad k = 0, 1, \ldots \tag{3.10}$$

Lemma (3.11)

Under (3.10) the LSE is unbiased.

Proof

The expectation of the last term in (3.7) can be rewritten as

$$E[E\{P_{n-1} \sum_{k=0}^{n-1} \phi_k w_{k+1} \mid \phi_s, s \leq n-1\}] =$$

$$E[P_{n-1} \sum_{k=0}^{n-1} \phi_k E\{w_{k+1} \mid \phi_s, s \leq n-1\}] = 0. \qquad \square$$

Note that (3.10) does not require that $\{w_k\}$ be iid.

Exercise (3.12)

Suppose that

$$E\{w_k w_j \mid \phi_s, s < \infty\} = \sigma^2 \delta_{kj} \tag{3.13}$$

but that σ^2 is unknown. (Here δ_{kj} is the Kronecker delta.) Let

$$\epsilon_k^n := y_k - \phi_{k-1}^T \hat{\theta}_n, \quad 1 \leq k \leq n,$$

be the *residual* at time n. Suppose that $\theta^\circ \in R^{p+1}$ as in (3.3). Show that

$$\frac{1}{n-p-1} \sum_{k=1}^{n} (\epsilon_k^n)^2 \tag{3.14}$$

is an unbiased estimate of σ^2.

[Hint: Note $\epsilon_k^n = w_k - \phi_{k-1}^T P_{n-1} \sum_{j=1}^{n} \phi_{j-1} w_j$, and evaluate $E\{(\epsilon_k^n)^2 \mid \phi_s, s \leq n-1, \sigma^2\}$.]

The covariance of the estimation error is

$$\Sigma(\hat{\theta}_n)$$

$$:= E\{(\hat{\theta}_n - \theta^\circ)(\hat{\theta}_n - \theta^\circ)^T \mid \phi_s, s \leq n-1\}$$

$$= P_{n-1} \sum_{k=0}^{n-1} \sum_{j=0}^{n-1} \phi_k E\{w_{k+1} w_{j+1} \mid \phi_s, s \leq n-1\} \phi_j^T P_{n-1}.$$

Suppose now that (3.13) holds. Then

$$\Sigma(\hat{\theta}_n) = \sigma^2 P_{n-1}, \tag{3.15}$$

so

$$E\left(||\hat{\theta}_n - \theta^\circ||^2\right) = \sigma^2 E\left(Tr\ P_{n-1}\right). \tag{3.16}$$

From (3.16) we obtain conditions for consistency of $\hat{\theta}_n$ [recall (9.4.8).]

Exercise (3.17)
Suppose (3.13) holds. Show that $\hat{\theta}_n$ is consistent—i.e., converges to θ° in probability—if $\lim\limits_{n\to\infty} E\left(Tr\ P_{n-1}\right) = 0$.

[Hint: Use Chebyshev's inequality, $P(|x| \geq \epsilon) \leq \dfrac{Ex^2}{\epsilon^2}$.]

Suppose that $\{u_k\}$ is a deterministic input sequence and

$$Ew_k w_j = \sigma^2 \delta_{kj}. \tag{3.18}$$

From (3.2) and (3.6) we have

$$\frac{1}{n} P_{n-1}^{-1} = \begin{bmatrix} M_n(0,0) & .. & M_n(0,p) \\ \vdots & & \vdots \\ M_n(p,0) & .. & M_n(p,p) \end{bmatrix},$$

where

$$M_n(i,j) := \frac{1}{n} \sum_{k=0}^{n-1} u_{k-i}\ u_{k-j}$$

is the *sample correlation* of the input sequence $\{u_k\}$.

Definition (3.19)
The input sequence $\{u_k\}$ is said to be **persistently exciting** or sufficiently rich of order $p+1$ if there is a positive definite $(p+1) \times (p+1)$ matrix U such that for all large n,

$$\frac{1}{n} P_{n-1}^{-1} \geq U.$$

Corollary (3.20)
If $\{w_k\}$ satisfies (3.18) and the input sequence is persistently exciting of order $p+1$, then $\hat{\theta}_n \to \theta^\circ$ in probability.

Lastly we study efficiency. Let $\{u_k\}$ be deterministic and suppose that

$$Ew_{k+1} = 0. \tag{3.21}$$

Then by Lemma (3.11) the LSE $\hat{\theta}_n$ is unbiased. We now examine whether $\hat{\theta}_n$ is efficient. Recall, from Theorem (8.4.4), that if $I(\theta^\circ)$ is the (Fisher) information matrix, then

$$E(\hat{\theta}_n - \theta^\circ)(\hat{\theta}_n - \theta^\circ)^T \geq [I(\theta^\circ)]^{-1}$$

for an unbiased estimate $\hat{\theta}_n$, and that it is efficient if equality holds. Assume that (3.18) also holds. Then from (3.15)

$$E(\hat{\theta}_n - \theta^\circ)(\hat{\theta}_n - \theta^\circ)^T = \sigma^2 P_{n-1}.$$

The information matrix $I(\theta)$ is

$$I_{ij}(\theta^\circ)$$

$$:= \int [\frac{\partial \ln p (y^n \mid \theta^\circ)}{\partial \theta_i} \ \frac{\partial \ln p (y^n \mid \theta^\circ)}{\partial \theta_j}] p (y^n \mid \theta^\circ) dy^n,$$

where $y^n := (y_1, \ldots, y_n)$. To evaluate this we must specify the density $p(y^n \mid \theta^\circ)$. Assume, in addition to (3.21) and (3.18), that

$$w_k \sim N(0, \sigma^2), k = 1, 2, \ldots \tag{3.22}$$

Then,

$$p (y^n \mid \theta^\circ) = \prod_{k=1}^{n} \frac{1}{\sqrt{2 \pi \sigma^2}} \exp [\frac{-(y_k - \phi_{k-1}^T \theta^\circ)^2}{2 \sigma^2}],$$

and so

$$I_{ij} (\theta^\circ) = E^\theta (\sum_{k=0}^{n-1} \frac{w_{k+1} u_{k-i}}{\sigma^2} \sum_{l=0}^{n-1} \frac{w_{l+1} u_{l-j}}{\sigma^2})$$

$$= \frac{1}{\sigma^2} \sum_{k=0}^{n-1} u_{k-i} u_{k-j}.$$

This shows that

$$I(\theta^\circ) = \frac{1}{\sigma^2} P_{n-1}^{-1}$$

and we have the next result.

Lemma $\hspace{10cm}$ (3.23)
If $\{u_k\}$ is a deterministic input sequence, and $\{w_k\}$ is Gaussian white noise, then the LSE $\hat{\theta}_n$ is efficient.

It should be noted that throughout this section we have assumed either (3.10) or (3.13) or sometimes both. These assumptions are restrictive, since they effectively rule out the case where u_k is generated by feedback from the past outputs $\{y_s, s \le k\}$. A second point to note is that we did not investigate whether the LSE $\hat{\theta}_n$ is strongly consistent. The next section examines both issues for the general ARX case.

4. Analysis of LSE: the general ARX case

We will treat the general single input, single output ARX model,

$$y_{k+1} = a_0 y_k + \ldots + a_p y_{k-p} + b_0 u_k + \ldots$$

$$+ b_0 u_{k-p} + w_{k+1}, \tag{4.1}$$

where the true parameter is $\theta^\circ := (a_0, \ldots, a_p, b_0, \ldots, b_p) \in R^{2p+2}$.

As before, the analysis is conducted under P^{θ°—i.e., when the true parameter is θ°, so for simplicity of notation, we write E for E^{θ°.

Let F_k be the σ-algebra generated by the past $\{y_s, u_s, w_s, s \le k\}$, and assume throughout that

$$E\{w_{k+1} \mid F_k\} = 0, \, k = 0, 1, \ldots, \tag{4.2}$$

$$E\{w_{k+1}^2 \mid F_k\} \le \sigma^2. \tag{4.3}$$

Note that F_k is also the σ-algebra generated by $\{y_s, \phi_s, s \le k\}$, since w_k can be deduced from y_k and ϕ_{k-1} through the relation $w_k = y_k - \phi_{k-1}^T \theta^\circ$. Hence (2.4) holds for θ°. We have already seen that if it holds for all E^θ, then the LSE coincides with the PEE. Moreover, if assumption (2.8) also holds with f being the Gaussian density function, then the LSE also coincides with the MLE.

Denote

$$\phi_k := (y_k, \ldots, y_{k-p}, u_k, \ldots, u_{k-p})^T, \tag{4.4}$$

so (4.1) can be rewritten as

$$y_{k+1} = \phi_k^T \theta^\circ + w_{k+1}. \tag{4.5}$$

The data at time n is

$$(\phi_0, \ldots, \phi_{n-1}, y_1, \ldots, y_n),$$

so the LSE is

$$\begin{aligned}
\hat{\theta}_n &= [\sum_{k=0}^{n-1} \phi_k \phi_k^T]^{-1} \sum_{k=0}^{n-1} \phi_k y_{k+1} \\
&= P_{n-1} \sum_{k=0}^{n-1} \phi_k y_{k+1},
\end{aligned}$$

where

$$P_{n-1}^{-1} := \sum_{k=0}^{n-1} \phi_k \phi_k^T. \tag{4.6}$$

From (4.5), we can also write

$$\hat{\theta}_n = \theta^\circ + P_{n-1} \sum_{k=0}^{n-1} \phi_k w_{k+1}, \tag{4.7}$$

so the behavior of the LSE depends crucially on that of the last term in (4.7).

To study this term requires the martingale convergence theorem.

Theorem (martingale convergence)   (4.8)

Let $\{m_k, F_k\}$ be a martingale—i.e., F_k is an increasing sequence of σ-algebras, m_k is F_k-measurable, and $E\{m_{k+1} \mid F_k\} = m_k$ a.s. for all k. If $\sup_k E \mid m_k \mid^p < \infty$ for some $p \geq 1$, then $\{m_k\}$ converges to an a.s. finite random variable. (When this holds with $p = 2$, the martingale is said to be square integrable.)

We also need the following result.

Theorem (Kronecker's lemma)   (4.9)

Let $\{x_k\}$ and $\{r_k\}$ be two real valued sequences satisfying

$$r_k > 0, \quad \lim_{k \to \infty} r_k = \infty, \quad \sum_{k=1}^{\infty} \frac{x_k}{r_k} < \infty. \qquad (4.10)$$

Then,

$$\lim_{N \to \infty} \frac{1}{r_N} \sum_{k=1}^{N} x_k = 0.$$

We can now study the term $\sum_{k=0}^{n-1} \phi_k w_{k+1}$ in (4.7). Let ϕ_k^i be the ith component of the vector ϕ_k, and for each i define $S_0^i := 1$, $S_{-1}^i := 1$, and

$$S_n^i := 1 + \sum_{k=0}^{n} (\phi_k^i)^2, \quad n = 1, 2, \dots$$

Clearly S_n^i is adapted to F_n. Also define $z_0^i := 0$, $i = 1, \dots, 2p + 2$ and

$$z_n^i = \sum_{k=0}^{n-1} \frac{\phi_k^i w_{k+1}}{S_k^i}, \quad n = 1, 2, \dots$$

Then z_n^i is also adapted to F_n and

$$E\{z_{n+1}^i \mid F_n\} = E\{z_n^i + \frac{\phi_n^i w_{n+1}}{S_n^i} \mid F_n\}$$

$$= z_n^i + \frac{\phi_n^i}{S_n^i} E\{w_{n+1} \mid F_n\} = z_n^i,$$

so $\{z_n^i, F_n\}$ is a martingale. Also

$$E(z_n^i)^2 = E \sum_{k=1}^{n} E\{(z_k^i)^2 - (z_{k-1}^i)^2 \mid F_{k-1}\}$$

$$= E \sum_{k=1}^{n} E\{(z_k^i - z_{k-1}^i)^2 \mid F_{k-1}\}$$

$$= E \sum_{k=1}^{n} E \{ (\frac{\phi_{k-1}^i w_{k+1}}{S_{k-1}^i})^2 \mid F_{k-1} \}$$

$$\leq \sigma^2 E \sum_{k=1}^{n} \frac{(\phi_{k-1}^i)^2}{(S_{k-1}^i)^2}$$

$$= \sigma^2 E \sum_{k=1}^{n} \frac{S_{k-1}^i - S_{k-2}^i}{(S_{k-1}^i)^2}$$

$$\leq \sigma^2 E \sum_{k=1}^{n} \frac{S_{k-1}^i - S_{k-2}^i}{S_{k-1}^i \ S_{k-2}^i}$$

$$= \sigma^2 E [\sum_{k=1}^{n} \frac{1}{S_{k-2}^i} - \frac{1}{S_{k-1}^i}]$$

$$= \sigma^2 E [\frac{1}{S_{-1}^i} - \frac{1}{S_{n-1}^i}]$$

$$\leq \sigma^2 E \frac{1}{S_{-1}^i} = \sigma^2.$$

This proves the next lemma.

Lemma   (4.11)
$\{z_n^i, F_n\}$ is a square integrable martingale if (4.2) and (4.3) hold.

By Theorem (4.8), $\{z_n^i\}$ converges a.s. for all i. Therefore, for $i = 1, \ldots, 2p + 2$,

$$\lim_n \sum_{k=0}^{n-1} \frac{\phi_k^i w_{k+1}}{S_k^i} \quad \text{exists and is finite } a.s. \qquad (4.12)$$

On the basic probability space (Ω, F, P) on which all the random variables are defined, let

$$\widetilde{\Omega} := \{ \omega \in \Omega \mid \lim_{n \to \infty} \lambda_{\min} (\sum_{k=0}^{n-1} \phi_k \phi_k^T) = + \infty \}$$

where $\lambda_{\min}$ denotes the smallest eigenvalue. Now

$$\lim_{n \to \infty} \lambda_{\min} (\sum_{k=0}^{n-1} \phi_k \phi_k^T) = + \infty$$

implies

$$\lim_{n \to \infty} S_{n-1}^i = 1 + \lim_{n \to \infty} \sum_{k=0}^{n-1} (\phi_k^i)^2 = + \infty.$$

Applying Kronecker's Lemma (4.9) to (4.12) then gives the next result.

Lemma (4.13)

If (4.2) and (4.3) hold, then

$$\lim_{n \to \infty} \frac{1}{S_{n-1}^i} \sum_{k=0}^{n-1} \phi_k^i \, w_{k+1} = 0 \quad i = 1, \ldots, 2p+2 \quad a.s. \text{ on } \widetilde{\Omega}.$$

Finally, since

$$S_{n-1}^i = 1 + \sum_{k=0}^{n-1} (\phi_k^i)^2 \leq 1 + Tr \sum_{k=0}^{n-1} \phi_k \phi_k^T, \quad i = 1, \ldots, 2p+2,$$

we note from Lemma (4.13) that

$$\lim_{n \to \infty} \frac{1}{Tr \sum_{k=0}^{n-1} \phi_k \phi_k^T} \sum_{k=0}^{n-1} \phi_k^i \, w_{k+1} = 0,$$

$$i = 1, \ldots, 2p+2 \ a.s. \text{ on } \widetilde{\Omega}. \tag{4.14}$$

Rewrite (4.7) as

$$\hat{\theta}_n = \theta^\circ + \left[\frac{1}{Tr \sum_{k=0}^{n-1} \phi_k \phi_k^T} \sum_{k=0}^{n-1} \phi_k \phi_k^T \right]^{-1}$$

$$\times \left[\frac{1}{Tr \sum_{k=0}^{n-1} \phi_k \phi_k^T} \sum_{k=0}^{n-1} \phi_k w_{k+1} \right]$$

Then we obtain the following result on strong consistency of the LSE.

Theorem (4.15)

Suppose (4.2) and (4.3) hold; then $\lim_{n \to \infty} \hat{\theta}_n = \theta^\circ$ $a.s.$ on the set of all ω for which $\lim_{n \to \infty} \lambda_{\min} (P_n^{-1}) = +\infty$ and $\dfrac{P_n^{-1}}{Tr \, P_n^{-1}} \geq \epsilon I$ for all large n and some $\epsilon > 0$.

The following corollary is sometimes useful, though it is a weaker statement.

Corollary (4.16)

Suppose (4.2) and (4.3) hold and there are positive definite matrices V, U and an integer N, all random, such that

$$V \geq \frac{1}{n} P_n^{-1} \geq U, \quad n \geq N \ a.s.$$

Then $\lim_{n \to \infty} \hat{\theta}_n = \theta^\circ$ $a.s.$

The condition $\frac{1}{n} P_n^{-1} \ge U$ $n \ge N$ is a persistency of excitation condition [see (3.19)]. The condition $\frac{1}{n} P_n^{-1} \le V$, $n \ge N$, is a stability condition stating that the input process $\{u_k\}$, where each u_k is determined possibly through feedback from $\{y_s, s \le k\}$, stabilizes the system (4.1). A central limit theorem can also be found, but we omit this.

To introduce the next section we consider bias of the LSE for the ARMAX system,

$$y_{k+1} = a_0 y_k + .. + a_p y_{k-p} + b_0 u_k + ..$$
$$+ b_0 u_{k-p} + c_0 w_k + .. + c_p w_{k-p} + w_{k+1}, \tag{4.17}$$

and we wish to estimate the parameters $(a_0, .., a_p, b_0, .., b_p, c_0, .., c_p)$. Then, as before, we rewrite the system as

$$y_{k+1} = \phi_k^T \theta^\circ + w_{k+1},$$

where

$$\theta^\circ := (a_0, .., a_p, b_0, .., b_p, c_0, .., c_p)^T,$$

$$\phi_k := (y_k, .., y_{k-p}, u_k, .., u_{k-p}, w_k, .., w_{k-p}).$$

The LSE is

$$\hat{\theta}_n = (\sum_{k=0}^{n-1} \phi_k \phi_k^T)^{-1} \sum_{k=0}^{n-1} \phi_k y_{k+1}.$$

However, since only the inputs and outputs are observed, the last $(p + 1)$ components of ϕ_k, namely, $w_k, .., w_{k-p}$, are not available as observations. Hence this LSE cannot be implemented.

Thus we cannot use the LSE to estimate all the parameters $(a_0, .., a_p, b_0, .., b_p, c_0, .., c_p)$. So we will try to estimate only the parameters $(a_0, .., a_p, b_0, .., b_p)$ by redefining

$$\theta^\circ := (a_0, .., a_p, b_0, .., b_p)^T,$$

$$\phi_k := (y_k, .., y_{k-p}, u_k, .., u_{k-p})^T,$$

and rewriting the system as

$$y_{k+1} = \phi_k^T \theta^\circ + (c_0 w_k + .. + c_p w_{k-p}) + w_{k+1}.$$

The estimate

$$\hat{\theta}_n = (\sum_{k=0}^{n-1} \phi_k \phi_k^T)^{-1} \sum_{k=0}^{n-1} \phi_k y_{k+1}$$

can be implemented based on the observations $\{u_s, y_s, s \le n\}$. However the noise process $\{c_0 w_k + c_1 w_{k-1} + ... + w_{k+1}\}$ is not white, and the estimate is biased unless $c_0, \ldots, c_p$ are 0, as we see in the next exercise.

Exercise   (4.18)

Consider the system $y_{k+1} = \theta° y_k + c_0 w_k + w_{k+1}$, where $\theta° = 0$, $c_0 \neq 0$.

Let $\{w_k\}$ be iid with $w_k \sim N(0, 1)$. Show that the LSE $\hat{\theta}_n = (\sum\limits_{k=0}^{n-1} y_k^2)^{-1}$

$\sum\limits_{k=0}^{n-1} y_k y_{k+1}$ converges to $\dfrac{c_0}{1 + c_0^2}$ a.s. Hence the LSE is asymptotically biased.

[Hint: Write

$$\hat{\theta}_n = [\frac{1}{n} \sum_{k=0}^{n-1} (c_0 w_{k-1} + w_k)^2]^{-1}$$

$$\times [\frac{1}{n} \sum_{k=0}^{n-1} (c_0 w_{k-1} + w_k)(c_0 w_k + w_{k+1})].$$

Then show that $\sum\limits_{k=0}^{n-1} \dfrac{w_k w_{k-m} - \delta_{m0}}{k}$ is a square integrable martingale, and

deduce using Kronecker's lemma that $\lim\limits_{n \to \infty} \dfrac{1}{n} \sum\limits_{k=0}^{n-1} w_k w_{k-m} = \delta_{m0}$ *a.s.*]

Thus the LSE of the parameters $(a_0, \ldots, a_p, b_0, \ldots, b_p)$ will generally be asymptotically biased unless $c_0 = \ldots = c_p = 0$. One must turn to other methods for estimating the parameters of a general ARMAX model such as (4.17).

5. Off-line methods for ARMAX systems: the PEE and the MLE

We consider other methods of estimating the parameters $(a_0, \ldots, a_p, b_0, \ldots, b_p, c_0, \ldots, c_p)$ of the ARMAX system

$$y_{k+1} = a_0 y_k + \ldots + a_p y_{k-p} + b_0 u_k + \ldots$$

$$+ b_0 u_{k-p} + c_0 w_k + \ldots + c_p w_{k-p} + w_{k+1}.$$

We start with the PEE. The first point to note is that, in general, there may be no recursive scheme for computing the estimates $\{\hat{\theta}_n\}$.

Example   (5.1)

[This is just Exercise (2.7) revisited.] Consider the system

$$y_{k+1} = b_0 u_k + c_0 w_k + w_{k+1}; \quad w_0 = 0, \quad k = 0, 1, \ldots$$

where $\{w_k\}$ is iid with mean 0 and variance 1. Let $\theta° := (b_0, c_0)^T$. Then the PEE $\hat{\theta}_n$ based on the data $\{u_s, y_s, s \leq n\}$ is given by

$$\hat{\theta}_n = Arg \min_{\theta = (\theta_1, \theta_2)} \sum_{k=0}^{n-1} [y_{k+1} - \theta_1 u_k$$

$$- \sum_{j=0}^{k-1} \theta_2^{k-j} (y_{j+1} - \theta_1 u_j)]^2.$$

The minimand contains terms involving powers such as $\theta_1^i \theta_2^j$ and is not easily minimized. Furthermore, there is no obvious alternative recursive scheme for generating $\hat{\theta}_{n+1}$ from $\hat{\theta}_n$.

Hence the PEE must be computed off-line. Nevertheless, since it serves as a benchmark, we analyze its behavior. For simplicity, we consider ARMA systems, since the analysis is similar even in the presence of an input $\{u_k\}$.

We first obtain the prediction $\hat{y}_{k+1 \mid k}$ for the ARMA model,

$$y_{k+1} = (y_k, \ldots, y_{k-p}, w_k, \ldots, w_{k-p}) \, \theta + w_{k+1}, \quad k = 0, 1, \ldots$$

where $\theta := (a_0(\theta), \ldots, a_p(\theta), c_0(\theta), \ldots, c_p(\theta))^T$. Assume that

$$w_{-p} = w_{-p+1} = \ldots = w_0 = 0, \quad \text{and}$$

$$y_{-p} = y_{-p+1} = \ldots = y_0 = 0, \tag{5.2}$$

$$E^\theta \{w_{k+1} \mid w_s, s \le k\} = 0, \tag{5.3}$$

$$E^\theta \{w_{k+1}^2 \mid w_s, s \le k\} = \sigma^2. \tag{5.4}$$

Define the polynomials

$$A(z, \theta) := 1 - a_0(\theta) z - \ldots - a_p(\theta) z^{p+1}, \tag{5.5}$$

$$C(z, \theta) := 1 - c_0(\theta) z - \ldots - cp(\theta) z^{p+1}, \tag{5.6}$$

and assume that $\Theta \subset R^{2p+2}$ satisfies the next assumption.

Assumption $\hfill (5.7)$

Θ is a compact set such that for every θ in Θ, all roots of the polynomials $A(z, \theta)$ and $C(z, \theta)$ are strictly outside the unit disc.

For $\theta \in \Theta$, let $\{h_k(\theta) z^k\}$ and $\{g_k(\theta) z^k\}$ be the power series

$$\sum_{k=0}^{\infty} h_k(\theta) z^k := \frac{C(z, \theta)}{A(z, \theta)}, \tag{5.8}$$

$$\sum_{k=0}^{\infty} g_k(\theta) z^k := \frac{A(z, \theta)}{C(z, \theta)}. \tag{5.9}$$

Note that $g_0(\theta) = h_0(\theta) = 0$.

Exercise $\hfill (5.10)$

Show that $\hat{y}_{k+1 \mid k}(\theta) := E^\theta \{y_{k+1} \mid y_s, s \le k\} = \sum_{j=0}^{k} h_{k+1-j}(\theta) w_j$. Hence

conclude that $\hat{y}_{k+1 \mid k}(\theta) = \sum_{j=0}^{k} -g_{k+1-j}(\theta) y_j$.

[Hint: Note that $w_n = \sum_{j=0}^{n} g_{n-j}(\theta) y_j$ can be obtained from y_s, $s \le n$. Now $y_n = \sum_{j=0}^{n} h_{n-j}(\theta) w_j$ and so $\hat{y}_{n \,|\, n-1}(\theta) = y_n - w_n$.]

We now analyze the behavior of the PEE assuming the true model is

$$y_{k+1} = (y_k, \ldots, y_{k-p'}, w_k, \ldots, w_{k-p'})\theta^\circ + w_{k+1}. \tag{5.11}$$

The analysis of the estimates is conducted under the assumption that the inputs and outputs are generated according to (5.11), i.e., under the probability measure P^{θ°. As before we write P^{θ° and E^{θ° as P and E, respectively. Assume that $\{w_k\}$ and $\{y_{-p'}, \ldots, y_0\}$ satisfy (5.2)-(5.4) (with $p = p'$), and also

$$E\{w_{k+1}^4 \mid w_s, s \le k\} = \delta. \tag{5.12}$$

Let

$$\theta^\circ := (a_0, \ldots, a_{p'}, c_0, \ldots, c_{p'}), \tag{5.13}$$

and suppose that the following assumption holds.

Assumption $\qquad\qquad\qquad\qquad\qquad\qquad\qquad\qquad\qquad\qquad\qquad\qquad$ (5.14)
All the roots of the polynomials

$$A(z) := 1 - a_0 z - \ldots - a_{p'} z^{p'+1} \quad \text{and}$$

$$C(z) := 1 + c_0 z + \ldots + c_{p'} z^{p'+1}$$

are strictly outside the closed unit disc.

For $\Theta \subset R^{2p+2}$ satisfying (5.7) the PEE is

$$\hat{\theta}_n = \text{Arg} \min_{\theta \in \Theta} \sum_{k=1}^{n} [y_k - \hat{y}_{k \,|\, k-1}(\theta)]^2. \tag{5.15}$$

It is not assumed that $\theta^\circ \in \Theta$ nor that p equals p' of (5.11). From Exercise (5.9),

$$\hat{y}_{k \,|\, k-1}(\theta) = \sum_{j=0}^{k-1} - g_{k-j}(\theta) y_j.$$

Hence

$$y_k - \hat{y}_{k \,|\, k-1}(\theta) = y_k + \sum_{j=0}^{k-1} g_{k-j}(\theta) y_j = \sum_{j=0}^{k} g_{k-j}(\theta) y_j$$

[noting that $g_0(\theta) = 1$ from (5.5), (5.6), and (5.9)]. Hence

$$\hat{\theta}_n = \text{Arg} \min_{\theta \in \Theta} \sum_{k=1}^{n} [\sum_{j=0}^{k} g_{k-j}(\theta) y_j]^2. \tag{5.16}$$

Let

$$V_n(\theta) := \frac{1}{n} \sum_{k=1}^{n} \left[\sum_{j=0}^{k} g_{k-j}(\theta) y_j\right]^2 . \tag{5.17}$$

We shall show that $V_n(\theta)$ converges a.s. to a constant $V(\theta)$ and that every limit point of the sequence $\{\hat{\theta}_n\}$ is a minimizer of $V(\theta)$ over $\theta \in \Theta$. This allows us to conclude that $\{\hat{\theta}_n\}$ converges to the set of minimizers of $V(\theta)$. We then examine the properties of the set of minimizers of $V(\theta)$.

Let $\{h_k z^k\}$ and $\{g_k z^k\}$ be the power series

$$\sum_{k=0}^{\infty} h_k z^k := \frac{C(z)}{A(z)} , \tag{5.18}$$

$$\sum_{k=0}^{\infty} g_k z^k := \frac{A(z)}{C(z)} . \tag{5.19}$$

Exercise (5.20)

Show that $\displaystyle\lim_{n\to\infty} \frac{1}{n} \sum_{\substack{m=1}}^{n} y_m\, y_{m-l} = \sigma^2 \sum_{k=1}^{\infty} h_k h_{k+l}$ a.s.

[Hint : Use $y_m = \displaystyle\sum_{j=0}^{m} h_j w_{m-j}$ to get

$$\frac{1}{n} \sum_{m=0}^{n} y_m\, y_{m-l} = \sum_{k=0}^{n-l} \sum_{j=0}^{n} h_j h_k$$

$$\times \left[\frac{1}{n} \sum_{m=\max(j,k+l)}^{n} w_{m-j}\, w_{m-l-k}\right].$$

Then show that $\displaystyle\sum_{k=1}^{n} \frac{w_k w_{k-m} - \sigma^2 \delta_{m0}}{k}$ is a square integrable martingale

and deduce using Kronecker's Lemma that $\displaystyle\lim_{n\to\infty} \frac{1}{n} \sum_{k=1}^{n} w_k w_{k-m} = \sigma^2 \delta_{m0}$

a.s. Now use the Schwarz inequality to show that

$$\left|\frac{1}{n} \sum_{m=\max(j,k+l)}^{n} w_{m-j}\, w_{m-l-k}\right| \le \frac{1}{n} \sum_{m=1}^{n} w_m^2$$

and hence that for some random constant M,

$$\left|\frac{1}{n} \sum_{m=\max(j,k+l)}^{n} w_{m-j}\, w_{m-l-k}\right| \le M.$$

Finally use (5.14) and (5.18) to deduce that $|h_j| \le \alpha \gamma^j$ for some $0 < \gamma < 1$.]

Exercise   (5.21)
Show that $V_n(\theta)$ converges uniformly on Θ to

$$V(\theta) := \sigma^2 \sum_{j=0}^{\infty} \sum_{k=0}^{\infty} \sum_{l=0}^{\infty} g_j(\theta) \, g_k(\theta) \, h_l h_{l+|j-k|} \quad a.s. \tag{5.22}$$

[Hint: Use the same procedure as in Exercise (5.20) with the additional fact that because of (5.7) there are constants α and $0 < \gamma < 1$ such that $|g_j(\theta)| \le \alpha \gamma^j$ for all $\theta \in \Theta$.]

    Thus

    1) $V_n(\theta)$ is continuous on $\Theta \subset R^{2p+2}$, and

    2) $V_n \to V$ uniformly on Θ a.s.   (5.23)

Exercise   (5.24)
Note that $\hat{\theta}_n$ minimizes $V_n(\theta)$ over Θ. Use (5.22) to show that every limit point θ^* of $\{\hat{\theta}_n\}$ is such that $V(\theta^*) \le V(\theta)$ for all $\theta \in \Theta$.

    Let $\bar{\Theta}$ be the set of minimizers of $V(\theta)$, i.e.,

$$\bar{\Theta} := \{\bar{\theta} \in \Theta \mid V(\bar{\theta}) \le V(\theta) \text{ for all } \theta \in \Theta\}. \tag{5.25}$$

We rewrite (5.24) as a lemma.

Lemma   (5.26)
Assume that the ARMA system (5.1) satisfies (5.2)-(5.4) and (5.12)-(5.14). Then the PEE $\{\hat{\theta}_n\}$ given by (5.16) is such that $\hat{\theta}_n \to \bar{\Theta}$ a.s.

    We now characterize $\bar{\Theta}$ more explicitly.

    First note that by (5.14), for some $0 < \gamma < 1$,

$$\frac{C(z)}{A(z)} \frac{C(z^{-1})}{A(z^{-1})}$$

$$= \sum_{m=0}^{\infty} h_m z^m \sum_{l=0}^{\infty} h_l z^{-l}, \quad \gamma < |z| < \gamma^{-1},$$

$$= \sum_{p=-\infty}^{\infty} z^p \sum_{l=\max(0,-p)}^{\infty} h_l h_{l+p}, \quad \gamma < |z| < \gamma^{-1}.$$

Hence, for $m \ge 0$,

$$\sum_{l=0}^{\infty} h_l h_{l+m} = \text{coefficient of } z^m \text{ in } \frac{C(z)}{A(z)} \frac{C(z^{-1})}{A(z^{-1})}$$

$$\text{for } \gamma < |z| < \gamma^{-1},$$

$$= \text{coefficient of } z^{-m} \text{ in } \frac{C(z)}{A(z)} \frac{C(z^{-1})}{A(z^{-1})}$$

for $\gamma < |z| < \gamma^{-1}$.

These coefficients can be evaluated by contour integration,

$$\sum_{l=0}^{\infty} h_l h_{l+m} = \frac{1}{2\pi i} \oint z^{-m} \frac{C(z)}{A(z)} \frac{C(z^{-1})}{A(z^{-1})} \frac{dz}{z}$$

$$= \frac{1}{2\pi i} \oint z^m \frac{C(z)}{A(z)} \frac{C(z^{-1})}{A(z^{-1})} \frac{dz}{z},$$

where the integral is over the unit circle traversed in the counterclockwise direction. Therefore we can rewrite

$V(\theta)$

$$= \sigma^2 \sum_{j=0}^{\infty} \sum_{k=0}^{\infty} g_j(\theta) g_k(\theta) \frac{1}{2\pi i} \oint z^{j-k} \frac{C(z)}{A(z)} \frac{C(z^{-1})}{A(z^{-1})} \frac{dz}{z}.$$

By (5.5)-(5.7) and (5.9), we get

$$V(\theta) = \frac{\sigma^2}{2\pi i} \oint \sum_{j=0}^{\infty} \sum_{k=0}^{\infty} g_j(\theta) g_k(\theta) z^{j-k} \frac{C(z)}{A(z)} \frac{C(z^{-1})}{A(z^{-1})} \frac{dz}{z}$$

$$= \frac{\sigma^2}{2\pi i} \oint \frac{A(z,\theta)}{C(z,\theta)} \frac{A(z^{-1},\theta)}{C(z^{-1},\theta)} \frac{C(z)}{A(z)} \frac{C(z^{-1})}{A(z^{-1})} \frac{dz}{z}. \qquad (5.27)$$

Theorem   (5.28)
Under the conditions of Lemma (5.26),

$$\hat{\theta}_n \to \bar{\Theta} \ a.s.,$$

where

$$\bar{\Theta} := \{\bar{\theta} \in \Theta \mid V(\bar{\theta}) \le V(\theta) \text{ for all } \theta \in \Theta\}$$

and $V(\theta)$ is as in (5.27).

There is a straightforward interpretation of (5.27). By (5.5.5), $V(\theta)$ can be regarded as the variance of the stationary process

$$z_{k+1}(\theta) = \frac{A(q^{-1},\theta)}{C(q^{-1},\theta)} \frac{C(q^{-1})}{A(q^{-1})} w_{k+1}, \qquad (5.29)$$

where $var(w_{k+1}) = \sigma^2$. We can view $\dfrac{C(q^{-1})}{A(q^{-1})} w_{k+1}$ as the stationary process

$$y_{k+1} = \frac{C(q^{-1})}{A(q^{-1})} w_{k+1}.$$

Hence (5.29) can be rewritten as

$$z_{k+1}(\theta) = y_{k+1} - [\frac{C(q^{-1},\theta) - A(q^{-1},\theta)}{C(q^{-1},\theta)}]y_{k+1}$$

and the last term can be interpreted as the steady-state predictor

$$\hat{y}_{k+1 \mid k}(\theta) = \frac{C(q^{-1},\theta) - A(q^{-1},\theta)}{C(q^{-1},\theta)} y_{k+1}.$$

Hence Theorem (5.28) has the interpretation that the PEE converges to the set of θ that provide the best steady-state predictors.

So far we have not assumed that $\theta^\circ \in \Theta$. If we assume this, then a stronger result can be obtained.

Exercise   (5.30)
Show that if $\theta^\circ \in \Theta$, then $\overline{\Theta}$ of Theorem (5.28) can be written as

$$\overline{\Theta} = \{\theta \in \Theta \mid \frac{C(z,\theta)}{A(z,\theta)} = \frac{C(z)}{A(z)} \}.$$

[Hint: The term

$$\frac{\sigma^2}{2\pi i}\oint[\frac{A(z,\theta)C(z)}{A(z)C(z,\theta)} - 1][\frac{A(z^{-1},\theta)C(z^{-1})}{A(z^{-1})C(z^{-1},\theta)} - 1]\frac{dz}{z}$$

is nonnegative. The cross terms in the integrand all integrate to σ^2 and so the integral equals $V(\theta) - \sigma^2$. Hence the minimum value of $V(\theta)$ is σ^2, which is attained when $A(z,\theta)C(z) = A(z)C(z,\theta)$.]

Thus if $\theta^\circ \in \Theta$, we can identify the transfer function $\frac{C(z)}{A(z)}$ from the noise w to the output y. Can we identify $C(z)$ and $A(z)$ individually?

Exercise   (5.31)
Suppose $\theta^\circ \in \Theta$, $p' = p, a_p \neq 0, c_p \neq 0$ and $A(z)$ and $C(z)$ have no common factors. Show that $\lim_{n \to \infty} \theta_n = \theta^\circ$ a.s.

Exercise   (5.32)
If $\{w_k\}$ is Gaussian white noise, show that the MLE estimate of θ° for the system (5.11) coincides with the PEE.

Exercise   (5.33)
Consider the state space model

$$x_{k+1} = A(\theta^\circ) x_k + B(\theta^\circ) u_k + w_{k+1},$$

$$y_k = C(\theta^\circ) x_k + v_k,$$

where $\{w_k\}$ is Gaussian white noise, with $w_k \sim N(0,Q)$, $\{v_k\}$ is Gaussian white noise, with $v_k \sim N(0,R)$, $x_0 \sim N(0,P_0)$, and $\{w_k\},\{v_k\}$, and x_0 are independent. Given $\{y_0, \ldots, y_n, u_0, \ldots, u_{n-1}\}$, how does one obtain the PEE for θ° ?

6. Closed loop identification

Sometimes one is faced with the task of estimating the parameters of a system operating under feedback control. So suppose that we have the ARMAX system

$$y_{k+1} = a_0 y_k + \ldots + a_p y_{k-p} + b_p u_k + \ldots$$
$$+ b_p u_{k-p} + c_0 w_k + \ldots + c_p w_{k-p} + w_{k+1},$$

which can be written more compactly as

$$A(q^{-1}) y_{k+1} = B(q^{-1}) u_{k+1} + C(q^{-1}) w_{k+1},$$

with $A(z) := 1 - a_0 z - \ldots -a_p z^{p+1}$, $B(z) := b_0 z + \ldots + b_p z^{p+1}$, $C(z) := 1 + c_0 z + \ldots + c_p z^{p+1}$. Let $\{w_k\}$ satisfy (5.3), (5.4), and (5.12). Suppose

$$y_k = u_k = w_k = 0, \quad k \le 0. \tag{6.1}$$

Our goal is to identify $\theta^\circ := (a_0, \ldots, a_p, b_0, \ldots b_p, c_0, \ldots, c_p)^T$.

Suppose we have a compact parameter set $\Theta \subset R^{3p+3}$. Then for $\theta = (a_0(\theta), \ldots, a_p(\theta), b_0(\theta), \ldots, b_p(\theta), c_0 (\theta), \ldots, c_p(\theta)) \in \Theta$, we can form the predictions

$$\hat{y}_{k+1 \mid k} (\theta) = \theta^T [y_k, \ldots, y_{k-p}, u_k, \ldots$$
$$u_{k-p}, y_k - \hat{y}_{k \mid k-1}(\theta), \ldots, y_{k-p} - \hat{y}_{k-p \mid k-p-1}(\theta)] \tag{6.2}$$

with initial conditions (6.1) and $\hat{y}_{j \mid j-1} (\theta) = 0$ for $j \le 0$. Let

$$\hat{\theta}_n := Arg \min_{\Theta} \sum_{k=1}^{n} [y_k - \hat{y}_{k \mid k-1}(\theta)]^2. \tag{6.3}$$

We study the behavior of $\{\hat{\theta}_n\}$ when the input $\{u_k\}$ is generated by the causal, linear feedback control law

$$u_k = \frac{F(q^{-1})}{G(q^{-1})} y_k$$

with $F(z) := f_0 + \ldots + f_l z^l$ and $G(z) := 1 + g_1 z + \ldots + g_l z^l$. By this we mean that u_k is generated by the difference equation

$$u_k = f_0 y_k + \ldots + f_l y_{k-l} - g_1 u_{k-1} - \ldots - g_l u_{k-l}, \tag{6.4}$$

with initial conditions (6.1). Note that the PEE $\hat{\theta}_n$ of (6.2) and (6.3) can be obtained without knowing how the sequence $\{u_k\}$ is generated.

Moreover, even if we did know that $\{u_k\}$ is generated by the causal feedback law (6.4), the predictions, and therefore the PEE, would remain unchanged, as the following exercise shows.

Exercise (6.5)

For $\theta = (a_0(\theta), \ldots, c_p(\theta))^T \in \Theta$, define $A(z,\theta) := 1 - a_0(\theta) z - \ldots - a_p(\theta)$ z^{p+1}, $B(z,\theta) := b_0(\theta)z + \ldots + b_p(\theta) z^{p+1}$, $C(z,\theta) := 1 + c_0(\theta)z + \ldots + c_p(\theta) z^{p+1}$ and $H(z,\theta) := A(z,\theta) G(z) - B(z,\theta) F(z)$, $K(z,\theta) := C(z,\theta)$ $G(z)$. Then if θ was the value of the parameter and the inputs satisfied (6.4), the system would be given by

$$H(z,\theta)y_{k+1} = K(z,\theta) w_{k+1}. \tag{6.6}$$

Let $\hat{y}_{k+1 \mid k}(\theta) := E\{y_{k+1} \mid y_s, s \leq k\}$. Show that $\hat{y}_{k+1 \mid k}(\theta)$ satisfies (6.2) with initial conditions (6.1) and $\hat{y}_{j \mid j-1}(\theta) = 0$ for $j \leq 0$.

Once we see that the estimate $\{\hat{\theta}_n\}$ is the PEE for the system (6.6) we can use the results obtained in Section 5 where we treated ARMA systems of the type (6.6). So assume that for each $\theta \in \Theta$, the polynomials $H(z,\theta)$ and $K(z,\theta)$ have all their roots strictly outside the closed unit disc and $\theta^\circ \in \Theta$.

Then we can immediately use the result of Exercise (5.29) to conclude that $\hat{\theta}_n \to \overline{\Theta}$ *a.s.*, where

$$\overline{\Theta} := \{\theta \in \Theta \mid \frac{K(z,\theta)}{H(z,\theta)} = \frac{K(z,\theta^\circ)}{H(z,\theta^\circ)} \}.$$

If we assume further that the coefficient of z^{p+l+1} in $K(z,\theta^\circ) \neq 0$, the coefficient of z^{p+l+1} in $H(z,\theta^\circ) \neq 0$, and $K(z,\theta^\circ)$ and $H(z,\theta^\circ)$ have no common factors, then, as in Exercise (5.31), we get the following result.

Theorem (6.7)

Under the assumptions above, $\hat{\theta}_n \to \overline{\Theta}$ *a.s.*, where $\overline{\Theta}$ is the set of all θ in Θ for which

$$A(z,\theta) - B(z,\theta)\frac{F(z)}{G(z)} = A(z) - B(z)\frac{F(z)}{G(z)} \quad \text{and}$$

$$C(z,\theta) = C(z).$$

Note that $A(z) - B(z)\dfrac{F(z)}{G(z)}$ is the closed loop transfer function. The result says that we can identify it as well as $C(z)$. However, we may not be able to identify $A(z)$ and $B(z)$ individually as the next exercise shows.

Example (6.8)

Consider the system $y_{k+1} = a_0 y_k + b_0 u_k + w_{k+1}$ with u_k selected by feedback according to $u_k = f_0 y_k$, with $|a_0 + b_0 f_0| \leq 1 - \epsilon$ and

$$\Theta := \{(a,b) \mid |a + b f_0| \leq 1 - \epsilon\} \quad \text{for some } \epsilon > 0.$$

According to Theorem (6.7) $\hat{\theta}_n \to \overline{\Theta}$ *a.s.* where

$$\bar{\Theta} := \{(a,b) \mid a + bf_0 = a_0 + b_0f_0\}.$$

Hence we can identify only $a_0 + b_0f_0$ and not a_0 and b_0 separately. This inability to identify the polynomials $A(z)$ and $B(z)$ gives rise to the *closed loop identification problem*. It should be kept in mind whenever the inputs $\{u_k\}$ are generated by feedback.

Note that the estimates $\{\hat{\theta}_n\}$ satisfying (6.2) and (6.3) coincide with the PEE for the system

$$A(z,\theta)y_{k+1} = B(z,\theta) \frac{F(z)}{G(z)} y_{k+1} + C(z,\theta)w_{k+1},$$

because the feedback $u_k = \dfrac{F(z)}{G(z)}$ is causal, i.e., u_k depends on $\{y_s,$ $s \le k\}$ [more generally, u_k is adapted to F_k and $\{w_k, F_k\}$ is a martingale difference sequence].

Exercise   (6.9)
Suppose $y_{k+1} = a_0y_k + b_0u_k + w_{k+1}$, where $u_k = f_0y_{k+1}$. Show that the estimates (6.2), (6.3), and the PEE for the system $(1 - b_0f_0)y_{k+1} = a_0y_k + w_{k+1}$ do not coincide.

Exercise   (6.10)
In Exercise (6.8), let $\phi_k := (y_k, u_k)^T$ and $P_n^{-1} := \displaystyle\sum_{k=0}^{n} \phi_k\phi_k^T$. Evaluate

$\displaystyle\lim_{n\to\infty} \frac{1}{n} P_n^{-1}$ and deduce that persistency of excitation does not hold.
[Hint: To evaluate the limit, use Exercise (5.20). In fact, P_n^{-1} is singular for every n, i.e., P_n does not exist.]

7. Recursive identification methods for ARMAX systems

For the ARMAX system

$$y_{k+1} = a_0y_k + \ldots + a_p y_{k-p} + b_0u_k + \ldots$$
$$+ b_p u_{k-p} + c_0w_k + \ldots + c_p w_{k-p} + w_{k+1},$$ (7.1)

we saw that

1.   The LSE for the parameters $(a_0, \ldots, a_p, b_0, \ldots, b_p)$ is asymptotically biased; see Exercise (4.18).

2.   The LSE for the parameters $(a_0, \ldots, a_p, b_0, \ldots, b_p, c_0, \ldots, c_p)$ might be a good estimate, but it cannot be implemented since the last $(p + 1)$ components of the vector $\phi_k := (y_k, \ldots, y_{k-p}, u_k, \ldots, u_{k-p}, w_k, \ldots, w_{k-p})$ used to generate the LSE cannot be observed.

3.   The PEE is a good estimate, but it cannot be recursively implemented; see Exercise (5.1).

So each procedure has some drawback. In this section we study some other recursive methods that can potentially overcome these drawbacks.

We begin with the instrumental variables method. Rewrite (7.1) as

$$y_{k+1} = \phi_k^T \theta^\circ + v_{k+1},$$

where

$$v_{k+1} := c_0 w_k + .. + c_p w_{k-p} + w_{k+1},$$

$$\phi_k^T := (y_k, .. , y_{k-p}, u_k, .. , u_{k-p}),$$

$$\theta^\circ := (a_0, .. , a_p, b_0, .. , b_p)^T.$$

Then the LSE for θ° is

$$(\sum_{k=0}^{n-1} \phi_k \phi_k^T)^{-1} \sum_{k=0}^{n-1} \phi_k y_{k+1},$$

which can be rewritten as

$$\theta^\circ + (\frac{1}{n} \sum_{k=0}^{n-1} \phi_k \phi_k^T)^{-1} (\frac{1}{n} \sum_{k=0}^{n-1} \phi_k v_{k+1}).$$

The basic problem is that ϕ_k is correlated with v_{k+1}, and so the term $\frac{1}{n} \sum_{k=0}^{n-1} \phi_k v_{k+1}$ need not converge to 0 [Exercise (4.18)].

This suggests an immediate remedy. Suppose we can find a random vector ψ_k such that:

1. ψ_k is uncorrelated with v_{k+1} so that $\frac{1}{n} \sum_{k=0}^{n-1} \psi_k v_{k+1} \to 0$, and

2. ψ_k is sufficiently correlated with ϕ_k^T so that $(\frac{1}{n} \sum_{k=0}^{n-1} \psi_k \phi_k^T)^{-1}$ remains

   bounded.

Then the estimate

$$\hat{\theta}_n = (\sum_{k=0}^{n-1} \psi_k \phi_k^T)^{-1} \sum_{k=0}^{n-1} \psi_k y_{k+1} \tag{7.2}$$

$$= \theta^\circ + (\frac{1}{n} \sum_{k=0}^{n-1} \psi_k \phi_k^T)^{-1} (\frac{1}{n} \sum_{k=0}^{n-1} \psi_k v_{k+1})$$

will be strongly consistent. Such a vector ψ_k, which can be found in some problems, is called an instrumental variable (IV) and $\hat{\theta}_n$ of (7.2) is an **instrumental variable estimate** (IVE).

Exercise   (7.3)

Let $y_{k+1} = a_0 y_k + c_0 w_k + w_{k+1}$, with $0 < |a_0| < 1$. Let $\psi_k := y_{k-1}$ be the IV and consider the IVE for a_0,

$$\hat{\theta}_n = \left(\sum_{k=0}^{n-1} \psi_k y_k \right)^{-1} \sum_{k=0}^{n-1} \psi_k y_{k+1}.$$

Assuming that $\{w_k\}$ is iid with $E\, w_k = 0$, $E\, w_k^2 = \sigma^2$, and $E\, w_k^4 = \delta$, show that $\lim_{n \to \infty} \hat{\theta}_n = a_0$.

[Hint: Use Exercise (5.20).]

Note that in Exercise (7.3), we can also use $\psi_k = y_{k-m}$ for $m \geq 1$, and so there is a large number of choices for the IVs. However, to keep the error variance $E\, (\hat{\theta}_n - \theta^\circ)^2$ small, lower values of m are generally better.

Exercise   (7.4)

Show that the IVE (7.2) can be recursively generated by

$$\hat{\theta}_{n+1} = \hat{\theta}_n + R_n^{-1} \psi_n (y_{n+1} - \phi_n^T \hat{\theta}_n),$$

$$R_{n+1} = R_n + \psi_{n+1} \phi_{n+1}^T.$$

Next we consider the pseudolinear regression method. Rewrite (7.1) as

$$y_{k+1} = \phi_k^T \theta^\circ + w_{k+1},$$ (7.5)

where

$$\phi_k^T := (y_k, \ldots, y_{k-p}, u_k, \ldots, u_{k-p}, w_k, \ldots, w_{k-p}),$$

$$\theta^\circ := (a_0, \ldots, a_p, b_0, \ldots, b_p, c_0, \ldots, c_p)^T.$$

The difficulty with the LSE

$$\hat{\theta}_{n+1} = \hat{\theta}_n + R_n^{-1} \phi_n (y_{n+1} - \phi_n^T \hat{\theta}_n),$$

$$R_{n+1} = R_n + \phi_{n+1} \phi_{n+1}^T,$$

is that the components $w_k, \ldots, w_{k-p}$ of ϕ_k are not available as observations. So we try and approximate them. From (7.5)

$$w_k = y_k - \phi_{k-1}^T \theta^\circ.$$

Moreover, at time k, if $\hat{\theta}_k$ is available, and if it is close to θ°, then

$$\epsilon_k = y_k - \phi_{k-1}^T \hat{\theta}_k$$

can serve as an approximation to w_k. This suggests the following recursive scheme:

$$\hat{\theta}_{n+1} = \hat{\theta}_n + R_n^{-1} \phi_n (y_{n+1} - \phi_n^T \hat{\theta}_n),$$ (7.6)

$$R_{n+1} = R_n + \phi_{n+1}\phi_{n+1}^T, \tag{7.7}$$

where

$$\phi_n := (y_n, \ldots, y_{n-p}, u_n, \ldots, u_{n-p}, \epsilon_n, \ldots, \epsilon_{n-p}) \tag{7.8}$$

$$\epsilon_n := y_n - \phi_{n-1}^T \hat{\theta}_n. \tag{7.9}$$

The recursive algorithm (7.6)-(7.9) is called a **pseudolinear regression** (PLR) algorithm, and the estimate $\hat{\theta}_n$ is called the pseudolinear regression estimate (PLRE). [PLR also goes by the names **extended least squares** (ELS) and **approximate maximum likelihood** (AML) estimates.]

The random variables $\{y_n - \phi_{n-1}^T\hat{\theta}_{n-1}\}$ and $\{y_n - \phi_{n-1}^T\hat{\theta}_n\}$ are residuals or prediction errors. The former are prior residuals or prior prediction errors, whereas the latter are posterior residuals or posterior prediction errors. We could obtain an algorithm that is slightly simpler to implement if we replace the ϵ_i in (7.8) and (7.9) by the prior residuals $y_i - \phi_{i-1}^T\hat{\theta}_{i-1}$. However, the convergence properties depend on which residuals are used.

The convergence properties of the PLRE $\hat{\theta}_n$ are difficult to analyze. The following result is available.

Theorem   (7.10)
Suppose that the noise sequence $\{w_k\}$ satisfies

$$E\{w_{k+1} \mid w_s, s \le k\} = 0,$$

$$E\{w_{k+1}^2 \mid w_s, s \le k\} = \sigma^2,$$

$$E\{w_{k+1}^4 \mid w_s, s \le k\} = \delta,$$

and the polynomials $1 - a_0 z - \ldots - a_p z^{p+1}$ and $C(z) := 1 + c_0 z + \ldots + c_p z^{p+1}$ have all their roots strictly outside the closed unit disc. Assume that

$$\text{Re}\left[\frac{1}{C(e^{i\omega})} - \frac{1}{2}\right] \ge 0 \quad \text{for all } \omega. \tag{7.11}$$

[Here Re means real part and $i := \sqrt{-1}$.] Assume that the input sequence $\{u_k\}$ is adapted to F_k and

$$\lim \frac{1}{n} \sum_{k=0}^{n-1} (y_k, \ldots, y_{k-p}, u_k, \ldots, u_{k-p}, w_k, \ldots, w_{k-p})$$

$$(y_k, \ldots, y_{k-p}, u_k, \ldots, u_{k-p}, w_k, \ldots, w_{k-p})^T = P,$$

where $P > 0$. Then $\lim_{n \to \infty} \hat{\theta}_n = \theta^\circ$ *a.s.*

The inequality (7.11) is called a **positive real condition**. It can equivalently be written as

$$| C(e^{i\omega}) - 1 | < 1 \quad \text{for all } \omega, \tag{7.12}$$

and can be interpreted as saying that the noise $\{w_{k+1} + c_0 w_k + .. + c_p w_{k-p}\}$ is not too different from white noise. In the next section we give an intuitive explanation for the condition (7.12).

The recursive prediction error method is based on the PEE. Recall that the PEE

$$\hat{\theta}_n := Arg \min \sum_{k=1}^{n} [y_k - \hat{y}_{k \mid k-1}(\theta)]^2,$$

is a good estimate of $\theta^\circ := (a_0, .., a_p, b_0, .., b_p, c_0, .., c_p)^T$ for the ARMAX model (7.1), but it is not recursively implementable. However, since it has good asymptotic properties, as we have seen in Section 5, we develop an approximation of the PEE that is recursively implementable.

First note that for the quadratic function

$$W(\theta) := \frac{1}{2}(\theta - \theta^*)^T Q (\theta - \theta^*) , Q > 0,$$

the algorithm

$$\theta_1 = \theta_0 - [W''(\theta_0)]^{-1} W'(\theta_0),$$

where W' is the gradient and W'' is the matrix of second partial derivatives, converges to the minimizing value of W in one step, starting from any θ_0. This suggests the Gauss-Newton scheme

$$\theta_n = \theta_{n-1} - \gamma_n [W''(\theta_{n-1})]^{-1} W'(\theta_{n-1})$$

to find the minimizer of a function that is locally quadratic near the minimizing value. Here γ_n determines the step size.

Thus for the PEE that minimizes

$$V_n(\theta) := \frac{1}{2} \sum_{k=1}^{n} [y_k - \hat{y}_{k \mid k-1}(\theta)]^2,$$

we consider the scheme

$$\theta_n = \theta_{n-1} - [V''_n(\theta_{n-1})]^{-1} V'_n(\theta_{n-1}).$$

We can approximate $V''_n(\theta_{n-1})$ and $V'_n(\theta_{n-1})$ as follows. Since

$$V_n(\theta) = V_{n-1}(\theta) + \frac{1}{2}[y_n - \hat{y}_{n \mid n-1}(\theta)]^2,$$

therefore

$$V'_n(\theta_{n-1}) = V'_{n-1}(\theta_{n-1})$$
$$- \hat{y}'_{n \mid n-1}(\theta_{n-1}) [y_n - \hat{y}_{n \mid n-1}(\theta_{n-1})],$$

$$V''_n(\theta_{n-1}) = V''_{n-1}(\theta_{n-1}) + \hat{y}'_{n\,|\,n-1}(\theta_{n-1})\hat{y}'_{n\,|\,n-1}(\theta_{n-1})^T$$
$$- \hat{y}''_{n\,|\,n-1}(\theta_{n-1})\,[y_n - \hat{y}_{n\,|\,n-1}(\theta_{n-1})],$$

where $\hat{y}'_{n\,|\,n-1}(\theta_{n-1})$ is the gradient and $\hat{y}''_{n\,|\,n-1}(\theta_{n-1})$ the matrix of second derivatives. If θ_{n-1} is close to minimizing $V_{n-1}(\theta)$, then $V'_{n-1}(\theta_{n-1}) \approx 0$, and so we approximate $V'_n(\theta_{n-1})$ by

$$V'_n(\theta_{n-1}) \approx -\hat{y}'_{n\,|\,n-1}(\theta_{n-1})[y_n - \hat{y}_{n\,|\,n-1}(\theta_{n-1})].$$

As regards $V''_n(\theta_{n-1})$ we drop the last term and just approximate it by R_{n-1}, where R_{n-1} is recursively generated by

$$R_{n-1} = R_{n-2} + \hat{y}'_{n\,|\,n-1}(\theta_{n-1})\hat{y}'^T_{n\,|\,n-1}(\theta_{n-1}).$$

We have thus arrived at the scheme

$$\theta_n = \theta_{n-1} + R^{-1}_{n-1}\,\hat{y}'_{n\,|\,n-1}(\theta_{n-1})\,[y_n - \hat{y}_{n\,|\,n-1}(\theta_{n-1})],$$
$$R_{n-1} = R_{n-2} + \hat{y}'_{n\,|\,n-1}(\theta_{n-1})\hat{y}'^T_{n\,|\,n-1}(\theta_{n-1}),$$

but we still need to approximate $\hat{y}'_{n\,|\,n-1}(\theta_{n-1})$, and that, too, recursively.

Example   (7.13)
Consider

$$y_k = a_0 y_{k-1} + b_0 u_{k-1} + c_0 w_{k-1} + w_k.$$

For $\theta := (a_0, b_0, c_0)^T$, the steady state predictor is

$$\hat{y}_{k\,|\,k-1}(\theta) = a_0 y_{k-1} + b_0 u_{k-1} + c_0(y_{k-1} - \hat{y}_{k-1\,|\,k-2}(\theta))$$
$$= [y_{k-1}, u_{k-1}, y_{k-1} - \hat{y}_{k-1|\,[\,|\,[k-2}(\theta)]^T\theta.$$

Hence the gradient is given by

$$\hat{y}'_{k\,|\,k-1}(\theta) = -c_0\hat{y}'_{k-1\,|\,k-2}(\theta)$$
$$+ [y_{k-1}, u_{k-1}, y_{k-1} - \hat{y}_{k-1\,|\,k-2}(\theta)]^T.$$

Similarly, for the ARMAX system

$$y_k = a_0 y_{k-1} + .\,. + a_p\,y_{k-P-1} + b_0 u_{k-1} + .\,.$$
$$+ b_p\,u_{k-p-1} + c_0 w_{k-1} .\,. + c_p\,w_{k-p-1} + w_k,$$

its steady-state predictor for $\theta := (a_0, .\,., a_p, b_0, .\,., b_p, c_0, .\,., c_p)$ is given by

$$\hat{y}_{k\,|\,k-1}(\theta) = [y_{k-1}, .\,., y_{k-p-1}, u_{k-1}, .\,., u_{k-p-1},$$
$$y_{k-1} - \hat{y}_{k-1\,|\,k-2}(\theta) .\,., y_{k-p-1} - \hat{y}_{k-p-1\,|\,k-p-2}(\theta)]^T\theta,$$

and the gradient by

$$\hat{y}'_{k\,|\,k-1}(\theta) = -c_0\hat{y}'_{k-1\,|\,k-2}(\theta) - .\,. - c_p\,\hat{y}'_{k-p-1\,|\,k-p-2}(\theta)$$

$$+ [y_{k-1}, \ldots, y_{k-p-1}, u_{k-1}, \ldots, u_{k-p-1}, y_{k-1} - \hat{y}_{k-1 \mid k-2}(\theta),$$

$$\ldots, y_{k-p-1} - \hat{y}_{k-p-1 \mid k-p-2}(\theta)]^T.$$

We need $\hat{y}_{k \mid k-1}(\theta_{k-1})$ and $\hat{y}'_{k \mid k-1}(\theta_{k-1})$. To obtain them recursively we introduce the approximation ϕ_{k-1} for $[y_{k-1}, \ldots, y_{k-p-1}, u_{k-1}, \ldots, u_{k-p-1}, y_{k-1} - \hat{y}_{k-1 \mid k-2}(\theta_{k-1}), \ldots, y_{k-p-1 \mid k-p-2}(\theta_{k-1})]^T$ where ϕ_{k-1} is generated by

$$\phi_{k-1} := (y_{k-1}, \ldots, y_{k-p-1}, u_{k-1}, \ldots, u_{k-p-1},$$

$$y_{k-1} - \phi_{k-2}^T \hat{\theta}_{k-1}, \ldots, y_{k-p-1} - \phi_{k-p-2}^T \hat{\theta}_{k-p-1})^T.$$

We also introduce the approximation ψ_{k-1} for $\hat{y}'_{k \mid k-1}(\theta_{k-1})$, where ψ_{k-1} is generated by

$$\psi_{k-1} = -\hat{c}_0(k-1)\psi_{k-2} - \ldots - \hat{c}_p(k-1)\psi_{k-p-2} + \phi_{k-1},$$

and $\hat{c}_0(k-1), \ldots, \hat{c}_p(k-1)$ are the last $(p+1)$ components of θ_{k-1}.

Finally we have the algorithm

$$\hat{\theta}_n = \hat{\theta}_{n-1} + R_{n-1}^{-1} \psi_{n-1} [y_n - \phi_{n-1}^T \hat{\theta}_{n-1}], \tag{7.14}$$

$$R_n = R_{n-1} + \psi_n \psi_n^T, \tag{7.15}$$

$$\phi_n := (y_n, \ldots, y_{n-p}, u_n, \ldots, u_{n-p}, \tag{7.16}$$

$$y_n - \phi_{n-1}^T \hat{\theta}_n, \ldots, y_{n-p} - \phi_{n-p-1}^T \hat{\theta}_{n-p})^T,$$

$$\psi_n := -\hat{c}_0(n)\psi_{n+1} - \ldots - \hat{c}_p(n)\psi_{n-p-1} + \phi_n, \tag{7.17}$$

$$(\hat{c}_0(n), \ldots, \hat{c}_p(n))$$

$$:= \text{last } (p+1) \text{ components of } \hat{\theta}_n. \tag{7.18}$$

The algorithm (7.14)-(7.18) is called the **recursive prediction error method** (RPEM). [It also goes by the name **recursive maximum likelihood** (RML) algorithm.]

The vector ψ_n in (7.17) can be seen as the output of a time varying filter driven by ϕ_n, where the instantaneous transfer function at time n (abusing notation) is $\dfrac{1}{C(q^{-1}, \hat{\theta}_n)}$, where q^{-1} is the shift operator and $C(q^{-1}, \hat{\theta}_n) := 1 + \hat{c}_0(n)q^{-1} + \ldots + \hat{c}_p(n)q^{-p-1}$. Since $C(q^{-1}, \hat{\theta}_n)$ is the estimate (at time n) of the polynomial $C(q^{-1})$ describing the moving average noise in the ARMAX system, we can say that the algorithm uses as approximation to the gradient of the predictor the vector $\{\phi_k\}$ filtered through a system with time-varying transfer function $\dfrac{1}{C(q^{-1}, \hat{\theta}_k)}$.

This points to a possible defect in this algorithm. Let $D \subset R^{3p+3}$ be the set of all $\theta \in R^{3p+3}$ such that $C(z, \theta)$ has all its roots strictly outside the closed unit disc and let D^C be its complement. For $\theta \in D^C$, the transfer function $\dfrac{1}{C(q^{-1},\theta)}$ is   unstable.   Hence   if $\hat{\theta}_k \approx \theta \in D^C$ for $n \le k \le m$ and if the time interval $m - n$ is large, then the   vector

$$\psi_k = \frac{1}{C(q^{-1},\hat{\theta}_k)}\ \phi_k$$

may blow up. Hence in practice, the algorithm is usually modified by replacing (7.14) by the following logic step.

Let $\hat{D}$ be a compact subset of D. Then set

$$\hat{\theta}_n = \hat{\theta}_{n-1} + R_{n-1}^{-1}\ \psi_{n-1}[y_n - \phi_{n-1}^T\hat{\theta}_{n-1}]$$

$$\text{if } \hat{\theta}_{n-1} + R_{n-1}^{-1}\ \psi_{n-1}[y_n - \phi_{n-1}^T\hat{\theta}_{n-1}] \in \hat{D} \qquad (7.19)$$

$$= \text{some point in } \hat{D}, \text{ otherwise.} \qquad (7.20)$$

The operation (7.20) is usually called a *projection facility*, and to implement it requires that the sequence $\{\hat{\theta}_{n-1} + R_{n-1}^{-1}\ \psi_{n-1}[y_n - \phi_{n-1}^T\hat{\theta}_{n-1}]\}$ be monitored to see when it leaves the set $\hat{D}$.

Use of the projection facility (7.19) and (7.20) leads to a sequence $\{\hat{\theta}_n\} \subset \hat{D}$, and so if $\{\hat{\theta}_n\}$ converges, it can converge only to points inside D. Thus $\{\hat{\theta}_n\}$ can only converge to a θ^* with $C(z,\theta^*)$ such that all its roots are strictly outside the closed unit disc. However, by the spectral factorization theorem (5.5.12), we know that it is not too restrictive to assume that for the moving average noise $\{w_{k+1} + c_0\ w_k + .. + c_p\ w_{k-P}\}$ the polynomial $1 + c_0z + .. + c_p z^{p+1}$ has all its roots strictly outside the closed unit disc.

There is an obvious connection between the RPEM algorithm (7.14)-(7.18) and the PLR algorithm (7.6)-(7.9). In the PLR algorithm the gradient ψ_n is replaced by ϕ_n, or to put it another way, we just filter $\{\phi_k\}$ through a system with transfer function 1. Hence we are replacing the filter $\dfrac{1}{C(q^{-1},\hat{\theta}_k)}$ with the filter 1. Thus if we believe that the RPEM is a rationally derived algorithm, then we should expect the PLR to be well behaved when $C(q^{-1},\theta^{\circ}) \approx 1$. We can therefore interpret the positive real condition (7.11) of Theorem (7.10) in this light.

A method to analyze the RPEM and related algorithms is treated in the next section.

8. The ODE method to analyze recursive stochastic algorithms

In previous sections we introduced several recursive algorithms such as the PLR and the RPEM algorithm. We now present a unified method for analyzing their asymptotic properties.

The algorithms we have seen so far, together with the ARMAX system, can be written in a uniform fashion in the state space form

$$x_{n+1} = A\,(\hat{\theta}_n)x_n + B\,(\hat{\theta}_n)w_{n+1}, \tag{8.1}$$

$$\begin{bmatrix} \hat{\theta}_{n+1} \\ \overline{R}_{n+1} \end{bmatrix} = \begin{bmatrix} \hat{\theta}_n \\ \overline{R}_n \end{bmatrix} + \frac{1}{n}\, f_n\,(\hat{\theta}_n, \overline{R}_n, x_{n+1}). \tag{8.2}$$

Typically, $\overline{R}_n = \dfrac{R_n}{n}$ for the algorithms considered earlier and $\{w_n\}$ is white noise.

Exercise (8.3)

Consider the RPEM (7.14)-(7.18) to estimate the parameter c_0 of the system $y_{k+1} = c_0 w_k + w_{k+1}$. Show that it can be written in the form

$$x_{n+1} = \begin{bmatrix} 0 & 0 & 0 & 0 \\ 1 & 0 & 0 & 0 \\ 0 & 1 - \hat{\theta}_n & -\hat{\theta}_n & -\hat{\theta}_n \\ 0 & 1 & 0 & -\hat{\theta}_n \end{bmatrix} x_n + \begin{bmatrix} c_0 \\ 1 \\ 0 \\ 0 \end{bmatrix} w_{n+1}, \tag{8.4}$$

$$\begin{bmatrix} \hat{\theta}_{n+1} \\ \overline{R}_{n+1} \end{bmatrix} = \begin{bmatrix} \hat{\theta}_n \\ \overline{R}_n \end{bmatrix} + \frac{1}{n} \begin{bmatrix} \overline{R}_n^{-1} x_{n+1}^{(3)}\,(x_{n+1}^{(2)} - x_{n+1}^{(4)}\,\hat{\theta}_n) \\ g_n\,(\hat{\theta}_n, \overline{R}_n, x_{n+1}) \end{bmatrix}, \tag{8.5}$$

where $x_{n+1} := (x_{n+1}^{(1)}, x_{n+2}^{(2)}, x_{n+1}^{(3)}, x_{n+1}^{(4)}) := (c_0 w_{n+1}, y_{n+1}, \psi_n, \phi_n)$, $\overline{R}_n := \dfrac{R_n}{n}$, and $g_n(\overline{\theta}_n, \overline{R}_n, x_{n+1}) = \dfrac{n}{n+1}\,(\psi_{n+1}^2 - \overline{R}_n)$.

The intuition underlying the method for the analysis of (8.1) and (8.2) is this. Assume that $A(\theta)$ is stable for all θ of interest to us; then (8.1) will have a steady-state distribution for x_n as $n \to \infty$ for each θ, if $\hat{\theta}_n$ is fixed at θ. Let P^θ and E^θ denote probabilities and expectations taken under this limiting steady state distribution. Now examine (8.2). Suppose that as in Exercise (8.3), $f_n(\theta, \overline{R}, x) \to f(\theta, \overline{R}, x)$ as $n \to \infty$ for some f. So we may approximate (8.2) as

$$\Delta \begin{bmatrix} \hat{\theta}_{n+1} \\ \overline{R}_{n+1} \end{bmatrix} \approx \frac{1}{n} f(\hat{\theta}_n, \overline{R}_n, x_{n+1}),$$

where $\Delta z_{n+1} := z_{n+1} - z_n$. When n is large, $\dfrac{1}{n}$ is small, and so $\dfrac{1}{n} f(\hat{\theta}_n, \overline{R}_n, x_{n+1})$ is also small. This suggests that for large n, the values of θ_n and $\overline{R}_n$ change very little. On the other hand x_{n+1} changes rapidly because of the noise term in (8.4), and so

$$\Delta \begin{bmatrix} \hat{\theta}_{n+1} \\ \bar{R}_{n+1} \end{bmatrix} \approx \frac{1}{n} E^{\hat{\theta}_n} f(\hat{\theta}_n, \bar{R}_n)$$

may be a good approximation to the behavior of (8.2). Here

$$E^{\hat{\theta}_n}(f(\hat{\theta}_n, \bar{R}_n)) := \int f(\hat{\theta}_n, \bar{R}_n, x_{n+1}) P^{\hat{\theta}_n}(dx_{n+1}).$$

For n large, the ordinary differential equation (ODE)

$$\frac{d}{dt} \begin{bmatrix} \hat{\theta}(t) \\ \bar{R}(t) \end{bmatrix} \approx \frac{1}{t} E^{\hat{\theta}(t)} f(\hat{\theta}(t), \bar{R}(t))$$

also suggests itself as an approximation in continuous time. Introduce now the change of time scale, $\tau = \tau(t) := \log t$ and we have the ODE

$$\frac{d}{d\tau} \begin{bmatrix} \hat{\theta}(\tau) \\ \bar{R}(\tau) \end{bmatrix} = E^{\hat{\theta}(\tau)}[f(\hat{\theta}(\tau), \bar{R}(\tau))] \tag{8.6}$$

which may asymptotically be a good approximation of the behavior of (8.1) and (8.2).

Indeed, under various assumptions on the noise $\{w_k\}$, the matrices A, B, and the functions f_n in (8.1), (8.2), the following result has been proved.

Result (8.7)

Let Ω_1 be the set of all ω for which $\hat{\theta}_n(\omega)$ and $\bar{R}_n(\omega)$ are bounded sequences. Let (θ^*, R^*) be a locally asymptotically stable equilibrium point of (8.5) with domain of attraction $D(\theta^*, R^*)$. For $\omega \in \Omega_1$, if $(\hat{\theta}_n(\omega), \bar{R}_n(\omega)) \in D(\theta^*, R^*)$ for infinitely many n, then

$$\lim_{n \to \infty} \hat{\theta}_n(\omega) = \theta^*, \quad \lim_{n \to \infty} \bar{R}_n(\omega) = R^*.$$

In applications, the projection facility (7.19), (7.20), which leads to a modification of the functions f_n and f of (8.2) and (8.6), is used to ensure the boundedness of $\{(\hat{\theta}_n, \bar{R}_n)\}$.

As an example consider the problem of estimating c_0 for the system

$$y_{k+1} = c_0 w_k + w_{k+1}; \quad |c_0| < 1, \tag{8.8}$$

where $\{w_k\}$ is white noise with mean 0 and variance 1. The RPEM for identifying c_0 gives the algorithm

$$\hat{c}_{n+1} = \hat{c}_n + \frac{1}{n} \bar{R}_n^{-1} \psi_n (y_{n+1} - \phi_n \hat{c}_n), \tag{8.9}$$

$$\bar{R}_{n+1} = \bar{R}_n + \frac{1}{n} \cdot \frac{n}{n+1} \, (\psi_{n+1}^2 - \bar{R}_n), \tag{8.10}$$

$$\phi_n = y_n - \phi_{n-1} \hat{c}_n, \tag{8.11}$$

$$\psi_n = y_n - \psi_{n-1} \hat{c}_n - \phi_{n-1} \hat{c}_n. \tag{8.12}$$

Thus the ODE (8.6) corresponding to this algorithm is

$$\frac{d}{d\tau} \hat{c}(\tau) = \bar{R}^{-1}(\tau) \, E^{\hat{c}(\tau)} [\psi_n y_{n+1} - \hat{c}(\tau) \psi_n \phi_n], \tag{8.13}$$

$$\frac{d}{d\tau} \bar{R}(\tau) = E^{\hat{c}(\tau)} (\psi_{n+1}^2) - \bar{R}(\tau), \tag{8.14}$$

where E^c denotes the steady-state expectation for the processes

$$y_n = (1 + c \, q^{-1}) w_n, \tag{8.15}$$

$$\phi_n = \frac{1}{1 + c \, q^{-1}} y_n, \tag{8.16}$$

$$\psi_n = \frac{1}{1 + c \, q^{-1}} \phi_n, \tag{8.17}$$

defined for $|c| < 1$.

Exercise (8.18)

Show that the ODE (8.13), (8.14) reduces to

$$\frac{d}{d\tau} \hat{c}(\tau) = \bar{R}^{-1}(\tau) \frac{(c_0 - \hat{c}(\tau))(1 - c_0 \hat{c}(\tau))}{(1 - \hat{c}^2(\tau))^2}, \tag{8.19}$$

$$\frac{d}{d\tau} \bar{R}(\tau) = \frac{(1 + c_0^2)(1 + \hat{c}^2(\tau)) - 4 c_0 \hat{c}(\tau)}{(1 - \hat{c}^2(\tau))^3} - \bar{R}(\tau). \tag{8.20}$$

[Hint: Use the fact that for $x_n = H(q^{-1}) w_n$, $v_n = F(q^{-1}) w_n$, H and F stable, $E \, x_n v_n = \frac{1}{2\pi i} \oint H(z) F(z^{-1}) \frac{dz}{z}$.]

There is only one equilibrium of (8.19), (8.20) in the region $|\hat{c}| < 1$, $\bar{R} > 0$, namely,

$$(\hat{c}, \bar{R}) = (c_0, \frac{1}{1 - c_0^2}).$$

Exercise (8.21)

For ϵ, δ positive and small enough show that the rectangle

$$\Gamma := \{ (\hat{c}, \bar{R}) \mid |\hat{c}| \le 1 - \epsilon,$$

$$\delta \le \bar{R}(\tau) \le \frac{(8 - 8\epsilon)(c_0 + 1)^2 + 4\epsilon^2(c_0^2 + 1)}{4\epsilon^3(2 - \epsilon)^3} + \delta \}$$

is an invariant set of the ODE (8.19), (8.20), i.e., if $(\hat{c}(0), \bar{R}(0)) \in \Gamma$, then

$(\hat{c}(\tau), \overline{R}(\tau)) \in \Gamma$ for all $\tau \geq 0$.
[Hint: Evaluate the right-hand sides of (8.19), (8.20) on the boundary of Γ and show that the derivatives point inside Γ.]

Exercise (8.22)
Show that for every solution of the ODE (8.19), (8.20) with initial conditions $(\hat{c}(0), \overline{R}(0)) \in \Gamma$,

$$\lim_{\tau \to \infty} \hat{c}(\tau) = c_0, \text{ and } \lim_{\tau \to \infty} \overline{R}(\tau) = \frac{1}{1 - c_0^2}.$$

[Hint: Consider $V(\hat{c}) := \frac{1}{2}(\hat{c} - c_0)^2$. Then

$$\frac{d}{d\tau} V(\hat{c}(\tau)) = -(\hat{c}(\tau) - c_0)^2 \left[\frac{1 - c_0 \hat{c}(\tau)}{\overline{R}(\tau)(1 - \hat{c}^2(\tau))} \right],$$

and so $\frac{d}{d\tau} V(\hat{c}(\tau)) \leq -\gamma V(c(\tau))$ for some $\gamma > 0$, and $\hat{c}(\tau) \to c_0$. Hence for every $\beta > 0$, there is T, such that

$$[\frac{1}{1 - c_0^2} - \overline{R}(\tau)] - \beta \leq \frac{d}{d\tau} \overline{R}(\tau)$$

$$\leq [\frac{1}{1 - c_0^2} - \overline{R}(\tau)] + \beta, \quad \tau > T.$$

Hence $\overline{R}(\tau) \to [\frac{1}{1 - c_0^2} - \beta, \frac{1}{1 - c_0^2} + \beta]$ for every $\beta > 0$.]

Thus if (8.7) is valid, then we can deduce that the RPEM (8.9)-(8.12) when used to estimate the coefficient c_0 of (8.8) has the property that

$$\lim_{n \to \infty} \hat{c}_n = c_0 \text{ and } \lim_{n \to \infty} \overline{R}_n = \frac{1}{1 - c_0^2}.$$

Some modification of the algorithm, such as the projection facility of (7.19), (7.20) may, however, be needed.

Indeed, a similar analysis can be conducted also for the behavior of the RPEM for general ARMAX systems; however the possibility of the existence of more than one locally asymptotically stable equilibrium point should be kept in mind when using (8.7).

Exercise (8.23)
For the off-line prediction error method for estimating c_0 in (8.8), let
$V(c) := \lim_{n \to \infty} \frac{1}{n} \sum_{k=1}^{n} [y_k - \hat{y}_{k \mid k-1}(c)]^2$ where $\hat{y}_{k \mid k-1}(c)$ is the prediction of y_k given $\{y_s, s \leq k-1\}$ if c is the value of the parameter. In Exercise (5.21) we saw that

$$V(c) = \frac{\sigma^2}{2\pi i} \oint \frac{1}{1+cz} \frac{1}{1+cz^{-1}} (1 + c_0 z)(1 + c_0 z^{-1}) \frac{dz}{z} .$$

Show that $\dfrac{d}{dc} V(c) = \dfrac{2\sigma^2(c_0 - c)(1 - cc_0)}{(1 - c^2)^2}$.

Thus for this example, the ODE (8.19) can also be written as

$$\frac{d}{d\tau} \hat{c}(\tau) = -\frac{\overline{R}^{-1}(\tau)}{2\sigma^2} \frac{d}{dc} V(\hat{c}(\tau)),$$

showing that its equilibrium points correspond to critical values (derivative = 0) of $V(c)$.

Moreover, since the right-hand side of (8.19) is the expected value of $(\hat{c}_{k+1} - \hat{c}_k)$, we see that this expected value has the same direction as the negative of the gradient $\dfrac{d}{dc} V(c_k)$. Thus, on the average, the RPEM takes steps in the direction of the negative gradient, which is reassuring since it shows that the RPEM does indeed have favorable properties with respect to the prediction error criterion.

If for general ARMAX systems, the equilibrium points of the ODE continue to correspond to critical points of $V(\theta)$ (and they do), then locally asymptotically stable equilibrium points of the ODE will correspond to local minima of $V(\theta)$. If there are several local minima of the limiting prediction error criterion $V(\theta)$, then the ODE method (8.7), if applicable, will show convergence to local minima of $V(\theta)$. This contrasts with the off-line PEM which converges to the global minimum of $V(\theta)$ as shown in Theorem (5.28). This, therefore, is the price to be paid for using the RPEM and it also shows the importance of examining the critical points of the asymptotic prediction error criterion $V(\theta)$.

9. Notes

1. To compute the LSE efficiently, one can use many of the algorithms developed for linear filtering in Chapter 7. For example, when fitting an AR model to data as in Section 3, $\phi_k := (y_k, \ldots, y_{k-p+1})^T$, and so equation (1.4) for determining the LSE is almost the same as the Yule-Walker equation (7.9.7). This leads to algorithms of the Levinson type; see (7.9.13). These issues are studied in detail in Ljung and Soderstrom (1983). Moreover, one can obtain adaptive versions of the filtering algorithms of Chapter 7. For example, if the reflection coefficients are unknown, then one can develop an adaptive lattice filter, where these coefficients are estimated on the basis of observed data. Such adaptive filtering algorithms are studied in Goodwin and Sin (1984).

2. The proof leading up to Theorem (4.15) is a slight modification of that of Ljung (1976a). Sharper results and central limit theorems, along

with references, are given in Solo (1982).

3. The proofs suggested in Exercises (5.20) and (5.21) are from Solo (1982). Exercise (5.30) is from Astrom and Soderstrom (1974), which also examines the question of local minima of the integral considered. Central limit theorems can also be obtained, see Ljung and Caines (1979) and Solo (1982); the latter also gives further references. The prediction error method is asymptotically efficient in the scalar case examined here.

4. A good reference for the closed loop identification problem of Section 6 is Gustavsson, Ljung and Soderstrom (1977).

5. Several choices of the instrumental variable ψ_k in (7.3) as well as references to the literature are given in Ljung and Soderstrom (1983).

6. Theorem (7.10) is proved in Solo (1979), see also You-Hong (1982). This is one of the few results for which a self-contained martingale-based proof is available.

7. The recursive prediction error method and the pseudolinear regression method can be used as the basis for unifying many identification algorithms which have been proposed in the literature; see Ljung and Soderstrom (1983).

8. The ODE approach of Section 8 was presented in Ljung (1977b) and for the full range of applications of this method also see Ljung and Soderstrom (1983). For more recent proofs of the ODE method, see Kushner and Clark (1978), Kushner (1984), Kushner and Schwartz (1984), and Metivier and Priouret (1984).

9. The attractive feature of the recursive prediction error method is that when it converges to the true parameter, the asymptotic variance of the parameter estimate error is the same as that obtained from the off-line prediction error method, see Ljung and Soderstrom (1983).

10. The martingale convergence theorem (4.8) and Kronecker's lemma (4.9) can be found in Stout (1974) as Theorem 2.7.2 and Lemma 3.2.3, respectively.

11. Solo (1984) has developed a different class of identification algorithms based on adaptive spectral factorization. Typically, these schemes use a method such as the IVE to estimate the AR coefficients. Then an estimate of the spectrum of the moving average noise is formed by subtracting out the estimated contribution of the AR part. Finally this estimated spectrum is factored by employing a recursive algorithm such as that developed by Wilson (1969) for spectral factorization of an MA process. These are the first recursive algorithms for identifying the coefficients of an ARMA process for which strong consistency of the estimates has been proved even when the colored noise does not satisfy a positive real condition.

10. References

[1]　K.J. Astrom and T. Soderstrom (1974), "Uniqueness of the maximum likelihood estimates of the parameters of an ARMA model," *IEEE Transactions on Automatic Control,* vol AC-19, 769-773.

[2]　G. Goodwin and K. S. Sin (1984), *Adaptive filtering prediction and control,* Prentice Hall, New Jersey.

[3]　I. Gustavsson, L. Ljung and T. Soderstrom (1977), "Adaptive processes in closed loop: identifiability and accuracy aspects," *Automatica,* vol 13, 59-79.

[4]　H. Kushner (1984), *Approximation and weak convergence methods for random processes,* MIT Press, Cambridge, MA.

[5]　H. Kushner and D. Clark (1978), *Stochastic approximation methods for constrained and unconstrained systems,* Applied Math. Science Series 26, Springer, Berlin.

[6]　H. Kushner and A. Schwartz (1984), "An invariant measure approach to the convergence of stochastic approximations with state dependent noise," *SIAM J. on Control and Optimization,* vol 22, 13-27.

[7]　L. Ljung (1976a). "Consistency of the least-squares identification method," *IEEE Transactions on Automatic Control,* vol AC-21, 779-781.

[8]　L. Ljung (1977b), "Analysis of recursive stochastic algorithms," *IEEE Transactions on Automatic Control,* vol AC-22, 551-575.

[9]　L. Ljung and P.E. Caines (1979), "Asymptotic normality of prediction error estimation for approximate system models," *Stochastics,* vol 3, 29-46.

[10]　L. Ljung and T. Soderstrom (1983), *Theory and practice of recursive identification,* MIT Press, Cambridge, MA.

[11]　M. Metivier and P. Priouret (1984), "Applications of a Kushner and Clark Lemma to general classes of stochastic algorithms," *IEEE Transactions on Information Theory,* vol IT-30, 140-151.

[12]　V. Solo (1979), "The convergence of AML," *IEEE Transactions on Automatic Control,* vol AC-24, 958-963.

[13]　V. Solo (1982), *Topics in advanced time series analysis,* Springer-Verlag Lecture Notes.

[14]　V. Solo (1984), "Adaptive spectral factorization," Preprint, Harvard University.

[15]　W. F. Stout (1974), *Almost sure convergence,* Academic Press, New York.

[16] G. Wilson (1969), "Factorization of the covariance generating function of a pure moving average process," *SIAM J. on Numerical Analysis,* vol. 6, 1-7.

[17] Z. You-Hong (1982), "Stochastic adaptive control and prediction based on a modified least squares — the general delay — colored noise case," *IEEE Transactions on Automatic Control,* vol AC-27, 1257-1260.

CHAPTER 11
BAYESIAN ADAPTIVE CONTROL

Adaptive control problems may be formulated in a non-Bayesian context, with no prior distribution for the unknown parameters, or in a Bayesian context, where there is such a prior distribution. In the former case we would attempt to design self-tuning or self-optimizing adaptive control laws. A simple illustrative example is provided to motivate the detailed treatment of such laws in Chapters 12 and 13. This chapter is devoted to Bayesian adaptive control problems. These can be viewed as optimal control problems of partially observed systems and studied using dynamic programming. A complete theory is presented for a special class of such problems known as bandit problems.

1. Introduction

In previous chapters we have separated the problems of control of known systems and identification of unknown systems. In controlling an unknown system, we could first use a method to identify the system and then control the identified system.

This procedure of identification followed by control may be unacceptable if it is important to maintain adequate control even during the identification phase. On the other hand, if the system changes slowly with time, an identification mechanism may have to be constantly in place. Thus there is need for a more wholistic approach to the control of unknown systems in which the tasks of identification and control are integrated together. This is the problem of adaptive control.

2. Non-Bayesian formulation: an example

Consider the system

$$y_{k+1} = \theta^\circ u_k + w_{k+1}, \tag{2.1}$$

where $\{w_k\}$ is white noise, θ° is a fixed but unknown parameter and no prior pd is given for θ°. Suppose the task of control is to keep

$$y_k \approx r \quad \text{for all } k = 0, 1, 2, \ldots,$$

where $r \neq 0$ is some desirable value for the output. If we knew θ°, we could simply choose

$$u_k = \frac{r}{\theta^\circ} , \tag{2.2}$$

and this will minimize both

$$E \lim_{N \to \infty} \sup \frac{1}{N} \sum_0^{N-1} (y_{k+1} - r)^2, \tag{2.3}$$

and

$$E \{(y_{k+1} - r)^2 \mid y_0, \ldots, y_k\}.$$

When θ° is not known we can proceed as follows. We first identify θ° using the LSE (10.1.4),

$$\hat{\theta}_k = \frac{\sum\limits_0^{k-1} y_{i+1} u_i}{\sum\limits_0^{k-1} u_i^2} , \quad k = 1, 2, \ldots, \tag{2.4}$$

and then choose

$$u_k = \frac{r}{\hat{\theta}_k} , \quad k = 1, 2, \ldots \tag{2.5}$$

In other words, at each time instant k we make an estimate of θ°, and then we choose the control as if the estimate were the true parameter. (This is sometimes called enforced certainty equivalence.)

The control law (2.4), (2.5) is an example of an **adaptive** control law. If $\hat{\theta}_k$ converges to the true parameter θ°, then

$$\lim_k u_k = \frac{r}{\theta^\circ} , \tag{2.6}$$

and the law (2.4), (2.5) asymptotically approaches the law (2.2) that would be used if θ° were known from the beginning. We shall refer to an adaptive control law with property (2.6) as **self-tuning.**

We now investigate when (2.4) and (2.5), applied to the system (2.1), is self-tuning. The following convergence result for martingales is useful.

Theorem (local convergence of martingales)   (2.7)
Let $\{w_n, F_n\}$ be a martingale difference sequence [see Theorem (8.5.26)], and suppose that

$$\sup_n E\{w_n^2 \mid F_n\} < \infty \ a.s. \tag{2.8}$$

Let $\{u_n\}$ be a process adapted to $\{F_n\}$, i.e., u_n is F_n-measurable for each n. Then:

1. $\sum_N u_{n-1} w_n$ converges a.s. on the set $\{\sum u_n^2 < \infty\}$.

2. $\sum_1^N u_{n-1} w_n = o(\sum_1^N u_{n-1}^2)$ a.s. on the set $\{\sum u_n^2 = \infty\}$.

To apply this result assume that $\{w_k, F_k\}$ is a martingale [see Theorem (10.4.8)] where F_k is the σ-field generated by $u_l, w_l, l \le k$, and assume that (2.8) holds. Substituting for y_{i+1} from (2.1), the LSE can be written as

$$\hat{\theta}_k = \theta^\circ + \frac{\sum_0^{k-1} u_i w_{i+1}}{\sum_0^{k-1} u_i^2}. \tag{2.10}$$

The following implications hold a.s.,

$$\sum u_i^2 < \infty \Rightarrow \sum u_i w_{i+1} \text{ converges}$$

$$\Rightarrow \frac{\sum u_i w_{i+1}}{\sum u_i^2} \text{ converges}$$

$$\Rightarrow \{\hat{\theta}_k\} \text{ converges}$$

$$\Rightarrow \lim \left| \frac{r}{\hat{\theta}_k} \right| > 0$$

$$\Rightarrow \lim |u_k| > 0 \Rightarrow \sum u_i^2 = \infty.$$

Hence $\sum u_i^2 = \infty$ a.s. From Theorem (2.7) it follows that $\lim \hat{\theta}_k = \theta^\circ$ a.s. and so (2.6) holds a.s. Therefore, the adaptive control law (2.4), (2.5) applied to the system (2.1) is self-tuning.

We study additional properties. First, the cost (2.3) when the adaptive control law is used is

$$\lim_N \sup \frac{1}{N} \sum_0^{N-1} (y_{k+1} - r)^2$$

$$= \lim_N \sup \frac{1}{N} \sum_0^{N-1} (\theta^\circ u_k + w_{k+1} - r)^2$$

$$= \lim_N \sup \frac{1}{N} [\sum_0^{N-1} (\theta^\circ u_k - r)^2 + w_{k+1}^2 - 2rw_{k+1} + 2\theta^\circ u_k w_{k+1}]$$

$$= \limsup_N \frac{1}{N} \sum_0^{N-1} w_{k+1}^2.$$

[The last equality follows because $(\theta^\circ u_k - r)^2 \to 0$, $\sum_0^{N-1} r\, w_{k+1} = o(N)$ by Theorem (2.7), and $\sum_0^{N-1} u_k w_{k+1} = o(\sum_0^{N-1} u_k^2)$, $\sum_0^{N-1} u_k^2 = O(N)$.]

On the other hand, if θ° is known, (2.3) is minimized by the control law (2.2), and the minimum cost is

$$\limsup_N \frac{1}{N} \sum_0^{N-1} (y_{k+1} - r)^2 = \limsup_N \frac{1}{N} \sum_0^{N-1} w_{k+1}^2,$$

so that the adaptive control law (2.4), (2.5) achieves the same cost as could be achieved if one knew θ°. This property will be called **self-optimizing.**

To summarize, the proposed adaptive control law is both self-tuning and self-optimizing.

Of course, to achieve a self-optimizing law it is necessary that the criterion is of the average cost type (2.3), since it only considers asymptotic behavior. In contrast, the discounted cost

$$E \sum_0^\infty \beta^k (y_{k+1} - r)^2, \tag{2.11}$$

is minimized only when

$$u_k = \frac{r}{\theta^\circ}, \; k = 0, 1, \dots,$$

which can be implemented only if θ° is known. Thus with respect to the discounted cost (2.11) the adaptive control (2.4), (2.5) is self-tuning, but not self-optimizing.

In conclusion, in a non-Bayesian formulation we can try to find self-tuning or self-optimizing control laws. In Chapters 12 and 13 we shall see that this can be surprisingly complicated.

3. Bayesian formulation: an example

Consider the same system

$$y_{k+1} = \theta^\circ u_k + w_{k+1},$$

and assume that $\theta^\circ \sim N(0, 1)$. With this prior distribution, we have the model

$$\theta_{k+1} = \theta_k, \; \theta_0 \equiv \theta^\circ \sim N(0, 1), \tag{3.1}$$

$$y_{k+1} = \theta_k u_k + w_{k+1}. \tag{3.2}$$

Assume that $\{w_k\}$ is white noise, $w_k \sim N(0, \sigma^2)$ and independent of θ°. Suppose we wish to minimize

$$E \sum_{k=1}^{N} (y_k - r)^2. \tag{3.3}$$

[Of course, the expectation in (3.3) is taken with respect to $\{w_k\}$ and θ_0.]

In the system (3.1), (3.2) we only observe y_k (but not θ_k) and so this is a stochastic optimal control problem with partial observations. For such problems we can define an information state $P_{k|k} (\cdot \mid y^k, u^{k-1})$, the conditional probability distribution of (θ_k, y_k) given the observed past (y^k, u^{k-1}). In Chapter 6 we saw that this information state can be regarded as the state of a new system with complete observations. We study this new system.

From Theorem (7.2.21), $P_{k|k}(\theta_k \mid y^k, u^{k-1}) \sim N(\hat{\theta}_k, \Sigma_k)$, where $\hat{\theta}_k$ and Σ_k are given by the Kalman filter,

$$\hat{\theta}_{k+1} = \hat{\theta}_k + \frac{\Sigma_k u_k}{u_k^2 \Sigma_k + \sigma^2} (y_{k+1} - u_k \hat{\theta}_k), \quad \hat{\theta}_0 = 0, \tag{3.4}$$

$$\Sigma_{k+1} = \Sigma_k - \frac{u_k^2 \Sigma_k^2}{u_k^2 \Sigma_k + \sigma^2}, \quad \Sigma_0 = 1. \tag{3.5}$$

Moreover, $y_{k+1} = \theta_k u_k + w_{k+1} = \hat{\theta}_k u_k + (\theta_k - \hat{\theta}_k) u_k + w_{k+1}$, and $(\theta_k - \hat{\theta}_k) \sim N(0, \Sigma_k)$ is independent of w_{k+1} and y^k. Denoting $(\theta_k - \hat{\theta}_k)u_k + w_{k+1} =: v_{k+1} \sim N(0, u_k^2 \Sigma_k + \sigma^2)$ we obtain

$$y_{k+1} = \hat{\theta}_k u_k + v_{k+1}. \tag{3.6}$$

Equations (3.4)-(3.6) characterize $P_{k|k}(\cdot \mid y^k, u^{k-1})$ and so $(\hat{\theta}_k, \Sigma_k, y_k)$ can be regarded as the new state. Thus (3.4)-(3.6) are the state equations for a completely observed system, and our goal is to choose $\{u_k\}$ to minimize (3.3). This problem can be solved by dynamic programming using Theorem (6.2.15). The state equations are nonlinear, and no explicit closed form for the optimal control law can be obtained.

However, some qualitative features are discernible. The choice of u_k influences Σ_{k+1}, the uncertainty in the value of θ°. Moreover, (3.5) shows that it might be advantageous to choose large values of $|u_k|$ since that will decrease Σ_{k+1}. However, this advantage must be balanced against the disadvantage of increasing $(y_{k+1} - r)^2$ and thereby the overall cost (3.3). Thus, the control serves a dual purpose (see Section 6.8).

In general, all Bayesian adaptive control problems can be viewed as stochastic control problems with partial observations. However, the resulting dynamic programming equations are computationally more complex. One interesting case where significant progress has been made is the so-called bandit problems, which we take up for consideration in the remainder of this chapter. Before that, note that if the criterion (3.3) is replaced by the average cost, then we can obtain optimal solutions quite easily by using a self-optimizing control law, as the following exercise shows.

Exercise   (3.7)
Consider the system

$$y_{k+1} = \theta^\circ u_k + w_{k+1},$$ (3.8)

where $\{w_k\}$ is iid with mean 0 and variance σ^2 and θ° is a random variable with prior distribution P. Show that

$$E\,[\limsup_N \frac{1}{N} \sum_1^N (y_k - r)^2]$$

is minimized by the law

$$u_k = \frac{r}{\hat{\theta}_k}, \quad \hat{\theta}_k = \frac{\sum_0^{k-1} y_{i+1} u_i}{\sum_0^{k-1} u_i^2}.$$ (3.9)

[Hint: If θ° were known, then the minimum cost is σ^2. Hence, irrespective of the prior distribution P, a lower bound on the optimal cost is σ^2. Now use the result of Section 2.]

4. Formulation of bandit problems

Suppose there are N stochastic processes $\{x_n(k)\}$, $k = 0, 1, .. , n = 1,.. ,N$, each taking values in the countable set $\{1, 2,.. \}$. At each time k, we choose a control input u_k from the control set $U = \{1, 2,.. ,N\}$. If $u_k = n$ is chosen, then the state $(x_1(k+1), .. , x_N(k+1))$ at time $k+1$ is given by

$$x_m(k+1) = x_m(k), \quad \text{for } m \neq n,$$ (4.1)

$$P(x_n(k+1) = j) = P_{x_n(k),j},$$ (4.2)

where $P = \{P_{i,j}, 1 \le i, j < \infty\}$ is a prespecified transition probability matrix. The state is observed, and the goal is to choose $\{u_k\}$ so that

$$\sum_{k=0}^{\infty} \beta^k R(x_{u(k)}(k))$$ (4.3)

is maximized, where R is a bounded reward function and $0 < \beta < 1$ is a discount factor.

To recapitulate, at each k, one of the N processes is selected to evolve in a Markovian fashion, whereas the remaining $N-1$ processes remain frozen. If process n is chosen at k, then the reward $R(x_n(k))$ is obtained. The goal is to maximize the expected total discounted reward.

By Exercise (8.2.26), the value function W is the unique solution of the dynamic programming equation

$$W(x_1, \ldots, x_N) = \max_{u \in \{1, \ldots, N\}} \{R(x_u)$$

$$+ \beta \sum_{j=1}^{\infty} P_{x_u, j} \, W(x_1, \ldots, x_{u-1}, j, x_{u+1}, \ldots, x_N)\}. \quad (4.4)$$

If $u = g(x_1, \ldots, x_N)$ attains the maximum, then g is an optimal stationary policy.

Before we begin the analysis of the bandit problem, we discuss two examples which can be modeled as bandit problems.

There are N slot machines, $M_1, \ldots, M_N$. For machine M_i, the probability of success on a play is θ_i, and the probability of failure is $(1 - \theta_i)$. A success yields 1 unit of reward, while a failure yields 0 units. The probabilities of success $\theta_1, \ldots, \theta_N$ are independent random variables with prior distributions $P_1(d\theta_1), \ldots, P_N(d\theta_N)$. At each k we can play one machine, denoted M_u. If r_k is the reward obtained at time k, the goal is to maximize

$$E \sum_{0}^{\infty} \beta^k \, r_k. \quad (4.5)$$

This Bayesian adaptive control problem gives rise to a bandit problem. Indeed, let $P_1^k, \ldots, P_N^k$ be the conditional probability distributions of the success probability of $M_1, \ldots, M_N$ given the observed past history up to time k. Suppose that at time k we choose M_n for a play. Then at time $k+1$, the conditional probabilities will be:

$$P_m^{k+1} = P_m^k, \quad \text{for } m \neq n, \quad (4.6)$$

$$P_n^{k+1}(d\theta_n)$$

$$= \frac{\theta_n P_n^k(d\theta_n)}{\int_0^1 \theta P_n^k(d\theta)} \quad \text{with prob } \int_0^1 \theta P_n^k(d\theta) \quad (4.7)$$

$$= \frac{(1-\theta_n) P_n^k(d\theta_n)}{\int_0^1 (1-\theta) P_n^k(d\theta)} \quad \text{with prob } 1 - \int_0^1 \theta P_n^k(d\theta). \quad (4.8)$$

If M_m is not played, its conditional probability distribution of the the success probability remains unchanged, implying (4.6). If M_n is played, it yields a success with probability $\int_0^1 \theta P_n(d\theta)$, and by Bayes' rule, its conditional probability distribution is updated according to (4.7). Similarly, a failure occurs on machine M_n with the complementary probability, and the corresponding updating rule is given by (4.8). Also, the expected reward for M_n is

$$R(P_n^k) := \int_0^1 \theta P_n^k(d\theta).$$

The goal is to maximize

$$E \sum_0^\infty \beta^k R(P_{u(k)}^k), \tag{4.9}$$

where $u(k) \in \{1,\ldots,N\}$ is the machine played at time k.

Once we identify P_i^k with $x_i(k)$, we see that the problem of maximizing (4.9) for the system (4.6)-(4.8) is similar to the problem (4.1)-(4.3) posed earlier. The precise correspondence is given in the next two exercises.

Exercise (4.10)
Show that given the prior pd P_n, the random variables P_n^k, $k = 0,1,\ldots$ take values in a countable set.
[Hint: Given P_n^k, P_n^{k+1} can assume one of only three possible values (4.6), (4.7), or (4.8).]

Exercise (4.11)
By Exercise (8.2.5), for every finite horizon T, the Bayesian adaptive control problem with reward criterion

$$E \sum_0^T \beta^k r_k,$$

is solved by the equations

$$W_0(P_1,\ldots,P_N) = 0,$$

$$W_{k+1}(P_1,\ldots,P_N) = \max_u \{R(P_u)$$

$$+ \beta \sum_j P_{P_u,j} W_k(P_1,\ldots,P_{u-1},j,P_{u+1},\ldots,P_N)\},$$

where the transition probability $P_{P_u,j}$ is as in (4.7) and (4.8). Show that the infinite horizon problem is solved by (4.4).
[Hint: Let $J^*(P_1,\ldots,P_N)$ be the supremum of (4.5) over all policies,

starting with the initial pds $(P_1, \ldots, P_N)$. Then $J^* \geq W_T$ for all T. Also $\{W_T\}$ is an increasing sequence in T which converges to the solution W of (4.4). Hence $J^* \geq W$. However, $J^* \leq W_T + \dfrac{\beta^T}{1-\beta}$ for all T. Hence $J^* = W$. Now show that the maximizing policy g for (4.4) is optimal by considering

$$E^g \sum_0^\infty \beta^k r_k = E^g \sum_0^\infty \beta^k E^g \{r_k \mid P_1^k, \ldots, P_N^k\}$$

$$= E^g \sum_0^\infty \beta^k R (P_{g(P_1^k, \ldots, P_N^k)})$$

and using the fact that $x_n(k)$ and P_n^k evolve in the same way under g, and so

$$E^g \sum_0^\infty \beta^k R(x_{u(k)}(k)) = E^g \sum_0^\infty \beta^k R(P_{u(k)}^k).]$$

The dual purpose of control is quite clear in this Bayesian adaptive control problem. By playing a slot machine, one acquires more knowledge about its success probability. This is valuable since one should play the machine with the largest success probability. However, to acquire this knowledge is costly, since one has to experiment with the various machines. Thus, the optimal policy must balance the benefits and costs of experimentation.

Bandit problems also arise naturally in the optimal scheduling of jobs to be done by a server. We begin with a simple case. There are N jobs that must be completed by a single server. Job i, $1 \leq i \leq N$, requires z_i (discrete) time units of service where z_i is a random variable with distribution function F_i, i.e., $F_i(n) := Prob(1 \leq z_i \leq n)$. If job i is completed at time k, a reward of $\beta^k r_i > 0$ is received. The problem is to schedule the single server among the n jobs so as to maximize the total expected reward.

Exercise   (4.12)
Show that this problem can be converted into a bandit problem.
[Hint: Let $x_n(k)$ be the number of times job n was worked on prior to k, if it is not completed at time k, and set $x_n(k) := \infty$, if it is completed by time k. Let $u(k) = n \in \{1, \ldots, N\}$ denote the job worked on at time k. Let

$$R(x_n(k)) = 0, \quad \text{if } x_n(k) = +\infty$$

$$= r_n \, P\{z_n = x_n(k)+1 \mid z_n > x_n(k)\},$$

$$\text{if } x_n(k) < +\infty$$

$$= r_n \frac{F_n(x_n(k)) - F_n(x_n(k) + 1)}{1 - F_n(x_n(k))} .]$$

We consider more elaborate scheduling problems. Suppose that there are J nodes in a queuing network. At each node j, $1 \le j \le J$, there are initially n_j waiting customers. A customer at node j requires a random number of units of service with distribution function F_j. If a customer completes his or her current service requirement at node j, the customer moves to node l with probability q_{jl} or departs from the network with probability $1 - \sum_{l=1}^{J} q_{jl}$. Every time a customer's service requirement at node j is completed at time k, a reward $\beta^k r_j > 0$ is received. Suppose that there is a single server who can work on one customer at a time, how should we schedule the server to maximize the total expected reward? It is assumed that all service times are independent and no new customers arrive.

Exercise   (4.13)

Show that the problem above can be coverted to a bandit problem. [Hint: Let $J + 1$ be a fictitious node to which all customers go when they depart from the network. For each customer i, $1 \le i \le \sum_{1}^{J} n_j$, let $x_i(k) :=$ (τ, j), if at time k the customer is at node j and has had τ units of service at node j in the latest visit. Let $u(k) \in U = \{1, 2, .., \sum_{1}^{J} n_j\}$, where $u(k) = i$ means that customer i is served at time k. Finally, let

$$R(x_i(k)) = 0, \quad \text{if } x_i(k) = (\tau, J + 1)$$

$$= r_j \frac{F_j(\tau) - F_j(\tau + 1)}{1 - F_j(\tau)} ,$$

$$\text{if } x_i(k) = (\tau, j) \text{ for } 1 \le j \le J.]$$

5. Deterministic bandit problems

To gain insight consider a bandit problem with two deterministic processes, $\{x_1(k)\}$ and $\{x_2(k)\}$, $k = 0, 1, ..$ with initial values $x_1(0) = 1$ and $x_2(0) = 0$. At each time k we choose a control input $u(k) \in \{1, 2\}$. If $u(k) = 1$, then

$$x_1(k + 1) = x_1(k) + 1, \quad \text{and } x_2(k + 1) = x_2(k).$$

whereas if $u(k) = 2$, then

$$x_1(k + 1) = x_1(k), \quad \text{and } x_2(k + 1) = x_2(k). \tag{5.1}$$

We are given a bounded reward function

$$R: \{0, 1, 2, ..\} \rightarrow (M, \infty),$$

where $M > 0$, and our goal is to choose $\{u(k)\}$ to maximize

$$\sum_0^\infty \beta^k R(x_{u(k)}(k)).$$

Suppose at some time k the state is $(x_1(k), x_2(k)) = (x_1, x_2)$ and control $u(k) = 2$ is optimal. Then at time $k + 1$, the state is again (x_1, x_2) by virtue of (5.1), and so $u(k + 1) = 2$ is again optimal. Continuing, we see that $u(k + l) = 2$ is optimal for $l \geq 0$. Thus, an optimal policy must be of the form

$$u(k) = 1 \quad \text{for } 0 \leq k \leq \tau - 1 \tag{5.2}$$

$$= 2 \quad \text{for } k \geq \tau \tag{5.3}$$

for some $0 \leq \tau \leq + \infty$, and so the maximum reward is

$$\max_{0 \leq \tau \leq +\infty} \{\sum_0^{\tau-1} \beta^k R(\bar{x}_1(k)) + \sum_\tau^\infty \beta^k R(x_2(0))\}, \tag{5.4}$$

where $\{\bar{x}_1(k)\}$ is the sequence $\{x_1(k)\}$ generated when $u(k) = 1$ is exclusively used for all k. Since $x_2(0) = 0$ and $\sum_\tau^\infty \beta^k R(x_2(0)) = \dfrac{\beta^\tau R(0)}{1 - \beta}$ the maximum reward (5.4) is

$$\max_{0 \leq \tau \leq +\infty} \{\sum_0^{\tau-1} \beta^k R(\bar{x}_1(k)) + \frac{\beta^\tau R(0)}{1 - \beta}\}.$$

The policy $u(k) = 2$ for $k \geq 0$, yields the reward $\dfrac{R(0)}{1 - \beta}$. Hence

$$\max_{0 \leq \tau \leq +\infty} \{\sum_0^{\tau-1} \beta^k R(\bar{x}_1(k)) + \frac{\beta^\tau R(0)}{1 - \beta}\}$$

$$> \frac{R(0)}{1 - \beta} \quad \Rightarrow \quad u(0) = 1 \quad \text{is optimal,} \tag{5.5}$$

$$= \frac{R(0)}{1 - \beta} \quad \Rightarrow \quad u(0) = 2 \quad \text{is optimal.} \tag{5.6}$$

Consider the family of problems obtained by varying $R(0)$. Clearly,

$$R(0) > \sup_k R(\bar{x}_1(k)) \quad \Rightarrow \quad (5.6) \text{ holds,} \tag{5.7}$$

$$\leq \inf_k R(\bar{x}_1(k)) \quad \Rightarrow \quad (5.5) \text{ holds.} \tag{5.8}$$

It is easy to check that

$$\rho(R(0))$$

$$:= \max_{0 \le \tau \le +\infty} \{ \sum_0^{\tau-1} \beta^k R(\bar{x}_1(k)) + \frac{\beta^\tau R(0)}{1-\beta} \} - \frac{R(0)}{1-\beta} \tag{5.9}$$

is a continuous, monotone decreasing function of $R(0)$. From (5.7) it follows that if $R(0) > \sup_k R(\bar{x}_1(k))$, then $\rho(R(0)) = 0$, whereas from (5.8) it follows that if $R(0) < \inf_k R(\bar{x}_1(k))$, then $\rho(R(0)) > 0$. Hence there is a critical value R^* such that

$$R(0) < R^* \implies \rho(R(0)) > 0 \text{ and } u(0) = 1 \text{ is optimal,}$$

$$\ge R^* \implies \rho(R(0)) = 0 \text{ and } u(0) = 2 \text{ is optimal.}$$

To determine R^* we proceed as follows. First,

$$\max_{0 \le \tau \le +\infty} \{ \sum_0^{\tau-1} \beta^k R(\bar{x}_1(k)) + \frac{\beta^\tau R^*}{1-\beta} \} = \frac{R^*}{1-\beta}, \tag{5.10}$$

and so

$$\max_{1 \le \tau \le +\infty} \{ \sum_0^{\tau-1} \beta^k R(\bar{x}_1(k)) + \frac{\beta^\tau R^*}{1-\beta} \} \le \frac{R^*}{1-\beta}. \tag{5.11}$$

On the other hand, for $R(0) < R^*$, the maximum in (5.9) cannot be attained for $\tau = 0$, and so

$$\max_{1 \le \tau \le +\infty} \{ \sum_0^{\tau-1} \beta^k R(\bar{x}_1(k)) + \frac{\beta^\tau R(0)}{1-\beta} \}$$

$$> \frac{R(0)}{1-\beta} \text{ for } R(0) < R^*. \tag{5.12}$$

Again, the difference between the left-and right-hand sides of (5.12) is a continuous, monotone decreasing function of $R(0)$, and by taking the limit as $R(0) \to R^*$, and using (5.11), we deduce that

$$\max_{1 \le \tau \le +\infty} \{ \sum_0^{\tau-1} \beta^k R(\bar{x}_1(k)) + \frac{\beta^\tau R^*}{1-\beta} \} = \frac{R^*}{1-\beta}. \tag{5.13}$$

Hence

$$\sum_0^{\tau-1} \beta^k R(\bar{x}_1(k)) + \frac{\beta^\tau R^*}{1-\beta} \le \frac{R^*}{1-\beta} \quad \text{for } 1 \le \tau \le +\infty,$$

and so

$$\frac{\sum_0^{\tau-1} \beta^k R(\bar{x}_1(k))}{[\frac{1-\beta^\tau}{1-\beta}]} \le R^* \quad \text{for } 1 \le \tau \le +\infty.$$

Therefore

$$\max_{1 \leq \tau \leq +\infty} \frac{\sum_{0}^{\tau-1} \beta^k R(\bar{x}_1(k))}{\sum_{0}^{\tau-1} \beta^k} \leq R^*.$$

On the other hand, the maximizing τ of (5.13) attains equality, and so

$$\max_{1 \leq \tau \leq +\infty} \frac{\sum_{0}^{\tau-1} \beta^k R(\bar{x}_1(k))}{\sum_{0}^{\tau-1} \beta^k} = R^*.$$

Thus we have a simple rule for selecting the value of $u(0)$,

$$\max_{1 \leq \tau \leq +\infty} \frac{\sum_{0}^{\tau-1} \beta^k R(\bar{x}_1(k))}{\sum_{0}^{\tau-1} \beta^k} > R(0) \implies u(0) = 1 \text{ is optimal}$$

$$\leq R(0) \implies u(0) = 2 \text{ is optimal}.$$

The quantity

$$\gamma(\{\bar{x}_1(k)\}) := \max_{1 \leq \tau \leq +\infty} \frac{\sum_{0}^{\tau-1} \beta^k R(\bar{x}_1(k))}{\sum_{0}^{\tau-1} \beta^k} \tag{5.14}$$

is called the **Gittins index** of the sequence $\{\bar{x}_1(k)\}_{k=0}^{\infty}$.

Similarly, let $\{\bar{x}_2(k)\}$ be the sequence generated when $u(k) = 2$ for all k. Since $\bar{x}_2(k) = R(0)$ for all k,

$$\gamma(\{\bar{x}_2(k)\}) = R(0).$$

Hence the optimal choice of $u(0)$ is given by the following rule, which merely compares the Gittins index of the two sequences:

$$\gamma(\{\bar{x}_1(k)\}) > \gamma(\{\bar{x}_2(k)\}) \implies u(0) = 1 \text{ is optimal} \tag{5.15}$$

$$\leq \gamma(\{\bar{x}_2(k)\}) \implies u(0) = 2 \text{ is optimal}.$$

In view of this form, the optimal policy described in (5.15) is said to be given by an **index rule**.

Note that the index of a process is defined without reference to the other process, and the index rule is to choose the process with the largest current index. In fact even more is true. Suppose that $\tau^* \geq 1$ maximizes (5.14) and $R(0) < R^*$. Then, since the same τ^* also maximizes (5.9), we know from (5.2) and (5.3), that the optimal policy will have

$$u(0) = .. = u(\tau^* - 1) = 1. \tag{5.16}$$

Thus when computing the index and using the index rule (5.15) we determine the optimal choice not only for $k = 0$ but also $0 \leq k \leq \tau^* - 1$.

We can also convince ourselves of the optimality of (5.16) by a different argument, as the following two exercises show.

Exercise (5.17)

Let $\tau^* \geq 1$ attain the maximum in (5.14). Show that

$$\sum_{\sigma}^{\tau^* - 1} \beta^{k - \sigma} R(\bar{x}_1(k)) \geq \gamma(\{\bar{x}_1(k)\}_{k=0}^{\infty}) \sum_{\sigma}^{\tau^* - 1} \beta^{k - \sigma}$$

for $0 \leq \sigma \leq \tau^* - 1$.

[Hint: Use the definition (5.14) and simple algebra.]

Exercise (5.18)

Let $\tau^* \geq 1$ attain the maximum in (5.14) and suppose that $R(0) \leq \gamma(\{\bar{x}_1(k)\}_{k=0}^{\infty})$. Show that (5.16) is part of an optimal policy by using the conclusion of Exercise (5.17).

[Hint: Shift the time origin to σ where $0 \leq \sigma \leq \tau^* - 1$. Then the sequence of rewards obtained by choosing the first process is $\{\bar{x}_1(\sigma), \bar{x}_1(\sigma+1), .. \}$ whose index is $\gamma(\{\bar{x}_1(k)\}_{k=\sigma}^{\infty})$. Since the index rule is optimal,

$$\gamma(\{\bar{x}_1(k)\}_{k=\sigma}^{\infty}) > R(0) \quad \Rightarrow \quad u(\sigma) = 1,$$

but this inequality follows from Exercise (5.17) since $\gamma(\{\bar{x}_1(k)\}_{k=0}^{\infty}) > R(0)$.]

The proof of the optimality of the index rule can be extended in a straightforward way to the case where the process $\{x_1(k)\}$ is random, as the following two exercises show.

Exercise (5.19)

Consider the system where, if $u(k) = 1$, the processes evolve according to

$$x_1(k+1) = j \quad \text{with probability } P_{x_1(k),j},$$

$$x_2(k+1) = x_2(k);$$

and if $u(k) = 2$,

$$x_1(k+1) = x_1(k),$$

$$x_2(k+1) = x_2(k).$$

The initial states are $(x_1(0), x_2(0))$, and $P = [P_{ij}]$ is a prespecified transition probability matrix. The goal is to maximize

$$E \sum_0^\infty \beta^k R(x_{u(k)}(k))$$

where R is a bounded reward function. Show that an index rule is optimal.

[Hint: Consider a process $\{\bar{x}_1(k)\}$ which is a Markov chain with transition probability matrix P. For the initial state $\bar{x}_1(0) = i$, define the index

$$\gamma(i) := \max_{1 \le \tau \le \infty} \frac{E \sum_0^{\tau-1} \beta^k R(\bar{x}_1(k))}{E \sum_0^{\tau-1} \beta^k}, \qquad (5.20)$$

where τ is a stopping time. (A nonnegative random variable τ is said to be a *stopping time* of the increasing sequence of σ-algebras $\{F_k\}$ if the event $\{\tau \le k\} \in F_k$ for every k.) Now use exactly the same arguments as above to show that

$$\gamma(\{\bar{x}_1(k)\}) > R(x_2(0)) \implies u(k) = 1 \text{ is optimal,}$$

$$\le R(x_2(0)) \implies u(k) = 2 \text{ is optimal.}]$$

Exercise (5.21)

In the preceding exercise $\{\bar{x}_1(k)\}$, the sequence obtained by $u(k) \equiv 1$, was Markovian. Suppose now that we drop the Markovian assumption and allow the evolution of $\{\bar{x}_1(k)\}$ to depend on the entire past according to:

If $u(k) = 1$, then $x_1(k+1) = j$

with probability $P(j \mid x_1(0), \ldots, x_1(k))$,

If $u(k) = 2$, then $x_1(k+1) = x_1(k)$,

where the conditional probability P is specified. Show that the index rule of Exercise (5.19) is still optimal, where the maximum in (5.20) is taken over stopping times τ of the sequence $\{F_k\}$ of σ-algebras, where $F_k := \sigma(\bar{x}_1(0), \ldots, \bar{x}_1(k))$.

[Hint: Simply define the new state $\hat{x}_1(k) := (x_1(0), \ldots, x_1(k))$, which lies in a countable state space and which evolves in a Markovian way.]

In the preceding development, the sequence $\{x_2(k)\}$ is constant. To generalize this, suppose that when $u(k) = 1$, the evolution of the state proceeds according to

$$x_1(k+1) = x_1(k) + 1,$$

$$x_2(k+1) = x_2(k),$$

and when $u(k) = 2$, the evolution proceeds according to

$$x_1(k+1) = x_1(k),$$

$$x_2(k+1) = x_2(k) - 1.$$

and the initial conditions are $(x_1(0), x_2(0))$ with $x_1(0) > x_2(0)$. The goal is again to maximize

$$\sum_0^\infty \beta^k R(x_{u(k)}(k))$$

where R is bounded. (The reader should note that this formulation is quite general.)

Let $\{\bar{x}_1(k)\} = \{x_1(0) + k\}$ and $\{\bar{x}_2(k)\} = \{x_2(0) - k\}$ be the sequences generated by choosing either the first or the second process exclusively. In view of the preceding results it seems natural to suspect that an optimal choice is given by the index rule

$$\gamma(\{\bar{x}_1(k)\}_{k=0}^\infty) > \gamma(\{\bar{x}_2(k)\}_{k=0}^\infty) \text{ then choose } u(0) = 1, \qquad (5.22)$$

$$\leq \gamma(\{\bar{x}_2(k)\}_{k=0}^\infty) \text{ then choose } u(0) = 2, \qquad (5.23)$$

where γ is defined in (5.14). Moreover, we also know from Exercise (5.17) that if (5.22) holds and if τ^* attains the maximum in (5.14), then

$$\gamma(\{\bar{x}_1(k)\}_{k=\sigma}^\infty) > \gamma(\{\bar{x}_2(k)\}_{k=0}^\infty) \quad \text{for } 0 \leq \sigma \leq \tau^* - 1$$

and so, by translating the time origin as in Exercise (5.18), it is natural to expect that

$$u(k) = 1 \quad \text{for } 0 \leq k \leq \tau^* - 1 \qquad (5.24)$$

is also optimal. We use this insight in fashioning a proof of optimality of the index rule (5.22) and (5.23). Suppose τ^* is as above and the inequality in (5.22) holds. We will show that (5.24) is optimal.

Evidently, any policy $\pi = \{u(k)\}$ will interweave the two sequences

$$\{R(\bar{x}_1(0)), R(\bar{x}_1(1)), R(\bar{x}_1(2)), \ldots \}, \qquad (5.25)$$

$$\{R(\bar{x}_2(0)), R(\bar{x}_2(1)), R(\bar{x}_2(2)), \ldots \},$$

into a sequence

$$\{R(z(0)), R(z(1)), R(z(2)), \ldots \}, \qquad (5.26)$$

where $z(k) := x_{u(k)}(k)$, and give the reward

$$V(\pi) := \sum_0^\infty \beta^k \, R(z(k)).$$

Let T be such that $z(T) = \bar{x}_1(\tau^*-1)$ [with $T = \infty$ if there is no such element in the sequence (5.26)]. Thus the sequence (5.26) is of the form

$$\{R(\bar{x}_2(0)), R(\bar{x}_2(1)), \ldots, R(\bar{x}_2(k_0-1)), R(\bar{x}_1(0)),$$

$$R(\bar{x}_2(k_0)), R(\bar{x}_2(k_0+1)), \ldots, R(\bar{x}_2(k_1-1)), R(\bar{x}_1(1)),$$

$$R(\bar{x}_2(k_1)), R(\bar{x}_2(k_1+1)), \ldots, R(\bar{x}_2(k_2-1)), R(\bar{x}_1(2)), \ldots,$$

$$R(\bar{x}_2(k_{\tau^*-1}-1)), R(\bar{x}_1(\tau^*-1)),$$

$$R(z(T+1)), R(z(T+2)), \ldots\}, \tag{5.27}$$

where k_i is the number of elements from the sequence (5.25) that precede $R(\bar{x}_1(i))$ in the sequence (5.26).

Consider now an alternative policy $\tilde{\pi}$ which yields the sequence of rewards,

$$\{R(\bar{x}_1(0)), R(\bar{x}_1(1)), \ldots, R(\bar{x}_1(\tau^*-1)), R(\bar{x}_2(0)),$$

$$R(\bar{x}_2(1)), \ldots, R(\bar{x}_2(k_{\tau^*-1}-1)),$$

$$R(z(T+1)), R(z(T+2)), \ldots\},$$

and let $V(\tilde{\pi})$ be the total discounted reward which results. This policy follows the index rule (5.24) for $k = 0, \ldots, \tau^*-1$ and so we expect that $V(\tilde{\pi}) \geq V(\pi)$, as we verify next. Clearly,

$$V(\pi) = \sum_0^{k_0-1} \beta^k \, R(\bar{x}_2(k)) + \beta^1 \sum_{k_0}^{k_1-1} \beta^k \, R(\bar{x}_2(k))$$

$$+ \beta^2 \sum_{k_1}^{k_2-1} \beta^k \, R(\bar{x}_2(k)) + \ldots$$

$$+ \beta^{\tau^*-1} \sum_{k_{\tau^*-2}}^{k_{\tau^*-1}-1} \beta^k \, R(\bar{x}_2(k))$$

$$+ \sum_0^{\tau^*-1} \beta^{k_t+t} \, R(\bar{x}_1(t)) + \sum_{T+1}^\infty \beta^k \, R(z(k)),$$

while

$$V(\tilde{\pi}) = \sum_0^{\tau^*-1} \beta^k \, R(\bar{x}_1(k)) + \beta^{\tau^*} \sum_0^{k_{\tau^*-1}-1} \beta^k \, R(\bar{x}_2(k))$$

$$+ \sum_{T+1}^\infty \beta^k \, R(z(k))$$

Hence $V(\tilde{\pi}) - V(\pi)$ can be written as

$$V(\tilde{\pi}) - V(\pi) = V_1 - V_2,$$

where

$$V_1 := \sum_{t=0}^{\tau^*-1} (\beta^t - \beta^{k_t+t}) \, R(\bar{x}_1(t))$$

$$V_2 := \sum_{k=0}^{k_0-1} (\beta^k - \beta^{\tau^*+k}) \, R(\bar{x}_2(k))$$

$$+ \sum_{k_0}^{k_1-1} (\beta^{k+1} - \beta^{\tau^*+k}) \, R(\bar{x}_2(k)) + \ldots$$

$$+ \sum_{k_{\tau^*-2}}^{k_{\tau^*-1}-1} (\beta^{k+\tau^*-1} - \beta^{\tau^*+k}) \, R(\bar{x}_2(k)).$$

Now,

$$V_1 = \sum_{0}^{\tau^*-1} (1 - \beta^{k_t}) \, \beta^t \, R(\bar{x}_1(t))$$

$$= \sum_{t=0}^{\tau^*-1} (\sum_{n=0}^{t} (\beta^{k_{n-1}} - \beta^{k_n}) \, \beta^t \, R(\bar{x}_1(t))) \quad \text{(with } k_{-1} := 0)$$

$$= \sum_{n=0}^{\tau^*-1} (\beta^{k_{n-1}} - \beta^{k_n}) \sum_{k=n}^{\tau^*-1} \beta^k \, R(\bar{x}_1(k))$$

$$\geq \sum_{n=0}^{\tau^*-1} (\beta^{k_{n-1}} - \beta^{k_n}) \sum_{k=n}^{\tau^*-1} \beta^k \, \gamma(\{\bar{x}_1(j)\}_{j=0}^{\infty}) \quad \text{[by (5.17)]}$$

$$= \gamma(\{\bar{x}_1(j)\}_{j=0}^{\infty}) \sum_{k=0}^{\tau^*-1} \beta^k \sum_{n=0}^{k} (\beta^{k_{n-1}} - \beta^{k_n})$$

$$= \gamma(\{\bar{x}_1(j)\}_{j=0}^{\infty}) \sum_{k=0}^{\tau^*-1} \beta^k \, (1 - \beta^{k_t}).$$

Also,

$$V_2 = (1 - \beta^{\tau^*}) \sum_{0}^{k_0-1} \beta^k \, R(\bar{x}_2(k))$$

$$+ (\beta - \beta^{\tau^*}) \sum_{k_0}^{k_1-1} \beta^k \, R(\bar{x}_2(k)) + \ldots$$

$$+ (\beta^{\tau^*-1} - \beta^{\tau^*}) \sum_{k_{\tau^*-2}}^{k_{\tau^*-1}-1} \beta^k \, R(\bar{x}_2(k))$$

$$= \sum_{n=0}^{\tau^*-1} (\beta^n - \beta^{\tau^*}) \sum_{k=k_{n-1}}^{k_n-1} \beta^k \, R(\bar{x}_2(k))$$

$$= \sum_{n=0}^{\tau^*-1} (\beta^n - \beta^{\tau^*}) \left[\sum_{k=0}^{k_n-1} \beta^k \, R(\bar{x}_2(k)) - \sum_{k=0}^{k_{n-1}-1} \beta^k \, R(\bar{x}_2(k)) \right]$$

$$= \sum_{n=0}^{\tau^*-1} (\beta^n - \beta^{n+1}) \sum_{k=0}^{k_n-1} \beta^k \, R(\bar{x}_2(k))$$

$$\leq \sum_{n=0}^{\tau^*-1} (\beta^n - \beta^{n+1}) \sum_{k=0}^{k_n-1} \beta^k \, \gamma(\{\bar{x}_2(j)\}_{j=0}^\infty)$$

$$= \gamma(\{\bar{x}_2(j)\}_{j=0}^\infty) \sum_{n=0}^{\tau^*-1} (\beta^n - \beta^{n+1}) \frac{(1 - \beta^{k_n})}{(1 - \beta)}$$

$$= \gamma(\{\bar{x}_2(j)\}_{j=0}^\infty) \sum_{n=0}^{\tau^*-1} \beta^n \, (1 - \beta^{k_n}).$$

[The last inequality is obtained from the fact that the index is defined as the maximum over the variable k_n of the ratio.] Hence

$$V_1 - V_2 \geq [\gamma(\{\bar{x}_1(j)\}_{j=0}^\infty) - \gamma(\{\bar{x}_2(j)\}_{j=0}^\infty)] \sum_{n=0}^{\tau^*-1} \beta^n \, (1 - \beta^{k_n})$$

$$\geq 0.$$

This shows that $V(\tilde{\pi}) \geq V(\pi)$. Hence any policy π can be improved by following the index rule up to time τ^*-1. Repeating the argument starting at τ^* and continuing in this manner, we conclude that the index rule is optimal. That is the following rule is optimal:

If $\gamma(\{x_1(k), x_1(k)+1, \ldots\}) > \gamma(\{x_2(k), x_2(k)-1, \ldots\})$,

then choose $u(k) = 1$

$\leq \gamma(\{x_2(k), x_2(k)-1, \ldots\})$,

then choose $u(k) = 2$.

6. Optimality of the index rule for the bandit problem

With the insights of the deterministic problem treated in the preceding section and Exercise (5.19), we are now ready to tackle the bandit problem posed in Section 4.

Theorem   (6.1)
(1) Let $\{\bar{x}(k)\}$ be a Markov chain with transition probabilities $\{P_{ij}\}$, $1 \le i, j \le \infty$. Define

$$\gamma(i) := \sup_{1 \le \tau \le \infty} \frac{E\{\sum_0^{\tau-1} \beta^k R(\bar{x}(0)) \mid \bar{x}(0) = i\}}{E\{\sum_0^{\tau-1} \beta^k \mid \bar{x}(0) = i\}}, \tag{6.2}$$

where the supremum is taken over stopping times τ of $\{F_k\}$ with $F_k := \sigma(\bar{x}(0), \ldots, \bar{x}(k))$. Then

$$\tau_i := \inf \{k > 0 \mid \gamma(\bar{x}(k)) < \gamma(i)\} \qquad (\inf \varnothing := \infty) \tag{6.3}$$

attains the supremum in (6.2).

(2) For the bandit problem the index rule policy that chooses the process with largest current index, i.e.,

$$u(k) = Arg \max_{1 \le m \le N} \gamma(x_m(k)), \tag{6.4}$$

is optimal.

Proof
The proof of Exercise (5.19) demonstrates that the stopping time of (6.3) maximizes (6.2), and so we need only prove (2). Let π be the index rule policy for which

$$u(k) = n \quad \text{if} \quad n = \min \{m \mid \gamma(x_m(k)) = \max_{1 \le k \le N} \gamma(x_k(k))\}$$

i.e., π satisfies (6.4) and breaks ties between competing maximizers by choosing the smallest one. We now show that π is optimal.

Suppose that π chooses process i at time $k = 0$ [i.e., $u(0) = i$]. Consider $j \ne i$ and let $\hat{\pi}$ be a nonstationary policy that chooses process j at time 0 and thereafter proceeds according to π. Under $\hat{\pi}$, process j will be chosen at times $0, 1, \ldots, \tau_j - 1$ where

$$\tau_j := \min \{k \ge 1 \mid \gamma(x_j(k)) \le \gamma(x_i(0))\} \quad \text{if} \quad j > i$$

$$:= \min \{k \ge 1 \mid \gamma(x_j(k)) < \gamma(x_i(0))\} \quad \text{if} \quad j < i.$$

Thereafter, under $\hat{\pi}$, process i will be chosen at least for times $k = \tau_j$, $\tau_j + 1, \ldots, \tau_j + \tau_i - 1$, where

$$\tau_j + \tau_i := \min \{k \ge \tau_j + 1 \mid \gamma(x_i(k)) \le \gamma(x_i(\tau_j)) = \gamma(x_i(0))\}.$$

Consider now yet another policy $\tilde{\pi}$, which chooses process i at times $k = 0, 1, \ldots, \tau_i - 1$, then chooses process j at times $k = \tau_i, \tau_i + 1, \ldots, \tau_i + \tau_j - 1$ and thereafter chooses the same processes as $\hat{\pi}$.

Denoting expectations under policies $\hat{\pi}$ and $\tilde{\pi}$ by $E^{\hat{\pi}}$ and $E^{\tilde{\pi}}$, respectively, we have

$$E^{\tilde{\pi}} \sum_{0}^{\infty} \beta^k \, R(x_{u(k)} \, (k)) - E^{\hat{\pi}} \sum_{0}^{\infty} \beta^k \, R(x_{u(k)}(k))$$

$$= E^{\tilde{\pi}} \, [\sum_{0}^{\tau_i - 1} \beta^k \, R(x_i(k)) + \sum_{\tau_i}^{\tau_i + \tau_j - 1} \beta^k \, R(x_j(k))]$$

$$- E^{\hat{\pi}} \, [\sum_{0}^{\tau_j - 1} \beta^k \, R(x_j(k)) + \sum_{\tau_k}^{\tau_k + \tau_i - 1} \beta^k \, R(x_i(k))]$$

$$= E^{\tilde{\pi}} \sum_{0}^{\tau_i - 1} \beta^k \, R(x_i(k)) + E^{\tilde{\pi}} \, [\beta^{\tau_i} \, E^{\hat{\pi}} \sum_{0}^{\tau_j - 1} \beta^k \, R(x_j(k))]$$

$$- E^{\hat{\pi}} \sum_{0}^{\tau_j - 1} \beta^k \, R(x_j(k)) - E^{\hat{\pi}} \, [\beta^{\tau_j} \, E^{\tilde{\pi}} \sum_{0}^{\tau_i - 1} R(x_i(k))]$$

$$= E^{\tilde{\pi}} \sum_{0}^{\tau_i - 1} \beta^k \, R(x_i(k)) \, [1 - E^{\hat{\pi}} \beta^{\tau_j}]$$

$$- E^{\hat{\pi}} \sum_{0}^{\tau_j - 1} \beta^k \, R(x_j(k)) \, [1 - E^{\tilde{\pi}} \beta^{\tau_i}]$$

$$\geq \gamma(x_i(0)) \, E^{\tilde{\pi}} \sum_{0}^{\tau_i - 1} \beta^k \, [1 - E^{\hat{\pi}} \beta^{\tau_j}]$$

$$- \gamma(x_j(0)) \, E^{\hat{\pi}} \sum_{0}^{\tau_j - 1} \beta^k \, [1 - E^{\tilde{\pi}} \beta^{\tau_i}]$$

$$= [\gamma(x_i(0)) - \gamma(x_j(0))] \, (1 - \beta) \, E^{\tilde{\pi}} \sum_{0}^{\tau_i - 1} \beta^k \, E^{\hat{\pi}} \sum_{0}^{\tau_j - 1} \beta^k \geq 0.$$

Therefore $\tilde{\pi}$, which coincides with π for $0 \leq k \leq \tau_i - 1$, is an improvement over $\hat{\pi}$. If at time τ_i, $\tilde{\pi}$ does not choose the process that π chooses, then by shifting the time origin to τ_i, we can repeat the argument above to obtain yet another improvement which coincides with π even at time τ_i. In this way we can obtain policies which coincide with π over arbitrarily large initial segments of time, and which are all improvements over $\hat{\pi}$. We conclude that π itself is better than $\hat{\pi}$.

This shows that to follow π at time $k = 0$ is optimal. By the fact that a stationary policy is optimal, we can deduce that π is an optimal policy.   □

7. An algorithm for computing the index

We present an algorithm for computing the Gittins index of a finite state Markov process. Consider the Markov chain $\{x(k)\}$ with transition probability matrix $P = \{P_{ij}\}$, $1 \le i, j \le I$. $R(i)$ is the reward obtained in state i. The Gittins index in state i is defined as

$$\gamma(i) := \max_{1 \le \tau \le \infty} \frac{E\{\sum_0^{\tau-1} \beta^k R(x(k)) \mid x(0) = i\}}{E\{\sum_0^{\tau-1} \beta^k \mid x(0) = i\}}. \qquad (7.1)$$

Suppose the states with the largest index (there may be more than one such state) are grouped into one set I_0, the states with the next largest index are grouped into set I_1, and so on into sets $I_0, I_1, \ldots, I_r$. From (6.3), we know that for $i \in I_s$,

$$\tau_i := \min \{k \ge 1 \mid x(k) \in \bigcap_{p=s}^r I_p\} \qquad (7.2)$$

attains the maximum in (7.1).

Exercise   (7.3)
Let $A \subset \{1, \ldots, I\}$ and let $\tau := \min \{k \ge 1 : x(k) \in A^C\}$. Define

$$W_j(A) := E\{\sum_0^{\tau-1} \beta^k R(x(k)) \mid x(0) = j\}.$$

Show that $\{W_j(A), j = 1, \ldots, I\}$ are the unique solutions of the equations,

$$W_j(A) = R(j) + \beta \sum_{k \in A} P_{jk} W_k(A), \quad j = 1, 2, \ldots, I,$$

or, in vector notation, with $W(A) := (W_1(A), \ldots, W_I(A))^T$, $R := (R(1), \ldots, R(I))^T$ and

$$P_{ij}^A := P_{ij} \quad \text{for } j \in A$$
$$:= 0 \quad \text{for } j \in A^C,$$

we get

$$W(A) = [I - \beta P^A]^{-1} R.$$

Similarly, if $V_j(A) := E\{\sum_0^{\tau-1} \beta^k \mid x(0) = j\}$, show that

$$V(A) = [I - \beta P^A]^{-1} (1, 1, \ldots, 1)^T.$$

Using the notation of this exercise, it follows from (7.1) and (7.2) that

$$\gamma(j) = \frac{W_j \left(\bigcup\limits_{p=0}^{s-1} I_p \right)}{V_j \left(\bigcup\limits_{p=0}^{s-1} I_p \right)} \quad \text{for } j \in I_s \tag{7.4}$$

$$\geq \frac{W_j \left(\bigcup\limits_{p=0}^{l} I_p \right)}{V_i \left(\bigcup\limits_{p=0}^{l} I_p \right)} \quad \text{for } l = 0,1,\ldots,r, \quad \text{for all } j. \tag{7.5}$$

Hence

$$j \in I_s \iff \frac{W_j \left(\bigcup\limits_{p=0}^{s-1} I_p \right)}{V_j \left(\bigcup\limits_{p=0}^{s-1} I_p \right)} = \gamma(j)$$

$$= \max_{i \in \bigcap\limits_{p=0}^{s-1} I_p^C} \gamma(i) = \max_{i \in \bigcap\limits_{p=0}^{s-1} I_p^C} \frac{W_i \left(\bigcup\limits_{p=0}^{s-1} I_p \right)}{V_j \left(\bigcup\limits_{p=0}^{s-1} I_p \right)}.$$

Thus we obtain the following recursive formula for determining I_s given $I_0, I_1, \ldots, I_{s-1}$,

$$I_s = \{ j \in \bigcap\limits_{p=0}^{s-1} I_p^C \mid$$

$$\frac{W_j \left(\bigcup\limits_{p=0}^{s-1} I_p \right)}{V_j \left(\bigcup\limits_{p=0}^{s-1} I_p \right)} = \max_{i \in \bigcap\limits_{p=0}^{s-1} I_p^C} \frac{W_i \left(\bigcup\limits_{p=0}^{s-1} I_p \right)}{V_i \left(\bigcup\limits_{p=0}^{s-1} I_p \right)} \}. \tag{7.6}$$

The reader can easily check that

$$I_0 = \{ j \mid R(j) = \max_{1 \leq i \leq I} R(i) \}. \tag{7.7}$$

The formulas (7.7) and (7.6), together with the computations of W_j and V_j in Exercise (7.3), give us a recursive algorithm for generating the Gittins indices of all the states.

8. A comment on Bayesian adaptive control

We show in this section that an adaptive control policy which is optimal for a Bayesian formulation need not be self-tuning. The converse, that self-tuning policies need not be Bayesian optimal, is clearly true, as can be seen from simple examples.

We will consider a simple bandit problem of the type discussed in Section 4. Suppose there are only two slot machines, M_1 and M_2. For slot machine M_i, the success probability is θ_i. Suppose that $\theta_2 > 0$ is known and θ_1 is uniformly distributed on [0, 1].

Exercise   (8.1)
Suppose that at time k, M_1 has been played a total of $(s_k + f_k)$ times, with s_k successes and f_k failures. Show that the posterior distribution of θ_1 is given by

$$P^k(B) := P(\theta_1 \in B \mid s_k, f_k)$$

$$= \frac{\Gamma(s_k + f_k + 2)}{\Gamma(s_k + 1)\,\Gamma(f_k + 1)} \int_B \theta^{s_k}\,(1 - \theta)^{f_k}\,d\theta, \tag{8.2}$$

where Γ is the Gamma function. Thus the conditional expectation and variance of θ_1, at time k, are

$$E\,\{\theta_1 \mid s_k, f_k\} = \frac{s_k + 1}{s_k + f_k + 2}\,, \tag{8.3}$$

$$\text{var}\,\{\theta_1 \mid s_k, f_k\} = \frac{(s_k + 1)\,(f_k + 1)}{(s_k + f_k + 2)^2\,(s_k + f_2 + 3)}\,. \tag{8.4}$$

Let $u(k) = 1$ or 2 according as M_1 or M_2 is played at time k. The optimal policy is given by the index rule:

$$\text{If }\ \gamma(P^k) > \theta_2, \quad \text{then choose}\ \ u(k) = 1 \tag{8.5}$$

$$\leq \theta_2, \quad \text{then choose}\ \ u(k) = 2,$$

where $\gamma(P^k)$ is the Gittins index of M_1 at time k, and θ_2 is the Gittins index of M_2 at all times (as is easily verified).

Since this problem reduces to the same one as in Exercise (5.19), the optimal policy is of the form

$$u(k) = 1 \quad \text{for}\ k \leq \tau - 1$$

$$= 2 \quad \text{for}\ k > \tau - 1,$$

for some stopping time τ. Therefore,

$$\lim_{k \to \infty}\ u(k)$$

exists a.s. as a random variable.

If $P(\lim_{k \to \infty} u_k = 1 \mid \theta_1 > \theta_2) = 1$ and $P(\lim_{k \to \infty} u_k = 2 \mid \theta_1 \leq \theta_2) = 1$, then the optimal policy obtained through the Bayesian formulations is self-tuning, since it asymptotically converges to the better machine.

However we will now show that $P(\lim_{k \to \infty} u(k) = 2 \mid \theta_1 > \theta_2) > 0$, indicating that the Bayesian optimal policy is not self-tuning.

From Exercise (8.1), it is clear that P^k depends on (s_k, f_k) and we shall therefore write it as $P^k = \pi(s_k, f_k)$ to exhibit this dependence. Let

$$\Pi := \{\pi(s, f) \mid \gamma(\pi(s, f)) \le \theta_2\};$$

then the optimal policy (8.5) is of the form

If $P^k \notin \Pi$,   then choose $u_k = 1$

  $\in \Pi$,   then chose $u_k = 2$.

For any s,

$$\lim_{f \to \infty} \frac{s+1}{s+f+2} = 0, \quad \lim_{f \to \infty} \frac{(s+1)(f+1)}{(s+f+2)^2 (s+f+3)} = 0,$$

and so by using Chebyshev's inequality, (8.3) and (8.4), $\lim_{f \to \infty} \pi(s, f)$ is a pd with unit mass at 0. It is therefore also reasonable to suppose that (we will not show this)

$$\lim_{f \to \infty} \gamma(\pi(s, f)) = 0.$$

Hence for any s,

$\pi(s, f) \in \Pi$ for all f large enough.

Now, for any value of $\theta_1 < 1$, including $\theta_1 > \theta_2$, if one plays M_1 continually, there is a positive probability obtaining a long string of failures, and so

$$P(\pi(s_k, f_k) \in \Pi \mid 0 < \theta_2 < \theta_1 < 1) > 0 \quad \text{for large } k.$$

Hence

$$P(\lim_{k \to \infty} u(k) = 2 \mid 0 < \theta_2 < \theta_1 < 1) > 0, \tag{8.6}$$

and the Bayesian optimal policy is not self-tuning.

The next exercise shows that Bayesian optimal policies are not self-tuning in the more general adaptive control problem introduced in Section 4.

Exercise   (8.7)
Let u_k be the slot machine played according to the (optimal) index rule. For machine n, let Ω_n be the sample paths for which $u_k = n$ for infinitely many values of k.
(1) Show that

$$\text{Prob } (\theta_m = \theta_n \mid \Omega_m \cap \Omega_n) = 1,$$

i.e., if the index rule plays both m and n infinitely often, then the two success probabilities must be the same.

(2) Suppose the prior distributions $P_1(d\theta_1), \ldots, P_N(d\theta_N)$, are continuous so that Prob $(\theta_m = \theta_n) = 0$ for $m \neq n$. As a corollary to (1) show that there exists a random time $T < \infty$ a.s. such that

$$u_k = u_T, \quad k > T, \ a.s.$$

i.e., after a finite (random) time T, the index rule plays the same machine forever.

(3) Let $n^* = Arg \max_{n} \theta_n$ be the machine with largest success probability. Under the assumptions of (2) show that

$$Prob \ (u_k \neq n^*, t > T) > 0,$$

i.e., the index rule is not self-tuning.

[Hint: First show that if a machine is played infinitely often, then its index converges to its success probability.]

9. Notes

1. Theorem (2.7) can be found in Lai and Wei (1982).

2. To show that in general, optimal control problems for systems with partial observations can be replaced by equivalent problems in a new completely observed state, which is the conditional distribution of the old state, we need to overcome technical measurability problems. These are examined in Bertsekas and Shreve (1978).

3. The fundamental index result for bandit problems is from Gittins and Jones (1972). See also Gittins (1979), Gittins and Glazebrook (1977), and Glazebrook (1983). Books that treat index results are Whittle (1982) and Ross (1983).

4. For references to the literature on applications of bandit problems, see Glazebrook (1983) and Whittle (1982). For queueing network applications, a variation of the bandit problem called the tax problem is useful; see Varaiya, Walrand and Buyukkoc (1983).

5. A nice expression for W of (4.4) can be found in Whittle (1982).

6. The bandit problem can be generalized to the situation where, besides choosing a process for evolution, one can also choose a control action to guide its evolution. These so-called superprocesses are examined in Whittle (1980) and Gittins (1979). Another generalization is to allow new bandits to arrive in the course of time; see Whittle (1982). These questions are also discussed in Varaiya, Walrand and Buyukkoc (1983).

7. As Exercise (5.21) shows in a simple context, the various bandit processes need not evolve in a Markovian way for an index result to hold. For a general non-Markovian treatment see Varaiya, Walrand and Buyukkoc (1983).

8. The algorithm of Section 7 for computing indices of a finite state Markov chain can be found in Varaiya, Walrand and Buyukkoc (1983).

9. The result (8.6) that Bayesian, optimal policies need not be self-tuning can be shown to hold in more general situations; see Roth-schild (1974) and Kelly (1981).

10. Kelly (1981) examines the bandit problems when the discount factor $\beta \to 1$, and also provides some very nice interpretations of the dual purposes of control in bandit problems.

11. A survey of Bayesian adaptive control can be found in Kumar (1985), which also provides many other references.

12. Instead of using the average cost criterion as a performance measure for comparing different policies, one can examine the rate of the growth of the total cost criterion. This gives rise to a finer sense of optimality than that measured by the average cost and is examined for the non-Bayesian bandit problem in Lai and Robbins (1984).

10. References

[1]   D.P. Bertsekas and S.E. Shreve (1978), *Stochastic optimal control: the discrete time case,* Academic Press, New York.

[2]   J.C. Gittins (1979), "Bandit processes and dynamic allocation indices," *Journal of the Royal Statistical Society,* vol 41B, 148-177.

[3]   J.C. Gittins and K.D. Glazebrook (1977), "On Bayesian models in stochastic scheduling," *Journal of Applied Probability,* vol 14, 556-565.

[4]   J.C. Gittins and D.M. Jones (1972), "A dynamic allocation index for the sequential design of experiments," in J. Gani, K. Sarkadi and I. Vincze (eds), *Colloquia Mathematica Societatis Janos Bolyai,* 9, Progress in Statistics, European Meeting Of Statisticians, 241-266, North Holland, London.

[5]   K.D. Glazebrook (1983), "Optimal strategies for families of alternative bandit processes," *IEEE Transactions on Automatic Control,* vol AC-28, 858-861.

[6]   F.P. Kelly (1981), "Multi-armed bandits with discount factor near one: the Bernoulli case," *Annals of Statistics,* vol 9, 987-1001.

[7]   P.R. Kumar (1985), "A survey of some results in stochastic adaptive control," *SIAM J. on Control and Optimization,* vol 23(3), 329-380.

[8]   T. L. Lai and H. Robbins (1984), "Asymptotically efficient adaptive allocation rules," *Advances in Applied Mathematics,* vol 5, 1-19.

[9]   T.L. Lai and C.Z. Wei (1982), "Least squares estimates in stochastic regression models with applications to identification and control of dynamic systems," *Annals of Statistics,* vol 10(1), 154-166.

[10] S. Ross (1983), *Introduction to stochastic dynamic programming,* John Wiley, New York.

[11] M. Rothschild (1974), "A two armed bandit theory of market pricing," *Journal of Economic Theory,* vol 9, 185-202.

[12] P. Varaiya, J. Walrand and C. Buyukkoc (1985), "Extensions of the multi-armed bandit problem: the discounted case," *IEEE Transactions on Automatic Control,* vol AC-30, 426-439.

[13] P. Whittle (1980), "Multi-armed bandits and the Gittins index," *Journal of the Royal Statistical Society,* vol 42B, 143-149.

[14] P. Whittle (1982), *Optimization over time: dynamic programming and stochastic control,* John Wiley, New York.

CHAPTER 12
NON-BAYESIAN ADAPTIVE CONTROL

An important objective in non-Bayesian adaptive control is the design of self-tuning and self-optimizing control laws. These typically consist of an identification part and a control part. A fundamental difficulty arises in trying to identify a system in closed loop. This difficulty, as well as methods to overcome it, are studied in this chapter in the context of adaptive control of Markov chains. Self-tuning minimum variance control for ARMAX systems is analyzed in Chapter 13.

1. Introduction

In Chapter 11 we treated the problem of adaptive control from a Bayesian viewpoint. We assumed that the unknown parameter $\theta°$ determining the system behavior was chosen according to a known prior probability distribution. In this chapter we treat the non-Bayesian formulation of the adaptive control problem where no such prior probability distribution is assumed. We merely suppose that $\theta°$ belongs to a known set Θ.

Recall the simple example of such a situation considered in Chapter 11. We saw that the system

$$y_{k+1} = \theta° u_k + w_{k+1},\tag{1.1}$$

with the adaptive control law

$$\hat{\theta}(k) = \frac{\sum\limits_{i=0}^{k-1} y_{i+1} u_i}{\sum\limits_{i=0}^{k-1} u_i^2},\tag{1.2}$$

$$u_k = \frac{r}{\hat{\theta}(k)},\tag{1.3}$$

has three properties:
(1) The parameter estimates are strongly consistent, i.e., $\lim\limits_{k \to \infty} \hat{\theta}(k) = \theta°$ *a.s.*
(2) It follows that $\lim\limits_{k \to \infty} [u(k) - \frac{r}{\hat{\theta}(k)}] = 0$ a.s. Since the control law $u(k)$

$= \dfrac{r}{\theta(k)}$ minimizes each of the following cost criteria,

$$\sum_{k=0}^{N} (y_k - r)^2 \quad \text{for any fixed } N < \infty,$$

$$\sum_{k=0}^{\infty} \beta^k (y_k - r)^2 \quad \text{for any fixed } 0 < \beta < 1,$$

$$\lim_{N \to \infty} \frac{1}{N} \sum_{k=0}^{N-1} (y_k - r)^2,$$

the adaptive control law is self-tuning with respect to each of these criteria.

(3) For the last criterion, the adaptive control law yields the same cost as the optimal control law that would be used (namely, $u(k) \equiv \dfrac{r}{\theta^\circ}$) if we knew the true value of the parameter θ°. Hence with respect to this criterion, the adaptive control law is self-optimizing.

In the rest of this chapter we consider the problems involved in designing self-tuning or self-optimizing control laws for controlled Markov chains.

2. Identification of Markov chains

We begin by considering the problem of identifying the transition probabilities of a Markov chain. In the next section we introduce controls.

Suppose we have a Markov chain on the finite state space $\{1, \ldots, I\}$ with transition probability matrix

$$P(\theta^\circ) := \{P_{ij}(\theta^\circ) \mid 1 \le i, j \le I\}. \tag{2.1}$$

The parameter θ° is known to belong to the set Θ, although its actual value is not known:

$$\theta^\circ \in \Theta. \tag{2.2}$$

Our goal is to identify θ°. For simplicity assume that

$$\Theta \text{ is a finite set,} \tag{2.3}$$

and that the parametrization is canonical, i.e.,

$$\theta_1 \ne \theta_2 \;\Rightarrow\; P(\theta_1) \ne P(\theta_2). \tag{2.4}$$

To identify the unknown parameter, we shall use the maximum likelihood estimate (MLE). Given the observed past history of the transitions of the Markov chain up to time k, $(x_0, \ldots, x_k)$, the likelihood function is

$$L_k(\theta, \omega) := \prod_{s=0}^{k-1} P_{x_s, x_{s+1}}(\theta). \tag{2.5}$$

[The argument ω is introduced to stress the fact that $L_k(\theta)$ is a random variable since the transitions are random.] The MLE at time k is then given by

$$\hat{\theta}_k := Arg \max \{L_k(\theta) \mid \theta \in \Theta\}.$$

Sometimes the likelihood function reaches its maximum at more than one value of θ. To ensure a unique choice for the MLE in such cases, it is assumed that the elements of the parameter set Θ are ordered, and if there is more than one maximizer the MLE is taken to be the maximizer which is highest in the order.

We now show that the MLE $\hat{\theta}_k$ converges to $\theta°$ almost surely. For simplicity of presentation assume that

$$P_{ij}(\theta) > 0 \quad \text{for all } i, j, \theta. \tag{2.6}$$

The proof is organized as follows. For any limit point θ^* of the MLE $\{\hat{\theta}_k\}$ we will show that

$$P_{ij}(\theta^*) = P_{ij}(\theta°) \quad \text{for all } i, j. \tag{2.7}$$

From (2.4) this implies $\theta^* = \theta°$, and since θ^* is any limit point of the MLE it follows that $\lim_{k \to \infty} \hat{\theta}_k = \theta°$.

To prove (2.7) we first show that for some finite (but random) T

$$P_{x_k, x_{k+1}}(\theta^*) = P_{x_k, x_{k+1}}(\theta°), \ k \geq T, \tag{2.8}$$

and also that

For every i, j the event $\{x_k = i, x_{k+1} = j\}$

occurs infinitely often. $\tag{2.9}$

Since (2.7) follows readily from (2.8) and (2.9) it remains to prove these. Since Θ is finite by assumption (2.3), (2.8) is equivalent to

$$\lim_{k \to \infty} \frac{P_{x_k, x_{k+1}}(\theta^*)}{P_{x_k, x_{k+1}}(\theta°)} = 1. \tag{2.10}$$

From (2.5), it is clear that (2.10) will hold if we can show that

$$\lim_{k \to \infty} \frac{L_k(\theta^*)}{L_k(\theta°)} > 0.$$

Thus a quantity of interest is the *likelihood ratio* function

$$\Lambda_k(\theta, \omega) := \frac{L_k(\theta)}{L_k(\theta°)}, \tag{2.11}$$

where, again, the argument ω is introduced to stress that $\Lambda_k(\theta)$ is random. The most important property of the likelihood ratio is given in Lemma (2.13) which is based on the following martingale convergence theorem.

Theorem   (2.12)
If $\{M_k, F_k\}$ is a positive martingale, then $\{M_k\}$ converges a.s. to some random variable denoted M_∞.

Lemma   (2.13)
$\{\Lambda_k(\theta), F_k\}$ is a positive martingale for every $\theta \in \Theta$. Here $F_k := \sigma(x_0, \ldots, x_k)$ is the σ-algebra generated by the observations up to k. Hence for each $\theta \in \Theta$, $\{\Lambda_k(\theta)\}$ converges a.s. to a nonnegative finite random variable $\Lambda_\infty(\theta)$.

Proof
Define

$$l_k(\theta, \omega) := \frac{P_{x_{k-1}, x_k}(\theta)}{P_{x_{k-1}, x_k}(\theta^\circ)} \ , \ \theta \in \Theta,$$

then

$$\Lambda_k(\theta) = \prod_{s=1}^{k} l_s(\theta)$$ (2.14)

and

$$E\{l_{k+1}(\theta) \mid F_k\} = E\{\frac{P_{x_k, x_{k+1}}(\theta)}{P_{x_k, x_{k+1}}(\theta^\circ)} \mid F_k\}$$

$$= \sum_{j=1}^{I} \frac{P_{x_k, j}(\theta)}{P_{x_k, j}(\theta^\circ)} P_{x_k, j}(\theta^\circ)$$

$$= \sum_{j=1}^{I} P_{x_k, j}(\theta) = 1 \ a.s. \quad \text{for all } \theta.$$

From this and (2.14) it follows that for each θ

$$E\{\Lambda_{k+1}(\theta) \mid F_k\} = E\{\prod_{s=1}^{k+1} l_s(\theta) \mid F_k\}$$

$$= \prod_{s=1}^{k} l_s(\theta) E\{l_{k+1}(\theta) \mid F_k\}$$

$$= \Lambda_k(\theta) \ a.s.,$$

so $\{\Lambda_k(\theta), F_k\}$ is a martingale. Moreover, it is positive since by assumption (2.6) both the numerator and denominator are strictly positive.

Hence by Theorem (2.12) it converges.   □

The next theorem carries out all these arguments.

Theorem   (2.15)
Assume (2.2)-(2.4), and (2.6). Then the MLE $\{\hat{\theta}_k\}$ is strongly consistent, i.e.,

$$\lim_{k \to \infty} \hat{\theta}_k = \theta° \quad a.s.$$

Proof
The denominator on the right hand side of (2.11) does not depend on θ. Hence, $\hat{\theta}_k$ also maximizes Λ_k, i.e.,

$$\Lambda_k(\hat{\theta}_k) \geq \Lambda_k(\theta) \quad \text{for all } \theta \in \Theta.$$

Since $\Lambda_k(\theta°) = 1$, and $\theta° \in \Theta$,

$$\Lambda_k(\hat{\theta}_k) \geq 1 \ a.s. \tag{2.16}$$

Let $\widetilde{\Omega}$ be the set of all ω such that for each θ, $\Lambda_k(\theta, \omega)$ converges as $k \to \infty$. By Lemma (2.13), $Prob(\widetilde{\Omega}) = 1$. Fix $\omega \in \widetilde{\Omega}$ and let θ^* be a limit point of $\{\hat{\theta}_k(\omega)\}$. Then,

$$\hat{\theta}_{k_n}(\omega) = \theta^* \quad \text{for all } n$$

for some subsequence of times $\{k_n\}$. From (2.16) it follows that

$$\Lambda_{k_n}(\theta^*, \omega) \geq 1 \quad \text{for all } n,$$

and since $\Lambda_k(\theta^*, \omega)$ converges, this inequality implies

$$\lim_{k \to \infty} \Lambda_k(\theta^*, \omega) \geq 1.$$

Since this limit is strictly positive,

$$\lim_{k \to \infty} \frac{P_{x_{k-1}, x_k}(\theta^*)}{P_{x_{k-1}, x_k}(\theta°)} = \lim_{k \to \infty} l_k(\theta^*, \omega) = \lim_{k \to \infty} \frac{\Lambda_k(\theta^*, \omega)}{\Lambda_{k-1}(\theta^*, \omega)} = 1.$$

Hence there exists $T(\omega) < \infty$ such that

$$P_{x_{k-1}(\omega), x_k(\omega)}(\theta^*) = P_{x_{k-1}(\omega), x_k(\omega)}(\theta^0), \quad k \geq T(\omega). \tag{2.17}$$

From (2.17) it will follow that $P_{ij}(\theta^*) = P_{ij}(\theta°)$ provided

The event $\{x_{k-1} = i, x_k = j\}$

occurs infinitely often a.s.   (2.18)

By assumption (2.6), for every i,

$$E\{1(x_{k-1} = i) \mid F_{k-2}\} > \epsilon > 0 \ a.s.$$

By the martingale stability theorem (8.5.26), for every i,

$$\lim_{N \to \infty} \frac{1}{N} \sum_{k=1}^{N} [1(x_{k-1} = i) - E\{1(x_{k-1} = i) \mid F_{k-2}\}] = 0 \quad a.s.$$

and so for every i,

$$\lim_{N \to \infty} \inf \frac{1}{N} \sum_{k=1}^{N} 1(x_{k-1} = i) > \epsilon > 0 \quad a.s. \tag{2.19}$$

Also from (2.6) we get, for every i, j,

$$E\{1(x_{k-1} = i, x_k = j) \mid F_{k-1}\} > \epsilon 1(x_{k-1} = i) \quad a.s. \tag{2.20}$$

Again, by the martingale stability theorem,

$$\lim_{N \to \infty} \frac{1}{N} \sum_{k=1}^{N} [1(x_{k-1} = i, x_k = j)$$

$$- E\{1(x_{k-1} = i, x_k = j) \mid F_{k-1}\}] = 0 \; a.s.$$

and so from (2.19) and (2.20) it follows that for every i, j

$$\lim_{N \to \infty} \inf \frac{1}{N} \sum_{k=1}^{N} 1(x_{k-1} = i, x_k = j) > \epsilon^2 > 0 \quad a.s. \tag{2.21}$$

This clearly implies (2.18), and hence $P(\theta^*) = P(\theta^\circ)$. From assumption (2.4) it follows that $\theta^* = \theta^\circ$. Since θ^* was an arbitrary limit point of $\{\hat{\theta}_k(\omega)\}$, $\lim_{k \to \infty} \hat{\theta}_k(\omega) = \theta^\circ$. Moreover, since this is true for almost all ω, the result is proved .   □

This technique of using the martingale convergence and martingale stability theorems to establish convergence of the maximum likelihood parameter estimates is used repeatedly in what follows.

3. Identification of controlled Markov chains

Consider the controlled Markov chain on the state space $\{1, \ldots, I\}$ with control set U and transition probability matrix

$$P(u, \theta^\circ) = \{P_{ij}(u, \theta^\circ) \mid 1 \leq i, j \leq I\}. \tag{3.1}$$

As before the unknown parameter θ° satisfies (2.2), (2.3). For simplicity it is also assumed that

$$U \text{ is a finite set} \tag{3.2}$$

and

$$P_{ij}(\theta, u) > 0 \quad \text{for all } i, j, u, \theta. \tag{3.3}$$

As in the previous section the MLE based on the observed history $\{x_0, u_0, \ldots, u_{k-1}, x_k\}$ is

$$\hat{\theta}_k := Arg \ \max \ \{L_k(\theta) \mid \theta \in \Theta\}, \tag{3.4}$$

where $L_k(\theta)$ is the likelihood function,

$$L_k(\theta) := \prod_{s=0}^{k-1} P_{x_s, x_{s+1}}(u_s, \theta). \tag{3.5}$$

As before, the elements of Θ are assumed to be ordered, and if the likelihood function is maximized at more than one value of θ, the highest one in the order is taken as the MLE.

In contrast to the previous section, the likelihood function depends on $\{u_k\}$, and so the behavior of the MLE (3.4) will depend on how $\{u_k\}$ is chosen.

Exercise (3.6)

Assume (2.2), (2.3), and (3.1)-(3.5). Suppose that u_k is chosen as a function of past observations, i.e., u_k is F_k measurable, where $F_k := \sigma(x_0, u_0, \ldots, x_k)$. Show that there is a random variable T, such that for almost all ω, $T(\omega) < \infty$, and for any $\theta^* \in \Theta$ that is a limit point of $\{\hat{\theta}_k(\omega)\}$,

$$P_{x_k(\omega), x_{k+1}(\omega)} (u_k(\omega), \theta^*) = P_{x_k(\omega), x_{k+1}(\omega)} (u_k(\omega), \theta^\circ),$$

$$k \geq T(\omega). \tag{3.7}$$

[Hint: Use exactly the same proof as Theorem (2.15).]

By using the priority ordering on Θ, this result can be improved to show that $\{\hat{\theta}_k(\omega)\}$ actually converges.

Theorem (3.8)

Assume (2.2), (2.3), and (3.1)-(3.5) with u_k chosen to be F_k measurable for each k. Then:
(1) $\{\hat{\theta}_k\}$ converges almost surely to a random variable θ^*.
(2) There is an a.s. finite random time T such that wp 1

$$P_{x_k(\omega), x_{k+1}(\omega)}(u_k(\omega), \theta^*(\omega)) = P_{x_k(\omega), x_{k+1}(\omega)}(u_k(\omega), \theta^\circ),$$

$$k \geq T(\omega). \tag{3.9}$$

Proof

As before, consider the likelihood ratio

$$\Lambda_k(\theta) := \frac{L_k(\theta)}{L_k(\theta^\circ)}.$$

Let $\widetilde{\Omega}$ be the set of all ω for which (3.7) holds. Then $Prob(\widetilde{\Omega}) = 1$. Suppose that $\omega \in \widetilde{\Omega}$ and that θ^* and θ^{**} are two different limit points of $\{\hat{\theta}_k\}$.

Then, by (3.7),

$$\Lambda_k(\theta^*, \omega) = \Lambda_{T(\omega)}(\theta^*, \omega), \quad k \geq T(\omega),$$

$$\Lambda_k(\theta^{**}, \omega) = \Lambda_{T(\omega)}(\theta^{**}, \omega), \quad k \geq T(\omega).$$

Since $\theta^*(\omega)$ is a limit point, there exists $k_1 > T(\omega)$ such that $\hat{\theta}_{k_1}(\omega) = \theta^*(\omega)$. Hence

$$\Lambda_{k_1}(\theta^*(\omega), \omega) = \max_{\theta \in \Theta} \{\Lambda_{k_1}(\theta, \omega)\} \geq \Lambda_{k_1}(\theta^{**}(\omega), \omega).$$

Similarly, for some $k_2 > T(\omega)$,

$$\Lambda_{k_2}(\theta^{**}(\omega), \omega) = \max_{\theta \in \Theta} \{\Lambda_{k_2}(\theta, \omega)\} \geq \Lambda_{k_2}(\theta^*(\omega), \omega).$$

But then

$$\begin{aligned}
\Lambda_{T(\omega)}(\theta^*(\omega), \omega) &= \Lambda_{k_1}(\theta^*(\omega), \omega) \\
&\geq \Lambda_{k_1}(\theta^{**}(\omega), \omega) \\
&= \Lambda_{T(\omega)}(\theta^{**}(\omega), \omega) \\
&= \Lambda_{k_2}(\theta^{**}(\omega), \omega) \\
&\geq \Lambda_{k_2}(\theta^*(\omega), \omega) \\
&= \Lambda_{T(\omega)}(\theta^*(\omega), \omega).
\end{aligned}$$

From (3.7) it then follows that

$$\Lambda_k(\theta^*(\omega), \omega) = \Lambda_k(\theta^{**}(\omega), \omega), \quad k \geq T(\omega).$$

But by the priority ordering, either $\theta^*(\omega)$ or $\theta^{**}(\omega)$ will be preferred over the other for $k \geq T(\omega)$, and so both cannot possibly be limit points of $\{\hat{\theta}_k(\omega)\}$. Hence $\{\hat{\theta}_k(\omega)\}$ has only one limit point, i.e., it converges. Since $\omega \in \Omega$ is arbitrary, the result is proved.   □

It should be emphasized that Theorem (3.8) does not assert that

$$P_{ij}(u, \theta^*) = P_{ij}(u, \theta^\circ) \quad \text{for all } i, j, u. \tag{3.10}$$

It guarantees only that one can identify those transition probabilities $P_{ij}(u, \theta)$ for which the event $\{x_k = i, x_{k+1} = j, u_k = u\}$ occurs infinitely often.

To illustrate this more clearly, consider the special situation where we use the stationary control law

$$u_k = g(x_k). \tag{3.11}$$

Clearly, then, the event $\{x_k = i, x_{k+1} = j, u_k = u\}$ will occur infinitely often if and only if $u = g(i)$. In this case one should expect to identify only the transition probabilities $\{P_{ij}(u, \theta^\circ) \mid u = g(i)\}$ and not all the values $\{P_{ij}(u, \theta°) \mid u \in U\}$. The following exercise suggests how a counterexample to (3.10) can be constructed.

Exercise (3.12)

Construct an example with a stationary feedback control law (3.11) such that (3.10) does not hold.

[Hint: Let the state space be $\{1, 2\}$, the control set be $U = \{a, b\}$, and the parameter set be $\Theta = \{\theta°, \theta^1\}$. Suppose that $P(a, \theta°) = P(a, \theta^1)$ but $P(b, \theta°) \neq P(b, \theta^1)$. Now choose g and the priority order on Θ carefully.]

In fact, when the stationary control law (3.11) is used, we can prove only the following.

Corollary (3.13)

Under the same conditions as Theorem (3.8) suppose that the control law (3.11) is used; then the parameter estimates converge almost surely to a random variable θ^* such that for all i, j

$$P_{ij}(g(i), \theta^*) = P_{ij}(g(i), \theta°) \quad a.s. \tag{3.14}$$

Proof

From (3.9) it follows that

$$P_{x_k, x_{k+1}}(g(x_k), \theta^*) = P_{x_k, x_{k+1}}(g(x_k), \theta°) \, a.s., \quad k \geq T.$$

Now we need to show only that the event $\{x_k = i, x_{k+1} = j\}$ occurs infinitely often, and this is done in the same way that (2.21) was proved in Theorem (2.15). □

4. The closed loop identification problem in adaptive control

Consider the same system as in the previous section. Suppose that for each $\theta \in \Theta$, $g_\theta: \{1, \ldots, I\} \rightarrow U$ is a prespecified control law if the true parameter is θ. (Presumably g_θ leads to a desirable system behavior if the true parameter is θ.) In particular,

$$u(k) = g_{\theta°}(x_k)$$

is the control law that we would use if we knew the true parameter. The difficulty is that $\theta°$ is not known. We shall therefore investigate the possibility of designing a self-tuning or self-optimizing adaptive control law.

As in the example at the beginning of the chapter, at each time k we make an estimate of the unknown parameter and then use the estimate to choose a control input.

So let $\hat{\theta}_k$ be the MLE, and choose the control according to

$$u_k = g_{\hat{\theta}_k}(x_k). \tag{4.1}$$

Through (3.4) and (4.1) this defines a certainty equivalent adaptive control law for controlled Markov chains in exactly the same way that (1.2) and (1.3) formed a certainty equivalent adaptive control law for the linear stochastic system (1.1). We now analyze the performance of this adaptive control scheme.

Since this choice of u_k is subsumed by Theorem (3.8), the parameter estimate will converge to some random variable θ^*. After some random time T

$$\hat{\theta}_k = \theta^*, \quad k \geq T, \tag{4.2}$$

and so

$$u_k = g_{\theta^*}(x_k), \quad k \geq T. \tag{4.3}$$

Hence asymptotically, at least, we are using the fixed control law (4.3). Thus from Corollary (3.13), we should only expect convergence to a value θ^* for which (3.14) holds with g replaced by g_{θ^*}, i.e.,

$$P_{ij}(g_{\theta^*}(i), \theta^*) = P_{ij}(g_{\theta^*}(i), \theta^\circ), \quad \text{for all } i, j. \tag{4.4}$$

This is the content of the next theorem.

Theorem $\hspace{12cm}$ (4.5)
Assume (2.2), (2.3), and (3.1)-3.3); then for the adaptive control law (3.4), (4.1), the parameter estimate $\{\hat{\theta}_k\}$ converges a.s. to a random variable θ^* that satisfies (4.4) a.s.

Proof
By Theorem (3.8), $\{\hat{\theta}_k\}$ converges a.s. to a random variable θ^* for which (3.9) and (4.2) hold. From (4.1) this implies

$$u_k = g_{\theta^*}(x_k), \quad k \geq T(\omega).$$

Substituting in (3.9) gives, for almost all ω, and $k \geq T(\omega)$,

$$P_{x_k(\omega), x_{k+1}(\omega)}(g_{\theta^*(\omega)}(x_k(\omega)), \theta^*(\omega))$$

$$= P_{x_k(\omega), x_{k+1}(\omega)}(g_{\theta^*(\omega)}(x_k(\omega)), \theta^\circ).$$

Moreover, just as in the proof of (2.21) in Theorem (2.15), the event $\{x_k = i, x_{k+1} = j\}$ occurs infinitely often and so the result follows. $\hspace{0.5cm}\square$

Thus the set of possible limit points of $\{\hat{\theta}_k\}$ is the set of parameters θ^* that satisfy (4.4). Clearly $\theta^* = \theta^\circ$ satisfies (4.4), and so θ° is also a possible limit point. Since our goal is to find a self-tuning control law, it is important to know whether

$$\hat{\theta}_k \to \theta^\circ, \tag{4.6}$$

for then we would have a self-tuning control, i.e.,

$$\lim_{k \to \infty} g_{\hat{\theta}_k} = g_{\theta^\circ}. \tag{4.7}$$

Note that (4.6) is a stronger requirement than (4.7), but it is a reasonable objective if our goal is simultaneously to identify and control the system. But, as the analysis of the stationary control law case in Corollary (3.13) indicates, (4.6) and (4.7) may not hold. The following is a counterexample.

Example   (4.8)
Consider the system with state space $\{1, 2\}$, control set $U = \{a, b\}$, and parameter set $\Theta = \{\theta^\circ, \theta^1\}$. The true parameter is θ°. Let the transition probabilities of the system under the two possible values of the parameters be [recall the notation $P_{ij}(u, \theta)$]:

$$P_{11}(b, \theta^\circ) = P_{11}(b, \theta^1) = 0.7$$

$$P_{11}(a, \theta^\circ) = 0.6$$

$$P_{11}(a, \theta^\circ) = 0.8$$

$$P_{21}(u, \theta) = 0.2 \quad \text{for all } u, \theta.$$

Suppose we are interested in the average cost per unit time and the one step cost is $c(i, u) = 1(i = 2)$. Then the optimal control for the two parameter values is

$$g_{\theta^\circ}(i) = a \quad \text{for } i = 1, 2, \tag{4.9}$$

$$g_{\theta^1}(i) = b \quad \text{for } i = 1, 2. \tag{4.10}$$

Let the initial state and initial control input be $x_0 = 1$ and $u_0 = a$.

Consider the adaptive control algorithm (3.4), (4.1) for this example. The state at time $k = 1$ can be either 1 or 2. We consider only the case $x_1 = 1$, which happens with probability 0.6.

Since $P_{11}(a, \theta^1) > P_{11}(a, \theta^\circ)$, the MLE at $k = 1$ is $\hat{\theta}_1 = \theta^1$, and so the control applied at time $k = 1$ is $u_1 = b$. But then since $P_{ij}(b, \theta^\circ) = P_{ij}(b, \theta^1)$ for all i, j, it follows that no matter what x_2 is,

$$L_2(\theta^\circ) = \prod_{s=0}^{1} P_{x_s, x_{s+1}}(u_s, \theta^\circ) < \prod_{s=0}^{1} P_{x_s, x_{s+1}}(u_s, \theta^1) = L_2(\theta^1),$$

and so the MLE at time $k = 2$ will again be $\hat{\theta}_2 = \theta^1$. The control applied at $k = 2$ will again be $u_2 = b$. By repeating the argument it is seen that

$$\hat{\theta}_k = \theta^1, \quad k \geq 1.$$

Note that the foregoing analysis was contingent on the event $x_1 = 1$, which occurs with probability 0.6, so self-tuning does not occur with probability at least 0.6.

In this example self-tuning fails because

$$P_{ij}(g_{\theta^1}(i), \theta^1) = P_{ij}(g_{\theta^1}(i), \theta^\circ) \quad \text{for all } i, j, \tag{4.11}$$

so that one cannot distinguish between the parameters θ^1 and θ° as long as the control law g_{θ^1} is used. This *closed loop identification problem* prevents the adaptive control (3.4), (4.1) from being self-tuning.

Evidently this problem can only arise if there is $\theta^* \neq \theta^\circ$ for which (4.4) holds. Hence if we assume that Θ is such that (4.4) is satisfied only when $\theta^* = \theta^\circ$, then we can guarantee the self-tuning property.

Corollary   (4.12)
Suppose that for all $\theta \in \Theta$,

$$\theta \neq \theta^\circ \quad \Rightarrow \quad P(g_\theta, \theta) \neq P(g_\theta, \theta^\circ), \tag{4.13}$$

where $P(g_\theta, \theta) := \{P_{ij}(g_\theta(i), \theta) \mid 1 \leq i, j \leq I\}$. Then $\lim\limits_{k \to \infty} \hat{\theta}_k = \theta^\circ$ a.s.

The condition (4.13) is an *identifiability* condition. It guarantees that one can identify θ° even in closed loop. It is a restrictive condition and in what follows we investigate the possibility of designing self-tuning or self-optimizing adaptive control laws without this condition.

5. Self-tuning control: forced choice schemes

From the previous section we see that the source of the difficulty is that the adaptive controller prematurely converges to a parameter value θ^* and thus after some time it uses only the control law $u_k = g_{\theta^*}(x_k)$ to the exclusion of other controls. Because of this we can identify only the sub-set $\{P_{ij}(u) \mid u = g_{\theta^*}(i)\}$ of all transition probabilities $\{P_{ij}(u) \mid u \in U\}$. This suggests that the closed loop identification problem can be avoided if the adaptive controller investigates the effect of using each $u \in U$ in each state i infinitely often.

In this section we study one such method to overcome the closed loop identification problem. Another method is presented in Section 7. Consider the set of times $k = 1, 10, 100, 1000, 10000, \ldots$ We reserve this set of times purely for experimentation, and use a certainty equivalent control at other times. More specifically, suppose that U consists of J values,

$$U = \{v_1, \ldots, v_J\},$$

and consider the control law

$$u_k = v_n \quad \text{if } k = 10^{sJ+n} \text{ for some } s, \tag{5.1}$$

$$= g_{\hat{\theta}_k}(x_k) \quad \text{for all other } k.$$

Thus at the set of reserved times $\{10^s, s \geq 0\}$ the control is selected cyclically from the set U for purposes of experimentation.

Theorem   (5.2)
Assume (2.2), (2.3), and (3.1)-(3.3) and that the parametrization is

canonical; $P(u, \theta_1) \neq P(u, \theta_2)$ if $\theta_1 \neq \theta_2$. If the control law (5.1) is used where $\hat{\theta}_k$ is given by (3.4), (3.5), then

$$\lim_{k \to \infty} \hat{\theta}_k = \theta^\circ \quad a.s.$$

Proof
From Theorem (2.15) it is clear that all that we need to show is that for each i, j, u, the event $\{x_k = i, x_{k+1} = j, u_k = u\}$ occurs infinitely often. If we now consider the sequence of random variables $1(x_{10^k + n} = i, x_{10^k + n + 1} = j, u_{10^k + n} = v_n)$, we can proceed just as we did in Theorem (2.15).   □

Note that because we do not always use $u_k = g_{\hat{\theta}_k}(x_k)$ in generating the controls through (5.1), Theorem (5.2) does not imply that

$$\lim_{k \to \infty} g_{\hat{\theta}_k} = g_{\theta^\circ}.$$

However, such convergence does occur in a slightly weaker sense.

Exercise   (5.3)
Assume the same conditions as in Theorem (5.2). Show that

$$\lim_{N \to \infty} \frac{1}{N} \sum_{k=0}^{N-1} 1(g_{\hat{\theta}_k} = g_{\theta^\circ}) = 1 \quad a.s. \tag{5.4}$$

[Hint: Use the fact that $\displaystyle \lim_{n \to \infty} \frac{n}{10^n} = 0$ implies $\displaystyle \lim_{N \to \infty} \frac{1}{N} \sum_{k=0}^{N-1} 1(k = 10^n$ for some $n) = 0$. Note that (5.4) means that in the sense of average frequency of occurrence, the adaptive control law $\{g_{\hat{\theta}_k}\}$ self-tunes to g_{θ°.]

From the preceding analysis it is clear that instead of the set of times $\{1, 10, 100, ..\}$, we may choose any infinite sequence of times $\{k_n\}$ provided

$$\lim_{n \to \infty} \frac{n}{k_n} = 0.$$

6. Self-optimization with respect to the average cost criterion

So far we made no assumption regarding the way in which the prespecified control laws $\{g_\theta \mid \theta \in \Theta\}$ were chosen. Suppose now that for each θ, g_θ minimizes the average cost

$$\lim_{N \to \infty} \sup \frac{1}{N} \sum_{k=0}^{N-1} c(x_k, u_k), \tag{6.1}$$

when θ is the true parameter. Here c is some specified one-step cost function. Since (6.1) measures only the average performance, it is reasonable to expect that convergence in an average frequency of occurrence sense, as in (5.4), is sufficient for the adaptive control law to be self-optimizing with respect to the average cost (6.1).

Definition (6.2)
Let G_{θ° be the collection of all stationary control laws that are optimal with respect to some specified cost criterion for the true system. (There may be more than one stationary optimal control law.) Also let $U_{\theta^\circ}(x)$ be the set of all control inputs arising from some stationary optimal control law when the system state is x, i.e.,

$$U_{\theta^\circ}(x) := \{u \mid u = g(x) \text{ for some } g \in G_{\theta^\circ}\}. \tag{6.3}$$

We say that an adaptive control law is **self-tuning in a Cesaro sense** if

$$\lim_{N \to \infty} \frac{1}{N} \sum_{k=0}^{N-1} 1(u_k \in U_{\theta^\circ}(x_k)) = 1 \quad a.s. \tag{6.4}$$

For example, (5.4) implies that the adaptive control law (5.1) is self-tuning in a Cesaro sense with respect to some cost criterion if each g_θ is chosen to be optimal with respect to that cost criterion when the true parameter is θ.

The following theorem shows that for the average cost criterion (6.1), any adaptive control law which is self-tuning in a Cesaro sense is also self-optimizing.

Theorem (6.5)
Consider a system satisfying (3.1)-(3.3). Consider any adaptive control law that chooses inputs based on past observations on the system, i.e., such that u_k is F_k measurable. If such an adaptive control law is self-tuning in a Cesaro sense with respect to the average cost criterion (6.1), then it is self-optimizing, i.e.,

$$\lim_{N \to \infty} \frac{1}{N} \sum_{k=0}^{N-1} c(x_k, u_k) = J(\theta^\circ) \quad a.s. \tag{6.6}$$

where $J(\theta^\circ)$ is the minimum average cost attainable when the true parameter is θ°.

Proof
This is just a property of the average cost problem studied in Section 8.5. By Theorem (8.5.21) there exists $w = (w(1), \ldots, w(I))^T$ such that for all i

$$J(\theta^\circ) + w(i) = \min_{u \in U} \{c(i, u) + \sum_{j=1}^{I} P_{ij}(u, \theta^\circ) w(j)\}. \tag{6.7}$$

Moreover, a stationary control law g is optimal if and only if

$g(i)$ attains the minimum in (6.7) for each i.   (6.8)

Denote

$$h(i, u) := c(i, u) - J(\theta^\circ) - w(i)$$

$$+ \sum_{j=1}^{I} P_{ij}(u, \theta^\circ) \, w(j).$$ (6.9)

From (6.7) it follows that

$$h(i, u) \geq 0 \quad \text{for all } i, u,$$ (6.10)

and from (6.8),

$$g \text{ is optimal if and only if } h(i, g(i)) = 0 \text{ for all } i.$$ (6.11)

Hence $u(k) \in U_{\theta^\circ}(x_k)$ if and only if $h(x_k, u_k) = 0$. Since the adaptive control law is self-tuning in a Cesaro sense, it follows that

$$\lim_{N \to \infty} \frac{1}{N} \sum_{k=0}^{N-1} h(x_k, u_k) = 0 \quad a.s.$$ (6.12)

Next, let

$$y_{k+1} := c(x_k, u_k) - J(\theta^\circ) + w(x_{k+1}) - w(x_k) - h(x_k, u_k).$$

Proceeding as in Exercise (8.5.28), it follows from (6.9) that

$$E\{y_{k+1} \mid F_k\} = 0 \ a.s.$$

Thus $\{y_k, F_k\}$ is a martingale difference sequence. By Theorem (8.5.26)

$$\lim_{N \to \infty} \frac{1}{N} \sum_{k=0}^{N-1} y_{k+1} = 0 \quad a.s.$$

Substituting for y_{k+1} gives

$$\lim_{N \to \infty} \frac{1}{N} \sum_{k=0}^{N-1} c(x_k, u_k) - \lim_{N \to \infty} \frac{1}{N} \sum_{k=0}^{N-1} h(x_k, u_k)$$

$$- J(\theta^\circ) + \lim_{N \to \infty} \frac{1}{N} [w(x_N) - w(x_0)] = 0 \ a.s.$$

The last term vanishes, since $\{w(x_k)\}$ is a bounded sequence. From (6.12) we then get the desired result.   □

It should be noted from Exercise (8.5.28) that no control law can give a strictly smaller cost than $J(\theta^\circ)$ on any set of sample paths of positive probability measure. Hence self-optimizing control laws are optimal in a strong sample path sense.

Corollary   (6.13)

For each θ, let g_θ be a stationary optimal control law for the average cost criterion (6.1) when the true parameter is θ. Then, under the conditions of Theorem (5.2), the adaptive control law based on the forced choice scheme (5.1) is self-optimizing.

7. Self-tuning by the biased maximum likelihood method

In this section we show that although the simple adaptive control law based on the MLE (3.4) together with the certainty equivalent control law (4.1) is not self-tuning, a simple modification of the MLE will give us an adaptive control law which is self-tuning in a Cesaro sense.

Suppose that we are given either a discounted cost criterion

$$E \sum_{k=0}^{\infty} \beta^k c(x_k, u_k), \tag{7.1}$$

or an average cost criterion (6.1), assuming for convenience in what follows that

$$c(i, u) > 0 \quad \text{for all } i, u.$$

For each $\theta \in \Theta$, let g_θ be a stationary optimal control law when θ is the true parameter. We know that with the adaptive control law (3.4), (4.1), the parameter estimates $\{\hat{\theta}_k\}$ will converge to a random variable θ^* satisfying (4.4). The next exercise examines an important property satisfied by such θ^*.

Exercise (7.2)
(1) Let $J(g, \theta)$ be the average cost (6.1) incurred by the stationary control law g when the true parameter is θ, and let $J(\theta) := \min_g \{J(g, \theta)\}$ be the minimal cost. If θ^* satisfies (4.4), show that

$$J(\theta^*) = J(g_{\theta^*}, \theta^*) = J(g_{\theta^*}, \theta^\circ) \geq J(\theta^\circ). \tag{7.3}$$

(2) In the case of the discounted cost criterion, let $V(g, \theta)(i)$ be the cost (7.1) incurred by the control law g when the system starts in the initial state i and the parameter is θ, and let $V(\theta)(i)$ be the minimum cost. Show that a similar set of inequalities as in (7.3) holds with each $J(\theta)$ replaced by $V(\theta)(i)$. Also let $J(g, \theta) := \sum_{i=1}^{I} V(g, \theta)(i)$, $J(\theta) := \sum_{i=1}^{I} V(\theta)(i)$, and show a similar set of inequalities.

[Hint: For the average cost criterion show that (4.4) implies $\pi(g(\theta^*), \theta^*) = \pi(g(\theta^*), \theta^\circ)$, where $\pi(g, \theta)$ is the steady state probability resulting when g is used and the true parameter is θ. Then deduce that $J(g_{\theta^*}, \theta^*) = J(g_{\theta^*}, \theta^\circ)$. For the discounted cost use (4.4) and the fact that $V(g_{\theta^*}, \theta^*)$ is the unique solution of

$$V(i) = c(i, g_{\theta^*}(i)) + \beta \sum_{j=1}^{I} P_{ij}(g_{\theta^*}(i), \theta^*) V(j) \quad \text{for all } i,$$

to show that $V(g_{\theta^*}, \theta^°)$ is also the solution of the equation above.]

Thus the parameter estimates converge to some θ^* with a minimal cost (or sum of minimal costs in the case of the discounted cost criterion) greater than or equal to $J(\theta^°)$. This suggests the following idea. Suppose we could somehow bias the MLE (3.4) in favor of parameters θ with small minimal costs $J(\theta)$, then, perhaps, we could force θ^* to satisfy

$$J(\theta^*) \leq J^*(\theta^°).   \tag{7.4}$$

Combined with the inequalities in (7.3), this would result in the equality

$$J(g_{\theta^*}, \theta^°) = J(\theta^°),   \tag{7.5}$$

and then g_{θ^*} would be an optimal control law for the true system. However, we have to do this biasing carefully so that we still preserve an equality such as (4.4). In what follows we exhibit a biasing scheme that results in parameter estimates $\{\hat{\theta}_k\}$ converging in a Cesaro sense to θ^* satisfying (4.4) and (7.5).

To understand how small the bias must be, we first examine the likelihood ratio $\Lambda_k(\theta)$. Recall that the MLE maximizes $\Lambda_k(\theta)$, and so also the log-likelihood ratio

$$\ln \Lambda_k(\theta) = \sum_{s=0}^{k-1} \ln \frac{P_{x_s,x_{s+1}}(u_s, \theta)}{P_{x_s,x_{s+1}}(u_s, \theta^°)} .$$

This sum grows like $O(k)$—i.e., linearly with k—and so if the bias is not to damage the good properties of the MLE, the bias should only grow like $o(k)$. This suggests choosing our estimates to maximize the biased log-likelihood ratio

$$\ln \Lambda_k^{bias}(\theta) := -o(k) \ln \frac{J(\theta)}{J(\theta^°)} + \ln \Lambda_k(\theta).   \tag{7.6}$$

[In the discounted cost case, interpret $J(\theta)$ as $\sum_{i=1}^{I} V(\theta)(i)$, as in Exercise (7.2).] We cannot, of course, evaluate (7.6) since $\theta^°$ is unknown. However, just as maximization of $\Lambda_k(\theta)$ was equivalent to maximization of $L_k(\theta)$, so also maximization of the biased log-likelihood ratio can be done by maximizing the biased likelihood function

$$L_k^{bias}(\theta) := J(\theta)^{-o(k)}L_k(\theta)   \tag{7.7}$$

since, as is obvious,

$$\Lambda_k^{bias} = \frac{L_k^{bias}(\theta)}{L_k^{bias}(\theta^°)} .$$

The following theorem shows that an adaptive control scheme with this modified parameter estimation procedure does lead to self-tuning in a Cesaro sense.

Theorem   (7.8)

Consider the system (2.2), (2.3), and (3.1)-(3.3). Let

$$\hat{\theta}_k := Arg \max_{\theta \in \Theta} L_k^{bias}(\theta) \quad \text{for } k = 2, 4, \ldots, 2n, \ldots \tag{7.9}$$

$$:= \hat{\theta}_{k-1} \quad \text{for } k = 1, 3, \ldots, 2n+1, \ldots$$

where $L_k^{bias}(\theta)$ is given by (7.7), and $o(k)$ is any positive sequence satisfying

$$\lim_{k \to \infty} o(k) = +\infty, \quad \lim_{k \to \infty} \frac{o(k)}{k} = 0. \tag{7.10}$$

Let $u(k)$ be as in (4.1) where g_θ is a stationary optimal control law with respect to either the discounted cost criterion (7.1) or the average cost criterion (6.1) when the true parameter is θ . Then:

(1) The parameter estimates converge a.s. in a Cesaro sense to a random variable θ^* i.e.,

$$\lim_{N \to \infty} \frac{1}{N} \sum_{k=0}^{N-1} 1(\hat{\theta}_k = \theta^*) = 1 \quad a.s. \tag{7.11}$$

(2) θ^* satisfies (4.4).

(3) The control law g_{θ^*} is optimal for the true system with parameter θ°.

(4) The adaptive control law is self-tuning in a Cesaro sense.

The proof of this theorem is divided into a series of lemmas and exercises. Throughout the following, let $\widetilde{\Omega}$ be a subset of Ω of with $P(\widetilde{\Omega}) = 1$. All the results will be for a fixed $\omega \in \widetilde{\Omega}$. The argument ω of the random variables will frequently be omitted.

First we show that the bias does achieve the crucial inequality (7.4).

Lemma   (7.12)

Let θ^* be any limit point of $\{\hat{\theta}_k(\omega)\}$. Then

$$J(\theta^*) \le J(\theta^\circ) \quad a.s. \tag{7.13}$$

(Note: In keeping with our convention, this means that there is a subset $\widetilde{\Omega}$ of Ω with $P(\widetilde{\Omega}) = 1$ such that if $\omega \in \widetilde{\Omega}$ and θ^* is a limit point of $\{\hat{\theta}_k(\omega)\}$, then $J(\theta^*) \le J(\theta^\circ)$. All the statements in this section should be similarly interpreted.)

Proof

Since

$$\Lambda_k^{bias}(\hat{\theta}_k) \ge \Lambda_k^{bias}(\theta) \quad \text{for all } \theta \in \Theta \ a.s. ,$$

therefore, $\Lambda_k^{bias}(\hat{\theta}_k) \geq 1$ *a.s.*, since $\Lambda_k^{bias}(\theta^\circ) = 1$. Hence

$$\liminf_{k \to \infty} \Lambda_k^{bias}(\hat{\theta}_k) \geq 1 \quad a.s.$$

If θ^* is a limit point of $\{\hat{\theta}_k\}$ with $\hat{\theta}_{k_n} = \theta^*$ for all n, then clearly

$$\liminf_{n \to \infty} \Lambda_{k_n}^{bias}(\theta^*) \geq 1 \quad a.s. \tag{7.14}$$

As in (2.16), $\{\Lambda_k(\theta)\}$ is a positive martingale and therefore converges. In particular, it is bounded a.s. Suppose now that (7.13) is not true, so that

$$J(\theta^*) > J(\theta^\circ).$$

Then, from (7.6) and (7.10),

$$\lim_{k \to \infty} \Lambda_k^{bias}(\theta^*) = 0 \quad a.s.$$

which contradicts (7.14). □

Thus every limit point satisfies (7.4), one of the two crucial relations. It remains to show that equality (4.4) also holds. We actually show that (4.4) holds for every θ^* which is a *frequent limit point*, i.e.,

$$\limsup_{N \to \infty} \frac{1}{N} \sum_{k=0}^{N-1} 1(\hat{\theta}_k = \theta^*) > 0. \tag{7.15}$$

The proof is developed in the following exercises.

Exercise (7.16)
Show that

$$\lim_{N \to \infty} \frac{1}{N} \sum_{k=0}^{N-1} l_k(\theta) = 1 \quad a.s. \quad \text{for all } \theta \in \Theta,$$

where $l_k(\theta) := \dfrac{P_{x_k, x_{k+1}}(u_k, \theta)}{P_{x_k, x_{k+1}}(u_k, \theta^\circ)}$.

[Hint: Use (2.19) and Theorem (8.5.26).]

Exercise (7.17)
Suppose θ^* is a limit point, $\hat{\theta}_{k_n} = \theta^*$ for all n. Show that

$$\lim_{n \to \infty} \frac{1}{k_n} \ln \Lambda_{k_n}(\theta^*) \geq 0 \quad a.s.$$

[Hint: Note that $\dfrac{1}{k_n} \ln \Lambda_{k_n}^{bias}(\theta^*) \geq 0$. Use $\lim_{n \to \infty} \dfrac{o(k_n)}{k_n} \dfrac{J(\theta^*)}{J(\theta^\circ)} = 0$ from (7.10) to deduce the result from (7.6).]

Exercise (7.18)
For θ^* as in the previous exercise, show that

$$\lim_{k \to \infty} \sup \frac{1}{k} \ln \Lambda_k(\theta^*) \leq 0 \quad a.s.$$

[Hint: From Exercise (7.16) it follows that $\frac{1}{k} \sum_{s=0}^{k-1} l_{s+1}(\theta) \leq 1 + \epsilon$ for all $k \geq T$. Now use the concavity of the logarithm to deduce $\frac{1}{k} \sum_{s=0}^{k-1} \ln l_{s+1}(\theta) \leq \ln (1 + \epsilon)$ for all $k \geq T$.]

Exercise (7.19)

Show that if θ^* is a frequent limit point, then for all i, j,

$$\lim_{N \to \infty} \sup \frac{1}{N} \sum_{k=0}^{N-1} 1(\hat{\theta}_k = \theta^*, x_k = i, x_{k+1} = j) > 0 \quad a.s. \tag{7.20}$$

[Hint: Note that

$$\lim_{N \to \infty} \sup \frac{1}{N} \sum_{k=0}^{N-1} 1(\hat{\theta}_k = \theta^*, x_k = \hat{i}, k \text{ is even}) > 0$$

for at least one $\hat{i} \in \{1, .., I\}$. Now from (3.3)

$$E\{1(\hat{\theta}_k = \theta^*, x_k = \hat{i}, x_{k+1} = i, k \text{ is even}) \mid F_k\}$$
$$> \epsilon 1(\hat{\theta}_k = \theta^*, x_k = \hat{i}, k \text{ is even}) \, a.s.$$

for every i. By (7.9), $\{\hat{\theta}_k = \theta^*, k \text{ is even}\}$ implies that $\{\hat{\theta}_{k+1} = \theta^*\}$. Use Theorem (8.5.28) to deduce that

$$\lim_{N \to \infty} \sup \frac{1}{N} \sum_{k=0}^{N-1} 1(\hat{\theta}_{k+1} = \theta^*, x_{k+1} = i) > 0 \quad a.s.$$

for every i. Finally, once again use Theorem (8.5.28) appropriately to deduce the desired result.]

Lemma (7.21)

Let $B = \{b_1, .., b_l\}$ be a set of l distinct positive numbers. Let $\{z_i\}$ be a sequence with each $z_i \in B$ and satisfying

$$\lim_{n \to \infty} \frac{1}{n} \sum_{i=0}^{n-1} z_i = 1, \tag{7.22}$$

$$\lim_{k \to \infty} \frac{1}{n_k} \sum_{i=0}^{n_k-1} \ln z_i = 0, \tag{7.23}$$

for some subsequence $\{n_k\}$. If

$$d_q := \lim_{k \to \infty} \frac{1}{n_k} \sum_{i=0}^{n_k-1} 1(z_i = b_q) > 0 \tag{7.24}$$

for some $q \in \{1, .., l\}$, then

$$b_q = 1. \tag{7.25}$$

Proof

Define $d_j^n := \dfrac{1}{n} \sum_{i=0}^{n-1} 1(z_i = b_j)$, so $\sum_{j=1}^{l} d_j^n = 1$. Using convexity of the exponential gives

$$
\begin{aligned}
1 &= \lim_{n\to\infty} \frac{1}{n} \sum_{i=0}^{n-1} z_i = \lim_{n\to\infty} \frac{1}{n} \sum_{i=0}^{n-1} \sum_{j=1}^{l} 1(z_i = b_j) b_j \\
&= \lim_{n\to\infty} \sum_{j=1}^{l} \left[\frac{1}{n} \sum_{i=0}^{n-1} 1(z_i = b_j)\right] b_j = \lim_{n\to\infty} \sum_{j=1}^{l} d_j^n b_j \\
&= \lim_{n\to\infty} \sum_{j=1}^{l} d_j^n \exp(\ln b_j) \\
&\geq \lim_{n\to\infty} \sup \exp\left[\sum_{j=1}^{l} d_j^n \ln b_j\right] \\
&\geq \lim_{k\to\infty} \sup \exp\left[\sum_{j=1}^{l} d_j^{n_k} \ln b_j\right] \\
&= \exp\left[\lim_{k\to\infty} \sup \sum_{j=1}^{l} d_j^{n_k} \ln b_j\right] \\
&= \exp\left[\lim_{k\to\infty} \sup \sum_{j=1}^{l} \frac{1}{n_k} \sum_{i=0}^{n_k-1} 1(z_i = b_j) \ln b_j\right] \\
&= \exp\left[\lim_{k\to\infty} \sup \frac{1}{n_k} \sum_{i=0}^{n_k-1} \left(\sum_{j=1}^{l} 1(z_i = b_j) \ln b_j\right)\right] \\
&= \exp\left[\lim_{k\to\infty} \sup \frac{1}{n_k} \sum_{i=0}^{n_k-1} \ln z_i\right] \\
&= \exp\left[\lim_{k\to\infty} \frac{1}{n_k} \sum_{i=0}^{n_k-1} \ln z_i\right] = \exp(0) = 1.
\end{aligned}
$$

Equality therefore holds throughout, and so

$$1 = \lim_{n\to\infty} \sum_{j=1}^{l} d_j^n b_j = \lim_{k\to\infty} \sup \exp\left[\sum_{j=1}^{l} d_j^{n_k} \ln b_j\right]. \tag{7.26}$$

Now choose a subsequence of $\{n_k\}$, relabeling it as $\{n_k\}$, such that $\lim_{k\to\infty} d_j^{n_k} =: d_j$ exists for each j. Clearly $\sum_{j=1}^{l} d_j = 1$ and $d_j \geq 0$. From (7.26) we obtain

$$1 = \sum_{j=1}^{l} d_j b_j = \exp[\sum_{j=1}^{l} d_j \ln b_j],$$

and, taking logarithms,

$$0 = \ln \sum_{j=1}^{l} d_j b_j = \sum_{j=1}^{l} d_j \ln b_j.$$

Because the logarithm is strictly concave, if $d_j > 0$, then $b_j = 1$. Since $d_q > 0$, the result follows. $\square$

The next lemma uses these results to show that (4.4) holds whenever θ^* is a frequent limit point.

Lemma   (7.27)
Under the same conditions as Theorem (7.8), suppose that θ^* is a frequent limit point. Then θ^* satisfies (4.4).

Proof
Let $z_k := l_k(\theta^*)$. We show that $\{z_k\}$ satisfies the conditions of Lemma (7.21). From Exercise (7.16) we know that (7.22) is satisfied. From Exercises (7.17) and (7.18) we know that (7.23) also holds. Moreover, from (7.20) we know that if

$$b_q := \frac{P_{ij}(g_{\theta^*}(i), \theta^*)}{P_{ij}(g_{\theta^*}(i), \theta^\circ)},$$

then d_q as defined in (7.24) satisfies $d_q > 0$. Hence Lemma (7.21) can be applied to show that $b_q = 1$. Since this is true for every i, j, the result follows. $\square$

We are now ready to complete the proof of Theorem (7.8).

Define

$$\Theta^* := \{\theta^* \in \Theta \mid \theta^* \text{ is a limit point of } \{\hat{\theta}_k\}\},$$

$$\bar{\Theta} := \{\theta^* \in \Theta \mid \theta^* \text{ satisfies (7.15)}\},$$

and note that $\bar{\Theta} \subset \Theta^*$. For $\theta^* \in \bar{\Theta}$ we know from Exercise (7.2) and Lemma (7.12) that $J(\theta^*) = J(\theta^\circ)$. Hence from definition (7.6) we have

$$\Lambda_k^{bias}(\theta^*) = \Lambda_k(\theta^*), \quad \theta^* \in \bar{\Theta}.$$

But since $\theta^* \in \Theta^*$, it follows that

$$\Lambda_k^{bias}(\theta^*) \geq 1 \quad \text{for infinitely many } k,$$

and so, since $\{\Lambda_k(\theta^*), F_k\}$ is a positive martingale, it follows that

$$\lim_{k \to \infty} \Lambda_k(\theta^*) \geq 1.$$

Hence

$$\lim_{k \to \infty} l_k(\theta^*) = 1,$$

i.e., there exists T such that

$$l_k(\theta^*) = 1 \quad \text{for all } k \geq T.$$

This shows that

$$\Lambda_k^{bias}(\theta^*) = \Lambda_k(\theta^*) = \Lambda_T^{bias}(\theta^*) = \Lambda_T(\theta^*) \quad \text{for all } k \geq T.$$

Clearly a similar result holds for all $\theta \in \bar{\Theta}$. By the ordering of the set Θ used to select between competing maximizers of the likelihood function, it follows that there can only be a single element in the set $\bar{\Theta}$. This proves (7.11) and hence (1). (2) follows from Lemma (7.27). From Exercise (7.2) and (7.13) of Lemma (7.12), we get (3). Finally, (1) and (3) imply (4) and Theorem (7.8) is proved.

In the case of the average cost criterion, Theorem (6.5) shows that any adaptive control law which is self-tuning in a Cesaro sense is also self-optimizing. Hence we have the following corollary.

Corollary (7.28)
Suppose the cost criterion is of the average cost type (6.1). Then the adaptive control law of Theorem (7.8) is self-optimizing, i.e.,

$$\lim_{N \to \infty} \frac{1}{N} \sum_{k=0}^{N-1} c(x_k, g_{\hat{\theta}_k}(x_k)) = J(\theta^\circ) \quad a.s.$$

We might wonder whether the conclusion of Cesaro convergence of $\{\hat{\theta}_k\}$ to θ^*, as in (7.11), can be strengthened to ordinary convergence in Theorem (7.8). The following exercise shows that we cannot.

Exercise (7.29)
Consider the system with state space $\{1, 2\}$, control set $U = \{1, 2\}$, and parameter set $\Theta = \{1, 2, 3\}$. The true parameter is $\theta^\circ = 1$. The transition probabilities under the three possible values of the parameters are:

$$P_{11}(1, 1) = P_{11}(1, 3) = 0.5$$
$$P_{11}(1, 2) = 0.9$$
$$P_{11}(2, 1) = P_{11}(2, 2) = 0.8$$
$$P_{11}(2, 3) = 0.2$$
$$P_{21}(u, \theta) = 1 \quad \text{for all } (u, \theta)$$

Suppose that the one step cost function is

$$c(i, j, u) = 3 + (2 - i)(7.8 - 0.3u - 6j),$$

and consider the average cost criterion

$$\limsup_{N \to \infty} \frac{1}{N} \sum_{k=0}^{N-1} c(x_k, x_{k+1}, u_k),$$

Show first that the following control laws are optimal for the three parameter values:

$$g_1(i) = 1 \quad \text{for } i = 1, 2,$$

$$g_2(i) = g_3(i) = 2 \quad \text{for } i = 1, 2.$$

Let the initial state and control input be $x_0 = 1$ and $u_0 = 1$. Show that:

(1) $\hat{\theta}_k = 3$ for infinitely many k a.s.

(2) $\displaystyle\sum_{k=0}^{N-1} \frac{1}{N} 1(\hat{\theta}_k = 3) = 0$ a.s.

(3) $\displaystyle\sum_{k=0}^{N-1} \frac{1}{N} 1(\hat{\theta}_k = 1) = 1$ a.s.

[Hint: First show that $J(1) = 2$, $J(2) = 3$, $J(3) = 1$ are the minimum costs. Hence, by virtue of (7.3) and (7.13), $\theta^* = 1$ is the Cesaro limit of $\hat{\theta}_k$. Now suppose to the contrary that $\lim_{k \to \infty} \hat{\theta}_k = 1$; then $\hat{\theta}_k = 1$ and also $u_k = 1$ for all $k \ge T$. Hence $P_{x_k, x_{k+1}}(u_k, 3) = P_{x_k, x_{k+1}}(u_k, 1)$ for all $k \ge T$, and so $\Lambda_k(3) = $ constant > 0 for all $k \ge T$. Since $J(1) > J(3)$, it follows from (7.10) that $\lim_{k \to \infty} \Lambda_k^{bias}(3) = +\infty$. But since we have assumed that $\hat{\theta}_k = 1$ for all $k \ge T$, it is also true that $1 = \Lambda_k^{bias}(1) \ge \Lambda_k^{bias}(3)$ for all $k \ge T$, which is a contradiction.]

8. Notes

1. The pioneering work on the adaptive control of Markov chains is due to Mandl (1974). Under assumptions such as (4.13), the strong consistency of the adaptive control law as well as the self-optimality is proved there.

2. The adaptive control problem without the assumption of an identifiability condition was studied by Borkar and Varaiya (1979). The result of Theorem (4.5) as well as a counterexample of the type in Example (4.8) are given there. The countable state space case is studied in Borkar and Varaiya (1982). Kumar (1982) shows that the parameter estimates need not even converge if the parameter set is compact but infinite.

3. Schemes of the forced choice type given in Section 5.1 were introduced by Robbins (1952). The biased maximum likelihood method was introduced by Kumar and Becker (1982). The case of an arbitrary parameter set is studied in Kumar and Lin (1982), and the case of a general state space is studied in Kumar (1983a). The closed loop

identifiability also arises in adaptive control of a linear Gaussian system with a standard quadratic cost criterion; it is examined in Kumar (1983b).

4. To avoid the closed loop identification problem, one can also use randomized control laws. These are examined in Doshi and Shreve (1980) and Borkar and Varaiya (1982). A survey of the field, along with several references can be found in Kumar (1985).

5. Theorem (2.12) is a consequence of Theorem 4.1s in Doob (1953).

9. References

[1]   V. Borkar and P. Varaiya (1979), "Adaptive control of Markov chains, I: finite parameter set," *IEEE Transactions on Automatic Control,* vol AC-24, 953-958.

[2]   V. Borkar and P. Varaiya (1982), "Identification and adaptive control of Markov chains," *SIAM J. on Control and Optimization,* vol 20, 470-489.

[3]   J.L. Doob (1953), *Stochastic processes,* John Wiley, New York.

[4]   B. Doshi and S. Shreve (1980), "Strong consistency of a modified maximum likelihood estimator for controlled Markov chains," *Journal of Applied Probability,* vol 17, 726-734.

[5]   P.R. Kumar (1982), "Adaptive control with a compact parameter set," *SIAM J. on Control and Optimization,* vol. 20, 9-13.

[6]   P.R. Kumar (1983a), "Simultaneous identification and adaptive control of unknown systems over finite parameter sets," *IEEE Transactions on Automatic Control,* vol AC-28, 68-76.

[7]   P.R. Kumar (1983b), "Optimal adaptive control of linear quadratic Gaussian systems," *SIAM J. on Control and Optimization,* vol 21, 163-178.

[8]   P.R. Kumar (1985), "A survey of some results in stochastic adaptive control," *SIAM J. on Control and Optimization,* vol 23, 329-380,

[9]   P.R. Kumar and A. Becker (1982), "A new family of optimal adaptive controllers for Markov chains," *IEEE Transactions on Automatic Control,* vol AC-27, 137-146.

[10] P.R. Kumar and W. Lin (1982), "Optimal adaptive controllers for Markov chains," *IEEE Transactions on Automatic Control,* vol AC-27, 765-774.

[11] P. Mandl (1974). "Estimation and control in Markov chains," *Advances in Applied Probability,* vol 6, 40-60.

[12] H. Robbins (1952), "Some aspects of the sequential design of experiments," *Bulletin of the American Mathematical Society,* vol 58, 527-537.

CHAPTER 13
SELF-TUNING REGULATORS FOR LINEAR SYSTEMS

In this chapter we study the adaptive control of linear systems. For minimum variance control, the closed loop identifiability problem does not prevent self-tuning because, fortuitously, every possible limit point of the parameter estimates leads to an optimal control law. Hence special methods are not needed to circumvent the closed loop identifiability problem. Thus self-tuning regulators can be designed which converge to a regulator that minimizes the variance of the output of an ARMAX process. The parameter estimates need not converge to the true parameter; nevertheless the adaptive regulator converges to the optimal one.

1. Linear systems viewed as controlled Markov processes

In Chapter 12 we considered the design of self-tuning and self-optimizing adaptive laws for controlled finite state Markov chains. We now consider some examples of linear systems.

We begin by reexamining the adaptive control of the simple linear system (12.1.1),

$$y_{k+1} = \theta^\circ u_k + w_{k+1},$$

using the control law (12.1.2) and (12.1.3),

$$\hat{\theta}_k = \frac{\sum\limits_{i=0}^{k-1} y_{i+1} u_i}{\sum\limits_{i=0}^{k-1} u_i^2}, \quad u_k = \frac{r}{\hat{\theta}_k}.$$

For simplicity, suppose that

$$\{w_k\} \text{ is iid,} \tag{1.1}$$

and Gaussian,

$$w_k \sim N(0, 1). \tag{1.2}$$

Because of (1.1) one may regard the system as a controlled Markov chain (with an infinite state space). Because of assumption (1.2), the LSE and MLE coincide. Neglecting the fact that the state space is infinite, the two assumptions together permit the application of the theory developed in Chapter 12.

Viewed as a controlled Markov process, the system is described by the transition probability density

$$p(dy_{k+1} \mid y_k, u_k) = 2\pi^{-\frac{1}{2}}$$
$$\times \exp\left[-\frac{1}{2}(y_{k+1} - \theta^\circ u_k)^2\right] dy_{k+1}, \qquad (1.3)$$

or,

$$p(y_{k+1} \mid y_k, u_k) \sim N(\theta^\circ u_k, 1).$$

Also, for each θ we have the prespecified control law $g_\theta(y_k) = \dfrac{r}{\theta}$. Theorem (12.4.5) suggests that the possible limit points of $\{\hat{\theta}_k\}$ is the set of all θ^* satisfying

$$2\pi^{-\frac{1}{2}} \exp\left[-\frac{1}{2}(y_{k+1} - \theta^* g_{\theta^*}(y_k))^2\right]$$
$$= 2\pi^{-\frac{1}{2}} \exp\left[-\frac{1}{2}(y_{k+1} - \theta^\circ g_{\theta^*}(y_k))^2\right]$$

for all y_{k+1}, y_k, or equivalently,

$$N(\theta^* g_{\theta^*}(y_k), 1) = N(\theta^\circ g_{\theta^*}(y_k), 1) \quad \text{for all } y_k.$$

Substituting for $g_\theta(y)$, this reduces to

$$\theta^* = \theta^\circ,$$

so the identifiability condition (12.4.13) is satisfied.

This fortuitous coincidence is the reason why we did not encounter the closed loop identifiability problem in our analysis of (12.1.1)-(12.1.3), and why the control law is self-tuning.

Consider now the first order system,

$$y_{k+1} = a_0 y_k + b_0 u_k + w_{k+1}.$$

We shall often revisit this first order system in order to expose new concepts in a simple setting. Suppose (1.1) and (1.2) hold, and

$$\theta^\circ := (a_0, b_0)^T$$

is unknown. For each $\theta := (a, b)^T$, the variance of the output process,

$$\lim_{N \to \infty} \frac{1}{N} \sum_{k=0}^{N-1} y_k^2, \qquad (1.4)$$

is minimized by the law

$$u_k = -\frac{a}{b} y_k. \qquad (1.5)$$

Since $\theta°$ is unknown, we employ the MLE (which is also the LSE since the noise is white Gaussian)

$$\hat{\theta}_k := [\sum_{i=0}^{k-1} \phi_i \phi_i^T]^{-1} \sum_{i=0}^{k-1} \phi_i y_{i+1}, \qquad (1.6)$$

where

$$\phi_i := (y_i, u_i)^T.$$

Then we can use the certainty equivalent control

$$u_k := -\frac{\hat{a}_k}{\hat{b}_k} y_k, \qquad (1.7)$$

where $\hat{\theta}_k =: (\hat{a}_k, \hat{b}_k)^T$. This gives us the adaptive control law (1.6), (1.7).

Suppose the theory developed in Chapter 12 is applicable here even though the state space is infinite. By (12.4.5), if θ^* is a limit point of $\{\hat{\theta}_k\}$, then

$$N(a^* y_k + b^* g_{\theta^*}(y_k), 1) = N(a_0 y_k + b_0 g_{\theta^*}(y_k), 1) \text{ for all } y_k. \quad (1.8)$$

Substituting for $g_\theta(y)$ from (1.5), (1.8) reduces to

$$\frac{a^*}{b^*} = \frac{a_0}{b_0}. \qquad (1.9)$$

Even though this does not imply

$$\theta^* = \theta°,$$

it does imply that

$$g_{\theta^*} = g_{\theta°}.$$

Hence although the parameter estimate $\hat{\theta}_k$ may not converge to $\theta°$, the adaptive control law $g_{\hat{\theta}_k}$ does converge to the optimal control law $g_{\theta°}$. Thus it is self-tuning. Moreover, since the cost criterion (1.3) is of the average type, it is also self-optimizing.

2. The stochastic gradient algorithm for the ARX system

In view of the possibility of obtaining self-tuning control laws, we shall examine the minimum variance adaptive control problem in greater detail.

Consider the ARX system [see Section (7.8)],

$$y_{k+1} = \sum_{i=0}^{p} a_i y_{k-i} + \sum_{i=0}^{p} b_i u_{k-i} + w_{k+1}, \quad b_0 \neq 0, \qquad (2.1)$$

where $\{w_k\}$ is white noise, or more generally, a martingale difference sequence,

$$E\{w_{k+1} \mid F_k\} = 0 \ a.s. \quad \text{for all } k, \tag{2.2a}$$

with

$$E\{w_{k+1}^2 \mid F_k\} = \sigma^2 > 0 \ a.s. \quad \text{for all } k, \tag{2.2b}$$

$$\sup_k E\{w_{k+1}^4 \mid F_k\} < +\infty \ \ a.s. \tag{2.2c}$$

where $F_k := \sigma(y_0, u_0, \ldots, u_k)$. Note that the delay between the application of a control input and its effect at the output is 1 time unit, since $b_0 \neq 0$. (We will not consider the general delay case.)

From (7.8.3) the minimum variance control law is

$$u_k = -\frac{1}{b_0} \left[\sum_{i=0}^{p} a_i y_{k-i} + \sum_{i=1}^{p} b_i u_{k-i} \right].$$

The true parameter

$$\theta^\circ := (a_0, \ldots, a_p, b_0, \ldots, b_p)^T,$$

can be estimated by the LSE using the recursion (10.1.8) and (10.1.9),

$$\hat{\theta}_{k+1} = \hat{\theta}_k + R_k^{-1} \phi_k (y_{k+1} - \phi_k^T \hat{\theta}_k), \tag{2.3}$$

$$R(k) = \sum_{i=0}^{k} \phi_i \phi_i^T, \tag{2.4}$$

where

$$\phi_k := (y_k, \ldots, y_{k-p}, u_k, \ldots, u_{k-p})^T. \tag{2.5}$$

Unfortunately, the resulting adaptive control law is difficult to analyze, and we consider instead the estimate which is generated by the simpler algorithm,

$$\hat{\theta}_{k+1} = \hat{\theta}_k + \frac{\mu \phi_k}{r_k} (y_{k+1} - \phi_k^T \hat{\theta}_k), \tag{2.6a}$$

where

$$r_k := \sum_{i=0}^{k} \phi_i^T \phi_i. \tag{2.6b}$$

The certainty equivalent control law is accordingly

$$u_k = -\frac{1}{\hat{b}_0(k)} \left[\sum_{i=0}^{p} \hat{a}_i(k) y_{k-i} + \sum_{i=1}^{p} \hat{b}_i(k) u_{k-i} \right], \tag{2.7}$$

with $\hat{\theta}_k =: (\hat{a}_0(k), \ldots, \hat{a}_p(k), \hat{b}_0(k), \ldots, \hat{b}_p(k))^T$.

The major difference between (2.3) and (2.6) is that the term R_k has been replaced by its trace r_k. Also $\mu > 0$ is just some constant. This algorithm may be regarded as a simplification of the least squares recursion (2.3), (2.4) since the scalar gain r_k is simpler to implement than the matrix gain R_k.

For a more revealing interpretation, recall that the LSE minimizes

$$V_k(\theta) := \sum_{i=0}^{k} (y_i - \phi_{i-1}^T \theta)^2 =: \sum_{i=0}^{k} v_i(\theta).$$

Since

$$\nabla_\theta v_i(\theta) = -2(y_i - \phi_{i-1}^T \theta)\phi_i,$$

the recursion (2.6) can also be written as

$$\hat{\theta}_{k+1} = \hat{\theta}_k - \frac{\mu}{2r_k} \nabla_\theta v_k(\hat{\theta}_k).$$

Hence at each step k, the algorithm takes a step in the direction $-\nabla_\theta v_k(\hat{\theta}_k)$, i.e., in the direction opposite to the gradient and so tends to reduce $v_k(\theta)$. The quantity $\dfrac{\mu}{2r_k}$ can then be regarded as the *step size*. Since it is reasonable to expect that $r_k \to \infty$, the step size will have a limit of zero, a necessary condition for the algorithm to converge. For this reason (2.6), (2.7) is called a **stochastic gradient (SG)** algorithm. It is also called a *stochastic approximation* algorithm.

3. Analysis by the ODE method

We can now begin the analysis of the SG based certainty equivalent adaptive control law (2.6) and (2.7) applied to (2.1). We do not assume that $\{w_k\}$ is iid, so the system is not a controlled Markov process; moreover, neither the LSE nor the SG estimate need coincide with the MLE. So we cannot apply the theory for controlled Markov chains developed in Chapter 12, and we turn to other methods.

One such method is the ordinary differential equation (ODE) method of analyzing the behavior of recursive stochastic algorithms introduced in Section (10.8). Define

$$\bar{r}_k := \frac{r_k}{k} ,$$

and rewrite (2.6), (2.7) as

$$\hat{\theta}_{k+1} = \hat{\theta}_k + \frac{1}{k} \frac{\mu\phi_k}{\bar{r}_k} (y_{k+1} - \phi_k^T \hat{\theta}_k), \tag{3.1}$$

$$\bar{r}_{k+1} = \bar{r}_k + \frac{1}{k+1} (\phi_k^T \phi_k - \bar{r}_k). \tag{3.2}$$

Recall the intuitive justification for the ODE method. Because of the terms $\frac{1}{k}$ and $\frac{1}{k+1}$ in (3.1) and (3.2), when k is large, the difference terms $(\hat{\theta}_{k+1} - \hat{\theta}_k)$ and $(\bar{r}_{k+1} - \bar{r}_k)$ are small. Hence $\hat{\theta}_k$ and $\bar{r}_k$ fluctuate very little for large k. On the other hand, due to the presence of the noise term w_k, the system (2.1) fluctuates very rapidly. Replacing the rapid fluctuations by their expected values, we get the ODE

$$\frac{d\theta(\tau)}{d\tau} = \frac{\mu}{\bar{r}(\tau)} E_{\theta(\tau)}[(y_{k+1} - \phi_k^T \theta(\tau))\phi_k], \tag{3.3}$$

$$\frac{d\bar{r}(\tau)}{d\tau} = E_{\theta(\tau)}[\phi_k^T \phi_k] - \bar{r}(\tau), \tag{3.4}$$

which can be regarded as good approximations to the behavior of the recursive stochastic equations (3.1) and (3.2). For large τ we interpret $\hat{\theta}_\tau \approx \theta(\log\tau)$ and $r_\tau \approx \bar{r}(\log\tau)$. In (3.4) $E_{\theta(\tau)}$ denotes the expectation in steady state with respect to the random variables $\{y_k\}$ and $\{\phi_k\}$ generated by (2.1) and (2.5) when the fixed control law

$$u_k = -\frac{1}{b_0(\tau)} \left[\sum_{i=0}^{p} a_i(\tau) y_{k-i} + \sum_{i=1}^{p} b_i(\tau) u_{k-i} \right] \tag{3.5}$$

is used, and

$$\theta(\tau) =: (a_0(\tau), \ldots, a_p(\tau), b_0(\tau), \ldots, b_p(\tau))^T.$$

Since there is no steady state when the control law (3.5) makes the system unstable, the domain of definition of the ODE is restricted to the set of $\theta(\tau)$ which yield a stabilizing control law.

This ODE is a valid approximation only under some other conditions [compare (10.8.7)]. However, for the time being we are interested only in gaining some understanding of the behavior of the adaptive algorithm, and we do not bother with these conditions.

We begin with the first order system,

$$y_{k+1} = a_0 y_k + b_0 u_k + w_{k+1}. \tag{3.6}$$

The stochastic gradient algorithm generates the estimates

$$\hat{a}_{k+1} = \hat{a}_k + \frac{\mu}{k} \frac{(y_{k+1} - \phi_k^T \hat{\theta}_k)}{r_k} y_k, \tag{3.7}$$

$$\hat{b}_{k+1} = \hat{b}_k + \frac{\mu}{k} \frac{(y_{k+1} - \phi_k^T \hat{\theta}_k)}{r_k} u_k, \tag{3.8}$$

where $\hat{\theta}_k =: (\hat{a}_k, \hat{b}_k)^T$ and $\phi_k := (y_k, u_k)^T$. The certainty equivalent control law is

$$u_k := -\frac{\hat{a}_k}{\hat{b}_k} y_k. \tag{3.9}$$

This control law is also given implicitly by

$$\phi_k^T \hat{\theta}_k = 0. \tag{3.10}$$

Using this, (3.7) and (3.8) simplify to

$$\hat{a}_{k+1} = \hat{a}_k + \frac{\mu}{k} \frac{y_{k+1}}{\bar{r}_k} y_k,$$

$$\hat{b}_{k+1} = \hat{b}_k + \frac{\mu}{k} \frac{y_{k+1}}{\bar{r}_k} u_k.$$

The ODE (3.3) is

$$\frac{da(\tau)}{d\tau} = \frac{\mu}{\bar{r}(\tau)} E_{\theta(\tau)}[y_{k+1}y_k],$$

$$\frac{db(\tau)}{d\tau} = \frac{\mu}{\bar{r}(\tau)} E_{\theta(\tau)}[y_{k+1}u_k].$$

Substituting $u_k = -\dfrac{a(\tau)}{b(\tau)} y_k$ gives

$$\frac{da(\tau)}{d\tau} = \frac{\mu}{\bar{r}(\tau)} (a_0 - \frac{b_0 a(\tau)}{b(\tau)}) E_{\theta(\tau)} y_k^2, \tag{3.11}$$

$$\frac{db(\tau)}{d\tau} = \frac{-\mu a(\tau)}{\bar{r}(\tau)b(\tau)} (a_0 - \frac{b_0 a(\tau)}{b(\tau)}) E_{\theta(\tau)} y_k^2. \tag{3.12}$$

Setting the right hand side to 0,

$$a_0 - \frac{b_0 a(\tau)}{b(\tau)} = 0,$$

gives the set of equilibrium points

$$\{\theta = (a, b)^T \mid \frac{a}{b} = \frac{a_0}{b_0}\}, \tag{3.13}$$

which coincides with the set of possible limit points (1.9) of the least squares scheme (1.6) and (1.7). Thus at least as far as the equilibrium or limit points are concerned, the SG scheme exhibits the same behavior as the least squares scheme.

We can go one step further and try to determine whether the trajectories of the ODE actually converge to the set of equilibrium points (3.13). Dividing (3.12) by (3.11) we get

$$\frac{db(\tau)}{da(\tau)} = -\frac{a(\tau)}{b(\tau)},$$

and this describes the solution of the ODE in the ab plane. Solving, we obtain the integral curves of the ODE,

$$a(\tau)^2 + b(\tau)^2 = \text{constant},$$

where the constant is determined by the initial conditions $a(0)$ and $b(0)$. Hence the trajectories are circles in the ab plane. To determine the direction in which these circles are traversed in the course of time we can examine the sign of the right-hand sides of (3.11) or (3.12). The figure sketches the solutions for the case $a_0 > 0$ and $b_0 > 0$.

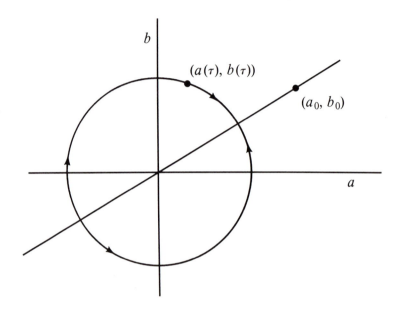

Thus the trajectories of the ODE converge to the set

$$\{\theta = (a, b)^T \mid \frac{a}{b} = \frac{a_0}{b_0} \text{ and } a > 0, b > 0\},$$

(when $a_0 > 0$, $b_0 > 0$). A similar analysis can be carried out when the parameter estimates are generated by the least squares method, as in the following exercise.

Exercise   (3.14)
Consider the first order system (3.6). Suppose the parameter estimates are generated by (2.3) and (2.4). The controls are chosen to satisfy (3.9) or (3.10). Show that the corresponding ODE is

$$\frac{d\theta(\tau)}{d\tau} = \bar{R}^{-1}(\tau) E_{\theta(\tau)}[\phi_k y_{k+1}],$$

$$\frac{d\bar{R}(\tau)}{d\tau} = E_{\theta(\tau)}[\phi_k \phi_k^T] - \bar{R}(\tau).$$

Now consider the Lyapunov function $V(\tau) := \theta(\tau)^T \bar{R}(\tau)\theta(\tau)$, and show that

$$\frac{dV(\tau)}{d\tau} = -V(\tau).$$

[However, $\lim_{\tau \to \infty} \overline{R}(\tau)$ may not be strictly positive definite.]

We return to the general ARX unit delay case. Due to (3.5), the control law is implicitly specified by

$$\phi_k^T \theta(\tau) = 0, \tag{3.15}$$

so the ODE (3.3) can be more simply expressed as

$$\frac{d\theta(\tau)}{d\tau} = \frac{\mu}{\overline{r}(\tau)} E_{\theta(\tau)}[\phi_k y_{k+1}]. \tag{3.16}$$

Drawing our intuition from the analysis of the first order system, we consider the quantity $||\theta(\tau)||^2$. From (3.15) and (3.16),

$$\frac{d ||\theta(\tau)||^2}{d\tau} = \frac{\mu}{\overline{r}(\tau)} \{ E_{\theta(\tau)}[\theta(\tau)^T \phi_k y_{k+1}]$$

$$+ E_{\theta(\tau)}[y_{k+1}\phi_k^T \theta(\tau)]\} = 0.$$

Hence, as before, the trajectories of (3.17) move along the surface of a sphere whose radius is determined by the initial condition. From the earlier analysis of the special case, we can expect $\theta(\tau)$ to converge to a positive multiple of $\theta°$. Since $||\theta(\tau)||$ stays constant, this positive multiple must be $\dfrac{||\theta(0)||}{||\theta°||}$, so we expect

$$\lim_{\tau \to \infty} \theta(\tau) = \frac{||\theta(0)||}{||\theta°||} \theta°.$$

To prove this we may consider the Lyapunov function

$$V(\tau) := || \theta(\tau) - \frac{||\theta(0)||}{||\theta°||} \theta° ||^2.$$

Using (3.16) and

$$E_{\theta(\tau)}[y_{k+1}\phi_k^T \theta°] = E_{\theta(\tau)}[y_{k+1}(y_{k+1} - w_{k+1})]$$

$$= E_{\theta(\tau)}[y_{k+1}^2 - \sigma^2],$$

gives

$$\frac{dV(\tau)}{d\tau} = -\frac{2\mu ||\theta(0)||}{\overline{r}(\tau)||\theta°||} E_{\theta(\tau)}[y_{k+1}^2 - \sigma^2] \le 0. \tag{3.17}$$

Clearly $E_{\theta(\tau)}y_k^2 = \sigma^2$ only when $\theta(\tau)$ gives a minimum variance control law. Moreover, (3.17) also shows that the solution of the ODE moves toward the point $\dfrac{||\theta(0)||}{||\theta°||}\ \theta°$. To prove that it actually converges to this point, we can continue the analysis in the same way as in Exercise (10.8.22).

4. Analysis using stochastic Lyapunov functions

In this section we prove that the SG based certainty equivalent adaptive control law for the ARX unit delay system is self-optimizing with respect to the average cost criterion

$$\lim_{N\to\infty}\frac{1}{N}\sum_{k=0}^{N-1} y_k^2.$$

The system under consideration is

$$y_{k+1} = \sum_{i=0}^{p} a_i y_{k-i} + \sum_{i=0}^{p} b_i u_{k-i} + w_{k+1},\ \ b_0 \neq 0,\tag{4.1}$$

and the adaptive control law is

$$u_k = -\frac{1}{\hat{b}_0(k)}\ [\sum_{i=0}^{p}\hat{a}_i(k)y_{k-i} + \sum_{i=1}^{p}\hat{b}_i(k)u_{k-i}],\tag{4.2}$$

with $\hat{\theta}_k =: (\hat{a}_0(k),\ldots,\hat{a}_p(k),\hat{b}_0(k),\ldots,\hat{b}_p(k))^T$. The estimate $\hat{\theta}_k$ is given by

$$\hat{\theta}_{k+1} = \hat{\theta}_k + \frac{\mu\phi_k}{r_k}\ (y_{k+1} - \phi_k^T\hat{\theta}_k),\tag{4.3}$$

$$r_k = 1 + \sum_{i=0}^{k}\phi_i^T\phi_i,\tag{4.4}$$

where

$$\phi_k := (y_k,\ldots,y_{k-p},u_k,\ldots,u_{k-p})^T.\tag{4.5}$$

It is assumed that $\{w_k\}$ satisfies (2.2) and

$$w_k \text{ has a strictly positive probability density,}\tag{4.6}$$

so that the event $\{\hat{b}_0(k) = 0\}$ has probability 0, and the control law (4.2) is well defined wp 1. The final assumption is that the constant μ in (4.3) satisfies

$$0 < \mu < 2.\tag{4.7}$$

(Later, in Exercise (4.27), this requirement will be eliminated.)

The proof of self-optimality centers around the quadratic form

$$V(k) := ||\widetilde{\theta}_k||^2,$$

where

$$\widetilde{\theta}_k := \hat{\theta}_k - \theta°$$

is the error in the estimate. In the last section, a similar quadratic form served as a Lyapunov function for proving convergence of the trajectories of a deterministic ODE. In the stochastic setting of this section, the analog of a Lyapunov function is a positive supermartingale. [$\{x_k, F_k\}$ is a *supermartingale* if $\{F_k\}$ is an increasing sequence of σ-algebras, x_k is F_k measurable, and for each k, $E\{x_{k+1} \mid F_k\} \le x_k$ *a.s.*]

Let $F_k := \sigma(y_s, u_s, s \le k)$. We calculate $E\{V(k+1) \mid F_k\}$. Observe that

$$\widetilde{\theta}_{k+1} = \widetilde{\theta}_k + \frac{\mu\phi_k}{r_k} (y_{k+1} - \phi_k^T\hat{\theta}_k),$$

and since the control law satisfies $\phi_k^T\hat{\theta}_k = 0$,

$$\widetilde{\theta}_{k+1} = \widetilde{\theta}_k + \frac{\mu\phi_k}{r_k} y_{k+1}.$$

Taking the square of the norm on both sides gives

$$E\{V(k+1) \mid F_k\} = V(k) + \frac{2\mu}{r_k} \phi_k^T\widetilde{\theta}_k E\{y_{k+1} \mid F_k\}$$

$$+ \frac{\mu^2}{r_k^2} \phi_k^T\phi_k E\{y_{k+1}^2 \mid F_k\}. \qquad (4.8)$$

From (4.1)

$$y_{k+1} = \phi_k^T\theta° + w_{k+1} = E\{y_{k+1} \mid F_k\} + w_{k+1}. \qquad (4.9)$$

Hence from (2.2),

$$E\{y_{k+1}^2 \mid F_k\} = (\phi_k^T\theta°)^2 + \sigma^2 = (E\{y_{k+1} \mid F_k\})^2 + \sigma^2.$$

Substituting in (4.8) gives

$$E\{V(k+1) \mid F_k\} = V(k) + \frac{2\mu}{r_k} \phi_k^T\widetilde{\theta}_k E\{y_{k+1} \mid F_k\}$$

$$+ \frac{\mu^2}{r_k^2} \phi_k^T\phi_k (E\{y_{k+1} \mid F_k\})^2$$

$$+ \frac{\mu^2}{r_k^2} \phi_k^T\phi_k \sigma^2.$$

Because the right-hand side may be larger than $V(k)$, $\{V(k), F_k\}$ is not a supermartingale. However, it is a positive *near supermartingale* to which the following convergence result applies.

Theorem (positive near supermartingale convergence)   (4.10)
Let $\{M_k\}$, $\{\alpha_k\}$, and $\{\beta_k\}$ be positive stochastic processes adapted to an increasing family of σ-algebras $\{F_k\}$ such that

$$E\{M_{k+1} \mid F_k\} \leq M_k - \alpha_k + \beta_k \ a.s., \quad \text{and} \ \sum_{k=1}^{\infty} \beta_k < \infty \ a.s.$$

Then $\{M_k\}$ converges a.s. and $\sum_{k=1}^{\infty} \alpha_k < \infty$ a.s.

We show that $V(k)$ is such a positive near supermartingale. Since $r_k \geq \phi_k^T \phi_k$,

$$E\{V(k+1) \mid F_k\}$$

$$\leq V(k) + \frac{2\mu}{r_k} \, \phi_k^T \tilde{\theta}_k E\{y_{k+1} \mid F_k\}$$

$$+ \frac{\mu^2}{r_k} (E\{y_{k+1} \mid F_k\})^2 + \frac{\mu^2}{r_k^2} \, \phi_k^T \phi_k \sigma^2$$

$$= V(k) - \frac{2\mu}{r_k} \, [-\phi_k^T \tilde{\theta}_k - \tfrac{1}{2}\mu E\{y_{k+1} \mid F_k\}] \, E\{y_{k+1} \mid F_k\}$$

$$+ \frac{\mu^2}{r_k^2} \, \phi_k^T \phi_k.$$

Since $\phi_k^T \hat{\theta}_k = 0$ and $\tilde{\theta}_k = \hat{\theta}_k - \theta°$, it follows that $\phi_k^T \tilde{\theta}_k = -\phi_k^T \theta° = -E\{y_{k+1} \mid F_k\}$. So

$$E\{V(k+1) \mid F_k\}$$

$$\leq V(k) - \frac{\mu(2-\mu)}{r_k}(E\{y_{k+1} \mid F_k\})^2 + \frac{\mu^2}{r_k^2}\phi_k^T \phi_k \sigma^2. \quad (4.11)$$

We can apply Theorem (4.10) provided

$$\sum_{k=1}^{\infty} \frac{\phi_k^T \phi_k}{r_k^2} < \infty.$$

But this is true since

$$\sum_{k=1}^{N} \frac{\phi_k^T \phi_k}{r_k^2} = \sum_{k=1}^{N} \frac{r_k - r_{k-1}}{r_k^2} \leq \sum_{k=1}^{N} \frac{r_k - r_{k-1}}{r_k r_{k-1}}$$

$$= \sum_{k=1}^{N} \frac{1}{r_{k-1}} - \frac{1}{r_k} \leq 1.$$

Hence from Theorem (4.10) it follows that   (4.12)

$$\{V(k)\} \text{ converges } a.s.,$$

and also

$$\sum_{k=1}^{\infty} \frac{(E\{y_{k+1} \mid F_k\})^2}{r_k} < \infty \quad a.s. \tag{4.13}$$

By Kronecker's lemma (10.4.9), if

$$r_k > 0 \quad \text{and} \quad \lim_{k \to \infty} r_k = \infty, \tag{4.14}$$

then

$$\lim_{N \to \infty} \frac{1}{r_N} \sum_{k=1}^{N} (E\{y_{k+1} \mid F_k\})^2 = 0 \quad a.s.$$

We need to verify that $\{r_k\}$ satisfies (4.14). Clearly if $\lim r_k < \infty$, then from (4.4) it follows that $\lim y_k^2 = \lim u_k^2 = 0$. From (4.1) it would then also follow that $\lim w_k = 0$. This can happen only on a set of zero probability measure because

$$\lim_{N \to \infty} \frac{1}{N} \sum_{k=0}^{N-1} w_k^2 = \sigma^2 \quad a.s. \tag{4.15}$$

by the martingale stability theorem (8.5.26). Hence

$$\lim_{N \to \infty} \frac{1}{r_N} \sum_{k=1}^{N} (E\{y_{k+1} \mid F_k\})^2 = 0 \quad a.s. \tag{4.16}$$

We want to show that

$$\lim_{N \to \infty} \frac{1}{N} \sum_{k=0}^{N-1} y_k^2 = \sigma^2 \quad a.s. \tag{4.17}$$

because then the adaptive control law would be self-optimizing. Now

$$\sum_{k=1}^{N} y_k^2 = \sum_{k=1}^{N} [(E\{y_{k+1} \mid F_k\})^2 + w_{k+1}^2$$
$$+ 2 w_{k+1} E\{y_{k+1} \mid F_k\}]. \tag{4.18}$$

By the Schwarz inequality

$$\sum_{k=1}^{N} w_{k+1} E\{y_{k+1} \mid F_k\} = \alpha(N) [\sum_{k=1}^{N} (E\{y_{k+1} \mid F_k\})^2]^{1/2}$$
$$\times [\sum_{k=1}^{N} w_{k+1}^2]^{1/2}$$

for some $|\alpha(N)| \le 1$. Hence if

$$\lim_{N \to \infty} \frac{1}{N} \sum_{k=0}^{N-1} (E\{y_{k+1} \mid F_k\})^2 = 0 \quad a.s., \tag{4.19}$$

then (4.17) will follow from (4.15) and (4.18). Suppose now that

$$\{\frac{r_N}{N}\} \text{ is bounded } a.s., \tag{4.20}$$

then clearly (4.16) would imply (4.19). Hence our goal is reduced to showing that (4.20) is true. We need to make another assumption.

Assumption (4.21)
The system (4.1) is minimum phase, i.e., all the roots of the polynomial $B(z) := b_0 + b_1 z + .. + b_p z^p$ are strictly outside the closed unit disk in the complex plane.

In Section (7.8) we saw that this assumption guarantees that the minimum variance control law leads to an internally stable system.

Minimum phase systems can be inverted to obtain $\{u_k\}$ as the output of an asymptotically stable linear system driven by $\{y_k\}$ and $\{w_k\}$. Hence there exist constants C_i such that

$$\frac{1}{N} \sum_{k=1}^{N} u_k^2 \leq \frac{C_1}{N} \sum_{k=1}^{N} y_{k+1}^2 + \frac{C_2}{N} \sum_{k=1}^{N} w_{k+1}^2 \quad \text{for all } N.$$

From (4.15) it follows that

$$\frac{1}{N} \sum_{k=1}^{N} u_k^2 \leq \frac{C_1}{N} \sum_{k=1}^{N} y_{k+1}^2 + C_3.$$

Therefore

$$\frac{r_N}{N} \leq \frac{C_4}{N} \sum_{k=1}^{N} y_{k+1}^2 + C_5.$$

In view of (4.9) and (4.15) it follows that

$$\frac{r_N}{N} \leq \frac{C_6}{N} \sum_{k=1}^{N} (E\{y_{k+1} \mid F_k\})^2 + C_7,$$

or, equivalently,

$$\frac{1}{r_N} \sum_{k=1}^{N} (E\{y_{k+1} \mid F_k\})^2 \geq \frac{r_N - NC_7}{C_6 r_N}$$

for some $C_6 > 0$. Suppose now that (4.20) is not true. Then taking the limit in the above along some subsequence $\{N_k\}$, we get

$$\lim_{k \to \infty} \frac{1}{r_{N_k}} \sum_{k=1}^{N_k} (E\{y_{k+1} \mid F_k\})^2 \geq \frac{1}{2} C_6 > 0, \tag{4.22}$$

which contradicts (4.16). This proves (4.20).

Theorem (4.23)
Consider the system (4.1) where the noise $\{w_k\}$ satisfies (2.2) and (4.6)
and assume that the system is minimum phase. If the stochastic gradient
based adaptive control law (4.2)-(4.5), (4.7) is used, then:
(1) The adaptive control law is self-optimizing with respect to the
minimum variance criterion, i.e.,

$$\lim_{N \to \infty} \frac{1}{N} \sum_{k=0}^{N-1} y_k^2 = \sigma^2 \ \text{a.s.}$$

(2) Moreover,

$$\sum_{k=1}^{\infty} \frac{(E\{y_{k+1} \mid F_k\})^2}{r_k} < \infty \tag{4.24}$$

$$\lim_{N \to \infty} \frac{1}{N} \sum_{k=0}^{N-1} (E\{y_{k+1} \mid F_k\})^2 = 0 \ \text{a.s.} \tag{4.25}$$

$$\lim_{N \to \infty} \sup \frac{1}{N} \sum_{k=0}^{N-1} u_k^2 < \infty \ \text{a.s.}$$

$$\|\tilde{\theta}_k\| \text{ converges } \ \text{a.s.} \tag{4.26}$$

Exercise (4.27)
Show that except possibly for (4.26), all the results of Theorem (4.23)
continue to hold for arbitrary $\mu \neq 0$.
[Hint: If $\hat{\theta}_0$ and μ are simultaneously scaled to $\alpha\hat{\theta}_0$ and $\alpha\mu$ respectively,
show by induction on k that $\hat{\theta}_k$ is then scaled to $\alpha\hat{\theta}_k$. Use the fact that
the control law (4.2) is invariant with respect to scaling of the parameter
estimate vector. Hence the resulting controls remain the same, and so the
new control law must still be self-optimizing.]

5. A geometric property of the SG algorithm

Starting with this section we prove the stronger result that the algorithm
is self-tuning.

Again we draw our intuition from the ODE analysis of Section 3.
We saw there that the solution $\theta(\tau)$ of the ODE moves toward a scalar
multiple of the true parameter. Here we establish such a convergence
result for the actual random sequence of estimates $\{\hat{\theta}_k\}$.

Recall that the solution of the ODE lies on the surface of a sphere
centered at the origin. We begin with a related result for $\{\hat{\theta}_k\}$.

Lemma (5.1)
$\hat{\theta}_{k+1} - \hat{\theta}_k$ is orthogonal to $\hat{\theta}_k$.

Proof
The recursion (4.3) for $\{\hat{\theta}_k\}$ can be written as

$$\hat{\theta}_{k+1} = \hat{\theta}_k + \gamma_k \phi_k,$$

where $\gamma_k := \dfrac{\mu y_{k+1}}{r_k}$ is a scalar. Hence $(\hat{\theta}_{k+1} - \hat{\theta}_k)$ is parallel to ϕ_k. On the other hand, since u_k satisfies $\phi_k^T \hat{\theta}_k = 0$, ϕ_k is orthogonal to $\hat{\theta}_k$ and the result follows. $\square$

The orthogonality relation is a property of the algorithm (4.2)-(4.5) and not the system being controlled by the use of this algorithm. It has several interesting consequences.

Lemma   (5.2)

(1) $||\hat{\theta}_k||^2 = ||\hat{\theta}_0||^2 + \sum\limits_{n=0}^{k} ||\hat{\theta}_n - \hat{\theta}_{n-1}||^2,$

(2) $||\hat{\theta}_k||$ increases with k.

(3) If $||\tilde{\theta}_k||$ converges, so does $||\hat{\theta}_k||$.

Proof
The proof is obvious since by Pythagoras' Theorem,

$$||\hat{\theta}_{k+1}||^2 = ||\hat{\theta}_k||^2 + ||\hat{\theta}_{k+1} - \hat{\theta}_k||^2.$$

For (3) simply note that if $||\tilde{\theta}_k||$ converges, then $||\hat{\theta}_k||$ is bounded, and since by (2) it is an increasing sequence, it converges. $\square$

By Theorem (4.23), $||\hat{\theta}_k - \theta^{\circ}||$ converges, so Lemma (5.2) applies and we have the following result.

Lemma   (5.3)
$||\hat{\theta}_k||$ and $||\hat{\theta}_k - \theta^{\circ}||$ both converge.

According to Lemma (5.3), $\hat{\theta}_k$ converges to the intersection of two spheres, one centered at the origin and the other centered at θ°. This has a nice geometric interpretation. First, exclude the uninteresting case $\theta^{\circ} = 0$. Two spheres in R^{2p+2} with different centers intersect in a hyper-sphere H of strictly smaller dimension. Hence $\hat{\theta}_k$ converges to H. To show that $\hat{\theta}_k$ converges to a point, we must show that H consists of exactly one point. That will only happen if the two spheres are tangential to each other. Moreover, if the spheres are indeed tangential, then their point of intersection lies on the straight line L passing through the centers of the spheres. $\hat{\theta}_k$ will then converge to a point on L. Recall now that our goal is not only to show that the sequence $\hat{\theta}_k$ converges, but also that it converges to a multiple of θ°. Since L consists precisely of all mul-tiples of θ°, we would have accomplished our goal.

Now H consists of only one point if and only if $H \cap L$ is not empty, since otherwise L will not intersect it. To show that H and L indeed have one point in common, it is enough to show there is a subsequence $\{\hat{\theta}_{k_n}\}$ which converges to L, because we already know that H contains all the limit points of $\{\hat{\theta}_k\}$. The following analysis is therefore aimed at finding a subsequence $\{k_n\}$ so that

$$\lim_{n \to \infty} \hat{\theta}_{k_n} = \gamma\theta° \text{ for some random scalar } \gamma. \tag{5.4}$$

The next lemma makes the geometric argument precise.

Lemma   (5.5)
Let $\theta°(i)$ and $\hat{\theta}_k(i)$ denote the ith components of the vectors $\theta°$ and $\hat{\theta}_k$ respectively. The following statements are equivalent.
(1) There is a subsequence $\{k_n\}$ such that (5.4) holds.
(2) There is a subsequence $\{k_n\}$ such that

$$\lim_{n \to \infty} \theta°(i)\,\hat{\theta}_{k_n}(p+2) - \theta°(p+2)\,\hat{\theta}_{k_n}(i) = 0 \tag{5.6}$$

for $i = 1, .., p + 2$.

(3) $\lim_{k \to \infty} \hat{\theta}_k = \gamma\theta°$ for some scalar γ.

Proof
Clearly (1) implies (2) and (3) implies (1). It remains to show that (2) implies (3). Since $\hat{\theta}_k$ is bounded, we can assume that $\lim_{n \to \infty} \hat{\theta}_{k_n} = \theta^*$ for some θ^*. By assumption (4.1), $\theta°(p+2) = b_0 \neq 0$, and so from (5.6),

$$\lim_{n \to \infty} \hat{\theta}_{k_n}(i) = \frac{\theta^*(p+2)}{\theta°(p+2)}\,\theta°(i), \quad i = 1, .., 2p, \tag{5.7}$$

i.e.,

$$\lim_{n \to \infty} \hat{\theta}_{k_n} = \gamma\theta°, \tag{5.8}$$

where $\gamma := \dfrac{\theta^*(p+2)}{\theta°(p+2)}$.

Since $||\tilde{\theta}_k||$ converges and $\tilde{\theta}_k = \hat{\theta}_k - \theta°$, $||\hat{\theta}_k||^2 - 2<\theta°, \tilde{\theta}_k> + ||\theta°||^2$ converges. Since $||\hat{\theta}_k||^2$ also converges, it follows that $<\theta°, \tilde{\theta}_k>$ converges. We can conclude that $\{||\hat{\theta}_k - \gamma\theta°||^2\}$ converges since $||\hat{\theta}_k - \gamma\theta°||^2 = ||\hat{\theta}_k||^2 - 2\gamma<\hat{\theta}_k, \theta°> + k^2||\theta°||^2$ and each term converges. From (5.8) it follows that

$$\lim_{k \to \infty} ||\hat{\theta}_k - \gamma\theta°||^2 = 0.$$

$\square$

If $\theta°$ is two-dimensional one need not even establish (5.6), as the following exercise shows.

Exercise   (5.9)
Suppose $\theta° \in R^2$ and $\theta° \neq 0$. Show that if $\hat{\theta}_{k+1} - \hat{\theta}_k$ is orthogonal to $\hat{\theta}_k$ and $\{\|\hat{\theta}_k - \theta°\|\}$ converges, then $\{\hat{\theta}_k\}$ converges. (However the limit need not be a multiple of $\theta°$.)
[Hint: Two circles with different centers in R^2 can intersect in at most two points. Now use the fact that $\{\|\hat{\theta}_k\|\}$ is an increasing sequence.]

In higher dimensions, however, the result above need not hold.

Exercise   (5.10)
Suppose the conditions of Exercise (5.9) hold except that $\theta° \in R^3$. Show that $\{\hat{\theta}_k\}$ need not converge.

[Hint: Consider $\hat{\theta}_k := (\rho_k \cos\alpha_k, \rho_k \sin\alpha_k, 1)^T$ where $\alpha_{k+1} := \alpha_k + \dfrac{1}{k}$ and $\rho_{k+1} := \rho_k \sec\dfrac{1}{k}$.]

6. The first order system: proof of self-tuning

Before taking up the proof of self-tuning in the general case, we study the first order case,

$$y_{k+1} = a_0 y_k + b_0 u_k + w_{k+1}, \quad b_0 \neq 0.$$

Let $(\hat{a}_k, \hat{b}_k)^T =: \hat{\theta}_k$, and note that by Lemma (5.5) our goal is to show that

$$\lim_{n \to \infty} a_0 \hat{b}_{k_n} - b_0 \hat{a}_{k_n} = 0 \quad a.s. \tag{6.1}$$

By (4.25)

$$\lim_{N \to \infty} \frac{1}{N} \sum_{k=0}^{N-1} (E\{y_{k+1} \mid F_k\})^2 = 0 \quad a.s.$$

Substituting $E\{y_{k+1} \mid F_k\} = a_0 y_k + b_0 u_k$ and $u_k = -\dfrac{\hat{a}_k}{\hat{b}_k} y_k$ gives

$$\lim_{N \to \infty} \frac{1}{N} \sum_{k=0}^{N-1} [(a_0 - b_0 \frac{\hat{a}_k}{\hat{b}_k}) y_k]^2 = 0 \quad a.s.$$

Since $\{\hat{b}_k\}$ is bounded, we can multiply the summand above by $\hat{b}_k^2$ and get

$$\lim_{N \to \infty} \frac{1}{N} \sum_{k=0}^{N-1} (a_0 \hat{b}_k - b_0 \hat{a}_k)^2 y_k^2 = 0 \quad a.s. \tag{6.2}$$

The idea behind the remainder of the proof is this: If we can replace the term y_k in (6.2) by 1, we get $\lim\limits_{N \to \infty} \dfrac{1}{N} \sum\limits_{k=0}^{N-1} (a_0 \hat{b}_k - b_0 \hat{a}_k)^2 = 0$. Then it will follow that there exists a subsequence satisfying (6.1). To establish this, we proceed as follows.

By Lemma (5.2), $||\hat{\theta}_N||^2 = \sum_{k=0}^{N} ||\hat{\theta}_k - \hat{\theta}_{k-1}||^2$, and, since $\{||\hat{\theta}_N||^2\}$ converges,

$$\lim_{k \to \infty} (\hat{\theta}_k - \hat{\theta}_{k-1}) = 0 \ \ a.s.$$

Exercise (6.3)

Suppose $\lim_{N \to \infty} \frac{1}{N} \sum_{k=0}^{N-1} \gamma_k y_k^2 = 0$ where $\{\gamma_k\}$ is a bounded sequence with $\lim_{k \to \infty} (\gamma_k - \gamma_{k-1}) = 0$. Use $\lim_{N \to \infty} \frac{1}{N} \sum_{k=0}^{N-1} y_k^2 = \sigma^2$ [by Theorem (4.23)], to show that

$$\lim_{N \to \infty} \frac{1}{N} \sum_{k=0}^{N-1} \gamma_{k-1} y_k^2 = 0.$$

Taking $\gamma_k := (a_0 \hat{b}_k - b_0 \hat{a}_k)^2$, Exercise (6.3) shows that in (6.2) we can replace $\hat{a}_k$ and $\hat{b}_k$ by $\hat{a}_{k-1}$ and $\hat{b}_{k-1}$; hence

$$\lim_{N \to \infty} \frac{1}{N} \sum_{k=0}^{N-1} (a_0 \hat{b}_k - b_0 \hat{a}_k)^2 y_{k+1}^2 = 0 \ \ a.s. \tag{6.4}$$

Since the system $y_{k+1} = a_0 y_k + b_0 u_k + w_{k+1}$ can also be written as

$$y_{k+1} = E\{y_{k+1} \mid F_k\} + w_{k+1},$$

substituting for y_{k+1} gives

$$\lim_{N \to \infty} \frac{1}{N} \sum_{k=0}^{N-1} (a_0 \hat{b}_k - b_0 \hat{a}_k)^2 [(E\{y_{k+1} \mid F_k\})^2$$
$$+ 2E\{y_{k+1} \mid F_k\} w_{k+1} + w_{k+1}^2] = 0 \ \ a.s. \tag{6.5}$$

We next analyze the cross term $\frac{1}{N} \sum_{k=1}^{N} (a_0 \hat{b}_k - b_0 \hat{a}_k)^2 E\{y_{k+1} \mid F_k\} w_{k+1}$

in (6.5). Let $\widetilde{\Omega} := \{\omega \mid \sum_{k=1}^{\infty} (a_0 \hat{b}_k - b_0 \hat{a}_k)^4 (E\{y_{k+1} \mid F_k\})^2 < \infty\}$ and consider two cases.

Case 1: $\omega \in \widetilde{\Omega}$.
By Theorem (11.2.7),

$$\sum_{k=1}^{\infty} (a_0 \hat{b}_k - b_0 \hat{a}_k)^2 E\{y_{k+1} \mid F_k\} w_{k+1} \text{ converges,}$$

and so

$$\lim_{N \to \infty} \frac{1}{N} \sum_{k=0}^{N-1} (a_0 \hat{b}_k - b_0 \hat{a}_k)^2 E\{y_{k+1} \mid F_k\} w_{k+1} = 0.$$

Hence (6.5) reduces to

$$\lim_{N \to \infty} \frac{1}{N} \sum_{k=0}^{N-1} (a_0 \hat{b}_k - b_0 \hat{a}_k)^2 [(E\{y_{k+1} \mid F_k\})^2 + w_{k+1}^2] = 0.$$

Since both terms in the summand are positive, this implies

$$\lim_{N \to \infty} \frac{1}{N} \sum_{k=0}^{N-1} (a_0 \hat{b}_k - b_0 \hat{a}_k)^2 (E\{y_{k+1} \mid F_k\})^2 = 0, \qquad (6.6)$$

and

$$\lim_{N \to \infty} \frac{1}{N} \sum_{k=0}^{N-1} (a_0 \hat{b}_k - b_0 \hat{a}_k)^2 w_{k+1}^2 = 0. \qquad (6.7)$$

Case 2: $\omega \notin \widetilde{\Omega}$.
By Theorem (11.2.7) again,

$$\sum_{k=1}^{N} (a_0 \hat{b}_k - b_0 \hat{a}_k)^2 E\{y_{k+1} \mid F_k\} w_{k+1}$$

$$= o[\sum_{k=1}^{N} (a_0 \hat{b}_k - b_0 \hat{a}_k)^4 (E\{y_{k+1} \mid F_k\})^2]$$

$$= o[\sum_{k=1}^{N} (a_0 \hat{b}_k - b_0 \hat{a}_k)^2 (E\{y_{k+1} \mid F_k\})^2],$$

since $\{(a_0 \hat{b}_k - b_0 \hat{a}_k)^2\}$ is bounded. Now we can rewrite (6.5) as

$$\lim_{N \to \infty} \frac{1}{N} \sum_{k=1}^{N} (a_0 \hat{b}_k - b_0 \hat{a}_k)^2 w_{k+1}^2$$

$$+ \frac{1}{N} \sum_{k=1}^{N} (a_0 \hat{b}_k - b_0 \hat{a}_k)^2 (E\{y_{k+1} \mid F_k\})^2$$

$$\times \{1 + \frac{o(\frac{1}{N} \sum_{k=1}^{N} (a_0 \hat{b}_k - b_0 \hat{a}_k)^2 (E\{y_{k+1} \mid F_k\})^2)}{\frac{1}{N} \sum_{k=1}^{N} (a_0 \hat{b}_k - b_0 \hat{a}_k)^2 (E\{y_{k+1} \mid F_k\})^2} \} = 0.$$

Since both terms above are positive for N sufficiently large, we can again conclude that (6.6) and (6.7) hold.

Thus (6.6) and (6.7) hold a.s. Let us focus on (6.7). Since $\{(a_0 \hat{b}_k - b_0 \hat{a}_k)^2\}$ is bounded, we can apply the martingale stability theorem (8.5.27) to (6.7) using the moment properties on the noise (2.2). This gives

$$\lim_{N\to\infty} \frac{1}{N} \sum_{k=0}^{N-1} (a_0 \hat{b}_k - b_0 \hat{a}_k)^2 = 0 \quad a.s. \tag{6.8}$$

and proves (5.7), since (6.8) implies that

$$\lim_k \inf (a_0 \hat{b}_k - b_0 \hat{a}_k)^2 = 0.$$

Hence Lemma (5.5) applies and the stochastic gradient based adaptive control algorithm is self-tuning for the first order system.

7. The ARX system: proof of self-tuning

To prove self-tuning in the general case, we follow the same argument as in the first order case, except for one additional feature.

Let $\theta = (\theta(1), \ldots, \theta(2p + 2))^T$ be the estimated parameter at time k. Then the control applied is

$$u_k = -\frac{\theta(1) + \theta(2)z + \ldots + \theta(p + 1)z^p}{\theta(p + 2) + \theta(p + 3)z + \ldots + \theta(2p + 2)z^p} \; y_k.$$

where z denotes the right (or backward) shift operator. Recall that the minimum variance control law for the system (2.1) is

$$u_k = -\frac{a_0 + a_1 z + \ldots + a_p z^p}{b_0 + b_1 z + \ldots + b_p z^p} \; y_k.$$

Definition (7.1)

A parameter θ produces a minimum variance control law if

$$\frac{\theta(1) + \theta(2)z + \ldots + \theta(p + 1)z^p}{\theta(p + 2) + \theta(p + 3)z + \ldots + \theta(2p + 2)z^p}$$
$$= \frac{a_0 + a_1 z + \ldots + a_p z^p}{b_0 + b_1 z + \ldots + b_p z^p} \; .$$

Let Θ_{MV} be the set of parameters that produce a minimum variance control law.

Definition (7.2)

The adaptive control law is **self-tuning in a Cesaro sense** if

$$\lim_{N\to\infty} \frac{1}{N} \sum_{k=0}^{N-1} 1(\hat{\theta}_k \in O) = 1 \quad a.s.$$

for every open set O containing Θ_{MV}.

We show later that any adaptive control law that is self-optimizing (as is the case for the SG based algorithm) is also self-tuning in a Cesaro sense.

Clearly $\Theta_{MV} \supset L := \{\gamma\theta^\circ \mid \gamma \in R\}$ (see Section 5), i.e., the minimum variance control law is unaffected if all coefficients are simultaneously scaled. We would like $\Theta_{MV} = L$; then from (7.2) we can show that there is a subsequence $\{\hat{\theta}_{k_n}\}$ which converges to L.

However, in general, Θ_{MV} may be larger than L. To see this, observe that any control law

$$u_k = -\frac{F(z)}{G(z)} y_k,$$

is also a minimum variance control law provided

$$\frac{F(z)}{G(z)} = \frac{a_0 + a_1 z + .. + a_p z^p}{b_0 + b_1 z + .. + b_p z^p}. \tag{7.3}$$

Hence one can obtain minimum variance control laws by the process of canceling common factors.

We say that the ARX unit delay system (2.1) does not have a reduced order minimum variance control law if there exist no polynomials $F(z)$ and $G(z)$, both of degree less than or equal to $p-1$ such that (7.3) holds. [Equivalently, $F(z)(b_0 + .. + b_p z^p) = G(z)(a_0 + .. + a_p z^p)$ implies $F(z) = G(z) = 0$.]

Exercise (7.4)
Show that the system has a reduced order minimum variance controller if and only if either
(1) $\deg(a_0 + .. + a_p z^p) \le p - 1$ and $\deg(b_0 + .. + b_p z^p) \le p - 1$, or
(2) $(a_0 + .. + a_p z^p)$ and $(b_0 + .. + b_p z^p)$ have a common factor.

Our proof proceeds as follows. We first establish self-tuning in a Cesaro sense; then we impose the assumption that the system has no reduced order minimum variance controllers and conclude that the adaptive controller is self-tuning.

The following results will find repeated use.

Exercise (7.5)
Let ρ_k be a sequence satisfying

$$\lim_{N\to\infty} \frac{1}{N} \sum_{k=0}^{N-1} \rho_k^2 = 0.$$

Then for any sequence $\{\psi_k\}$,

$$\lim_{N\to\infty} \frac{1}{N} \sum_{k=0}^{N-1} \psi_k^2 = 0$$

if and only if

$$\lim_{N\to\infty} \frac{1}{N} \sum_{k=0}^{N-1} (\rho_k + \psi_k)^2 = 0.$$

[Hint: Use the inequalities $(\rho_k + \psi_k)^2 \leq 2\rho_k^2 + 2\psi_k^2$ and $\psi_k^2 \leq 2\rho_k^2 + 2(\rho_k + \psi_k)^2$.]

Exercise (7.6)

Let $\{\pi_k\}$ and $\{s_k\}$ be stochastic processes such that

$$\pi_k \text{ and } s_k \text{ are } F_k \text{ measurable,} \tag{7.7}$$

$$\{\pi_k\} \text{ is bounded,} \tag{7.8}$$

$$\lim_{N\to\infty} \frac{1}{N} \sum_{k=0}^{N-1} (\pi_k w_{k+1} + s_k)^2 = 0 \quad a.s. \tag{7.9}$$

Show that

$$\lim_{N\to\infty} \frac{1}{N} \sum_{k=0}^{N-1} \pi_k^2 = 0 \quad a.s. \text{ and } \lim_{N\to\infty} \frac{1}{N} \sum_{k=0}^{N-1} s_k^2 = 0 \quad a.s. \tag{7.10}$$

[Hint: Rewrite (7.9) as

$$\lim_{N\to\infty} \frac{1}{N} \sum_{k=0}^{N-1} (s_k^2 + 2s_k \pi_k w_{k+1} + \pi_k^2 w_{k+1}^2) = 0.$$

Now analyze the cross term $\{s_k \pi_k w_{k+1}\}$ in the same way as the cross term of (6.5) was analyzed, to deduce as in (6.6) and (6.7) that

$$\lim_{N\to\infty} \frac{1}{N} \sum_{k=0}^{N-1} \pi_k^2 w_{k+1}^2 = 0 \text{ and } \lim_{N\to\infty} \frac{1}{N} \sum_{k=0}^{N-1} s_k^2 = 0.$$

Then use the boundedness of $\{\pi_k\}$ to apply the martingale stability theorem and conclude that $\lim_{N\to\infty} \frac{1}{N} \sum_{k=0}^{N-1} \pi_k^2 = 0$.]

Exercise (7.11)

Suppose $\{\eta_k\}$ is another bounded F_k-adapted sequence, while $\{\pi_k\}$ and $\{s_k\}$ satisfy (7.7) and (7.8). Suppose that

$$\lim_{N\to\infty} \frac{1}{N} \sum_{k=0}^{N-1} (\eta_k y_{k+1} + \pi_k w_{k+1} + s_k)^2 = 0 \quad a.s. \tag{7.12}$$

Show that (7.10) holds and

$$\lim_{N\to\infty} \frac{1}{N} \sum_{k=0}^{N-1} (\eta_k + \pi_k)^2 = 0 \quad a.s. \tag{7.13}$$

[Hint: Note that $y_{k+1} = E\{y_{k+1} \mid F_k\} + w_{k+1}$; hence (7.12) can be rewritten as

$$\lim_{N\to\infty} \frac{1}{N} \sum_{k=0}^{N-1} [(\eta_k + \pi_k) w_{k+1} + (\eta_k E\{y_{k+1} \mid F_k\} + s_k)]^2 = 0.$$

This is in a form that meets the conditions of the preceding exercise, and so (7.13) follows immediately. Also,

$$\lim_{N\to\infty} \frac{1}{N} \sum_{k=0}^{N-1} (\eta_k E\{y_{k+1} \mid F_k\} + s_k)^2 = 0.$$

Now use the boundedness of $\{\eta_k\}$ and (4.25) to get

$$\lim_{N\to\infty} \frac{1}{N} \sum_{k=0}^{N-1} (\eta_k E\{y_{k+1} \mid F_k\})^2 = 0.$$

Then use Lemma (7.5) to conclude (7.10).]

We are now ready to start the proof. As in the first order case begin by considering (4.25). By substituting for $E\{y_{k+1} | F_k\}$ from (4.9) we have

$$\lim_{N\to\infty} \frac{1}{N} \sum_{k=0}^{N-1} (a_0 y_k + \ldots + a_p y_{k-p} + b_0 u_k + \ldots + b_p u_{k-p})^2$$

$$= 0 \ a.s. \tag{7.14}$$

Substituting for u_k from (4.2) gives

$$\lim_{N\to\infty} \frac{1}{N} \sum_{k=0}^{N-1} [(a_0 - b_0 \frac{\hat{a}_0(k)}{\hat{b}_0(k)}) \, y_k + \ldots$$

$$+ (a_p - b_0 \frac{\hat{a}_p(k)}{\hat{b}_0(k)}) \, y_{k-p}$$

$$+ (b_1 - b_0 \frac{\hat{b}_1(k)}{\hat{b}_0(k)}) \, u_{k-1} + \ldots$$

$$+ (b_p - b_0 \frac{\hat{b}_p(k)}{\hat{b}_0}) \, u_{k-p}]^2 = 0 \ a.s.$$

Since $\{\hat{b}_0(k)\}$ is bounded by (4.26), we can multiply by it and get

$$\lim_{N\to\infty} \frac{1}{N} \sum_{k=0}^{N-1} (\alpha_0(k)y_k + \ldots + \alpha_p(k)y_{k-p}$$

$$+ \beta_1(k)u_{k-1} + \ldots + \beta_p(k)u_{k-p})^2 = 0 \ a.s., \tag{7.15}$$

where

$$\alpha_i(k) := a_i \hat{b}_0(k) - b_0 \hat{a}_i(k), \tag{7.16}$$

$$\beta_i(k) := b_i \hat{b}_0(k) - b_0 \hat{b}_i(k). \tag{7.17}$$

By Lemmas (5.2) and (5.3), $\sum_{k=0}^{N} ||\hat{\theta}_k - \hat{\theta}_{k-1}||^2 = ||\hat{\theta}_N||^2$ converges, hence $\lim_{k\to\infty} \hat{\theta}_k - \hat{\theta}_{k-l} = 0$ for every l, so

$$\lim_{k \to \infty} \alpha_i(k) - \alpha_i(k - l) = 0 \quad \text{and}$$

$$\lim_{k \to \infty} \beta_i(k) - \beta_i(k - l) = 0, \ l \geq 1.$$

The following exercise shows that one can replace $\alpha_i(k)$ and $\beta_i(k)$ by $\alpha_i(k - l)$ and $\beta_i(k - l)$ in (7.15).

Exercise   (7.18)

Suppose $\{x_k\}$, $\{v_k\}$, and $\{z_k\}$ are sequences such that

$$\lim_{N \to \infty} \sup \frac{1}{N} \sum_{k=0}^{N-1} v_k^2 < \infty,$$

$$\lim_{k \to \infty} (x_k - x_{k-1}) = 0,$$

$$\lim_{N \to \infty} \frac{1}{N} \sum_{k=0}^{N-1} (x_k v_k + z_k)^2 = 0.$$

Show that

$$\lim_{N \to \infty} \frac{1}{N} \sum_{k=0}^{N-1} (x_{k-1} v_k + v_k)^2 = 0.$$

[Hint: Note that $\lim_{N \to \infty} \frac{1}{N} \sum_{k=0}^{N-1} [(x_k - x_{k-1}) v_k]^2 = 0$. Now use Lemma (7.5).]

Exercise   (7.19)

Show that for every $l \geq 1$,

$$\lim_{N \to \infty} \frac{1}{N} \sum_{k=0}^{N-1} [\alpha_0(k - l) y_k + \ldots + \alpha_p(k - l) y_{k-p}$$

$$+ \beta_1(k - l) u_{k-1} + \ldots + \beta_p(k - l) u_{k-p}]^2 = 0 \ a.s. \quad (7.20)$$

[Hint: Use Exercise (7.18) on (7.15). Identify v_k with y_{k-1} or u_{k-l} and x_k with $\alpha_l(k)$ or $\beta_l(k)$.]

We can apply Exercise (7.11) to (7.20). Let l be a large integer. Identifying η_k with $\alpha_0(k - l)$ and setting $\pi_k = 0$, gives

$$\lim_{N \to \infty} \frac{1}{N} \sum_{k=0}^{N-1} (\alpha_0(k - l))^2 = 0 \ a.s., \quad (7.21)$$

and also

$$\lim_{N \to \infty} \frac{1}{N} \sum_{k=0}^{N-1} s_k^2 = 0 \ a.s.,$$

where

$$s_k := \sum_{i=1}^{p} [\alpha_i(k-l)y_{k-i} + \beta_i(k-l)u_{k-i}].$$

Using Lemma (7.5) we have

$$\lim_{N \to \infty} \frac{1}{N} \sum_{k=0}^{N-1} \left(\sum_{j=0}^{p} b_j s_{k-j} \right)^2 = 0 \quad a.s.$$

or, equivalently, substituting for s_{k-j},

$$\lim_{N \to \infty} \frac{1}{N} \sum_{k=0}^{N-1} [\sum_{j=0}^{p} b_j \sum_{i=1}^{p} \alpha_i(k-j-l)y_{k-j-i}$$
$$+ \beta_i(k-j-l)u_{k-j-i}]^2 = 0 \quad a.s.$$

As before we can change $\alpha_i(k-j-l)$ and $\beta_i(k-j-l)$ to $\alpha_i(k-l)$ and $\beta_i(k-l)$ and get

$$\lim_{N \to \infty} \frac{1}{N} \sum_{k=0}^{N-1} [\sum_{j=0}^{p} b_j \sum_{i=1}^{p} \alpha_i(k-l)y_{k-j-i} + \beta_i(k-l)u_{k-j-i}]^2$$
$$= 0 \quad a.s. \tag{7.22}$$

Now recall (4.25),

$$\lim_{N \to \infty} \frac{1}{N} \sum_{k=0}^{N-1} (E\{y_{k+1} \mid F_k\})^2 = 0 \quad a.s.$$

Since $\{\beta_i(k-l)\}$ is bounded,

$$\lim_{N \to \infty} \frac{1}{N} \sum_{k=0}^{N-1} [\sum_{i=1}^{p} \beta_i(k-l)E\{y_{k-i+1} \mid F_{k-i}\}]^2 = 0 \quad a.s.$$

Substituting $E\{y_{k-i+1} \mid F_{k-i}\} = \sum_{j=0}^{p} a_j y_{k-i-j} + b_j u_{k-i-j}$ we get

$$\lim_{N \to \infty} \frac{1}{N} \sum_{k=0}^{N-1} [\sum_{i=1}^{p} \beta_i(k-l) \sum_{j=0}^{p} a_j y_{k-i-j} + b_j u_{k-i-j}]^2$$
$$= 0 \quad a.s. \tag{7.23}$$

Using Lemma (7.5) we can subtract the square root of the summand in (7.23) from the square root of the summand in (7.22) to get

$$\lim_{N \to \infty} \frac{1}{N} \sum_{k=0}^{N-1} [\sum_{i=1}^{p} \sum_{j=0}^{p} (b_j \alpha_i(k-l) - a_j \beta_i(k-l))y_{k-i-j}]^2$$
$$= 0 \quad a.s.$$

By setting $m := i + j$ this can be rewritten as

$$\lim_{N\to\infty} \frac{1}{N} \sum_{k=0}^{N-1} [\sum_{j=0}^{p} \sum_{m=j+1}^{j+p} (b_j\alpha_{m-j}(k-l)$$

$$- a_j\beta_{m-j}(k-l))y_{k-m}]^2 = 0 \ a.s.$$

Reversing the order of summation gives

$$\lim_{N\to\infty} \frac{1}{N} \sum_{k=0}^{N-1} [\sum_{m=1}^{p} \sum_{j=0}^{m-1} (b_j\alpha_{m-j}(k-l) - a_j\beta_{m-j}(k-l))y_{k-m}$$

$$+ \sum_{m=p+1}^{2p} \sum_{j=m-p}^{p} (b_j\alpha_{m-j}(k-l) - a_j\beta_{m-j}(k-l))y_{k-m}]^2$$

$$= 0 \ a.s.$$

Now let $l > 2p + 1$. By identifying η with $(b_j\alpha_1 - a_j\beta_1)$ and setting $\pi_k = 0$, we can use Lemma (7.11) to extract the $m = 1$ (and hence $j = 0$) term and get

$$\lim_{N\to\infty} \frac{1}{N} \sum_{k=0}^{N-1} [b_0\alpha_m(k-l) - a_0\beta_1(k-l)]^2 = 0 \ a.s.$$

and also

$$\lim_{N\to\infty} \frac{1}{N} \sum_{k=0}^{N-1} [\sum_{m=2}^{p} \sum_{j=0}^{k-1} (b_j\alpha_{m-j}(k-l) - a_j\beta_{m-j}(k-l))y_{k-m}$$

$$+ \sum_{m=p+1}^{2p} \sum_{j=m-p}^{p} (b_j\alpha_{m-j}(k-l) - a_j\beta_{k-m}(k-l))y_{k-m}]^2$$

$$= 0 \ a.s.$$

Now identify η with $\sum_{j=0}^{1}(b_j\alpha_{2-j} - a_j\beta_{2-j})$, set $\pi_k = 0$, and again use Lemma (7.11) to extract the $m = 1$ (and thus the $j = 0$ or $j = 1$) term. Hence,

$$\lim_{N\to\infty} \frac{1}{N} \sum_{k=0}^{N-1} [\sum_{j=0}^{1} b_j\alpha_{2-j} - a_j\beta_{2-j}]^2 = 0 \ a.s.$$

and also

$$\lim_{N\to\infty} \frac{1}{N} \sum_{k=0}^{N-1} [\sum_{m=3}^{p} \sum_{j=0}^{m-1} (b_j\alpha_{m-j}(k-l) - a_j\beta_{m-j}(k-l))y_{k-m}$$

$$+ \sum_{m=p+1}^{2p} \sum_{j=m-p}^{p} (b_j\alpha_{m-j}(k-l) - a_j\beta_{m-j}(k-l))y_{k-m}]^2$$

$$= 0 \ a.s.$$

Continuing in this way we can repeatedly apply Lemma (7.8) $2p$ times, once for each index $m = 1, \ldots, 2p$ and get

$$\lim_{N \to \infty} \frac{1}{N} \sum_{k=0}^{N-1} [\sum_{j=0}^{m-1} b_j \alpha_{m-j}(k-l) - a_j \beta_{m-j}(k-l)]^2 = 0 \ a.s.,$$

$$m = 1, \ldots, p, \tag{7.24}$$

$$\lim_{N \to \infty} \frac{1}{N} \sum_{k=0}^{N-1} [\sum_{j=m-p}^{p} b_j \alpha_{m-j}(k-l) - a_j \beta_{m-j}(k-l)]^2 = 0 \ a.s.,$$

$$m = p+1, \ldots, 2p. \tag{7.25}$$

We now use (7.21), (7.24) and (7.25). By multiplying (7.24) and (7.25) by z^{m-1} and adding [using Lemma (7.5)], we get

$$\lim_{N \to \infty} \frac{1}{N} \sum_{k=0}^{N-1} [\sum_{m=1}^{p} z^{m-1} \sum_{j=0}^{m-1} (b_j \alpha_{m-j}(k) - a_j \beta_{m-j}(k))$$

$$+ \sum_{m=p+1}^{2p} z^{m-1} (\sum_{j=m-p}^{p} b_j \alpha_{m-j}(k) - a_j \beta_{m-j}(k))]^2$$

$$= 0 \ a.s. \tag{7.26}$$

For any parameter $\theta := (\theta_1, \ldots, \theta_{2p+2})$, define the polynomials

$$F(\theta, z) := \sum_{i=1}^{p} (a_i \theta_{p+2} - b_0 \theta_{i+1}) \, z^{i-1}, \tag{7.27}$$

$$G(\theta, z) := \sum_{i=1}^{p} (b_i \theta_{p+2} - b_0 \theta_{p+2+i}) \, z^{i-1}. \tag{7.28}$$

From the definitions of $\alpha_i(k)$ and $\beta_i(k)$ in (7.16) and (7.17) it is now clear that (7.26) has the following nice interpretation

$$\lim_{N \to \infty} \frac{1}{N} \sum_{k=0}^{N-1} [F(\hat{\theta}_k, z) \sum_{i=0}^{p} b_i z^i - G(\hat{\theta}_k, z) \sum_{i=0}^{p} a_i z^i]^2$$

$$= 0 \ a.s. \text{ for all } z. \tag{7.29}$$

Also recall (7.21),

$$\lim_{N \to \infty} \frac{1}{N} \sum_{k=0}^{N-1} (a_0 \hat{b}_k - b_0 \hat{a}_k)^2 = 0 \ a.s. \tag{7.30}$$

The preceding results are sufficient to prove that the adaptive control law is self-tuning in the Cesaro sense (7.2), as the following exercise shows.

Exercise (7.31)

Show that the stochastic gradient based certainty equivalent adaptive control law is self-tuning in the Cesaro sense (7.2).

[Hint: Let $D := \{\theta \mid F(\theta, z)\sum_{i=0}^{p} b_i z^i = G(\theta, z)\sum_{i=0}^{p} a_i z^i$ and $a_0\theta(p+2) = b_0\theta(1)\}$. Then every $\theta \in D$ gives an optimal control law, i.e., D is contained in Θ_{MV}. Let

$$d(\theta) := \sum_{k=0}^{2p-1} q_k^2 + (a_0\theta_{p+2} - b_0\theta_1)^2,$$

where

$$\sum_{k=0}^{2p-1} q_i z^i := F(\theta, z)\sum_{i=0}^{p} b_i z^i - G(\theta, z)\sum_{i=0}^{p} a_i z^i,$$

and note that $\theta \in D$ if and only if $d(\theta) = 0$. By (7.29) and (7.30)

$$\lim_{N\to\infty} \frac{1}{N} \sum_{k=0}^{N-1} d(\hat{\theta}_k) = 0 \text{ a.s.}$$

Since $\{\hat{\theta}_k\}$ is bounded, we know that $\hat{\theta}_k \in K$ for all k, where K is some compact random set. Now note that

$$\min_{\theta \in O^c \cap K} d(\theta) > 0,$$

since $O^c \cap K$ is compact. If

$$\lim_{N\to\infty} \sup \frac{1}{N} \sum_{k=0}^{N-1} 1(\hat{\theta}_k \in O^c \cap K) > 0,$$

this contradicts $\lim_{N\to\infty} \frac{1}{N} \sum_{k=0}^{N-1} d(\hat{\theta}_k) = 0.$]

Now that we have proved that the adaptive controller is self-tuning in a Cesaro sense, we impose the assumption that the system has no reduced order minimum variance controllers.

Theorem (7.32)

Consider the system (4.1) where the noise $\{w_k\}$ satisfies (2.2) and (4.6). Suppose that the system is minimum phase and has no reduced order minimum variance controllers. If the stochastic gradient based adaptive control algorithm (4.2)-(4.5), (4.7) is applied, then:

(1) The parameter estimate $\{\hat{\theta}_k\}$ converges to some random multiple of $\theta°$, i.e.,

$$\lim_{k \to \infty} \hat{\theta}_k = \gamma \theta^\circ \ \ a.s. \text{ for some scalar random variable } \gamma.$$

(2) the adaptive control law is self-tuning, i.e.,

$$\lim_{k \to \infty} (u_k - \frac{\sum_{i=0}^{p} a_i z^i}{\sum_{i=0}^{p} b_i z^i} \ y_k) = 0 \ \ a.s.$$

Proof
From (7.29) and (7.30) it follows that there is a subsequence k_n such that

$$\lim_{n \to \infty} F(\hat{\theta}_{k_n}, z) \sum_{i=0}^{p} b_i z^i - G(\hat{\theta}_{k_n}, z) \sum_{i=0}^{p} a_i z^i = 0,$$

$$\lim_{n \to \infty} \alpha_0 (k_n - l) = 0 \ \ a.s.$$

Since $\{\hat{\theta}_k\}$ is bounded, we can assume without loss of generality that

$$\lim_{n \to \infty} \hat{\theta}_{k_n} = \theta^*$$

for some θ^*. So

$$F(\theta^*, z) \sum_{i=0}^{p} b_i z^i - G(\theta^*, z) \sum_{i=0}^{p} a_i z^i = 0 \quad \text{for all } z, \tag{7.33}$$

and

$$a_0 \theta^* (p + 2) - b_0 \theta^* (1) = 0. \tag{7.34}$$

However, by assumption there are no reduced order minimum variance controllers, and so

$$F(\theta^*, z) = G(\theta^*, z) = 0. \tag{7.35}$$

Recalling the definitions of $F(\theta, z)$ and $G(\theta, z)$ in (7.27) and (7.28), it follows from (7.33)-(7.35) that the condition (5.7) of Lemma (5.6) is satisfied, and so part (1) follows. Part (2) is a simple consequence. $\square$

   The parameter estimates $\{\hat{\theta}_k\}$ need not converge to the true parameter θ°. For example, if the initial estimate is chosen too large, i.e., $||\hat{\theta}_0|| = ||\theta^\circ|| + r$ and $r > 0$, then from Lemma (5.2) it follows that $||\hat{\theta}_k|| \geq ||\theta^\circ|| + r$, for all k, and then $\hat{\theta}_k$ cannot converge to θ°.

   The following exercise shows that just as in Exercise (4.27), the requirement that $0 < \mu < 2$ can be eliminated.

Exercise   (7.36)
Show that Theorem (7.32) continues to hold as long as $\mu \neq 0$.

8. Colored noise: the ARMAX system and direct adaptive control

In (4.1) the noise entering into the system was white. We now turn to the general case where the noise is a moving average of white noise, i.e., the ARMAX case. We retain the restriction that the delay between the application of an input and its affect at the output is 1 time unit, $(b_0 \neq 0)$,

$$y_{k+1} = \sum_{i=0}^{p} a_i y_{k-i} + \sum_{i=0}^{p} b_i u_{k-i} + \sum_{i=0}^{p} c_i w_{k-i} + w_{k+1}. \tag{8.1}$$

From Section (7.8) we know that the control law

$$u_k = -\frac{1}{b_0} [\sum_{i=0}^{p}(a_i + c_i)y_{k-i} + \sum_{i=1}^{p} b_i u_{k-i}] \tag{8.2}$$

minimizes the average variance

$$\lim_{N \to \infty} \frac{1}{N} \sum_{k=0}^{N-1} y_k^2.$$

In (8.2) the coefficients a_i and c_i never appear separately but only together as $a_i + c_i$. Hence to implement the minimum variance control law (8.2) it is sufficient to know only $\{a_i + c_i, b_i\}$. With a slight abuse of notation, redefine the unknown parameter as

$$\theta := (a_0 + c_0, \ldots, a_p + c_p, b_0, \ldots, b_p)^T, \tag{8.3}$$

and let θ° be the true parameter value. Then θ° is of dimension $2p + 2$, whereas $(a_0, \ldots, b_0, \ldots, b_p, c_0, \ldots, c_p)$ is of dimension $3p + 3$.

If the optimal control (8.2) were used, the output of the system (8.1) would be close to the noise, at least asymptotically [recall the discussion in Section (7.8)], i.e.,

$$y_k = w_k.$$

Substituting $y_{k-i} = w_{k-i}$ in (8.1) we see that the closed loop system evolves according to

$$y_{k+1} = \sum_{i=0}^{p}(a_i + c_i)y_{k-i} + \sum_{i=0}^{p} b_i u_{k-i} + w_{k+1}. \tag{8.4}$$

Thus if the system (8.1) is under optimal control, it is reasonable to try and estimate the parameter vector θ° by the least squares estimate,

$$\hat{\theta}_k := [\sum_{s=0}^{k-1} \phi_s \phi_s^T]^{-1} \sum_{s=0}^{k-1} \phi_s y_{s+1}, \tag{8.5}$$

where

$$\phi_k := (y_k, \ldots, y_{k-p}, u_k, \ldots, u_{k-p})^T. \tag{8.6}$$

Since $\hat{\theta}_k$ is now an estimate of θ°, one can use it in place of θ° in the control law (8.2)and get the certainty equivalent control law

$$u_k = -\frac{1}{\hat{\theta}_k(p+2)}[\hat{\theta}_k(1)y_k + \ldots + \hat{\theta}_k(p+1)y_{k-p}$$

$$+ \hat{\theta}_k(p+3)u_{k-1} + \ldots + \hat{\theta}_k(2p+2)u_{k-p}], \tag{8.7}$$

where

$$\hat{\theta}_k =: (\hat{\theta}_k(1), \ldots, \hat{\theta}_k(2p+2))^T.$$

The adaptive control law (8.6)-(8.7) is called a **direct adaptive control** scheme because instead of first estimating all the parameters describing the model and then reconstructing an optimal control law based on the estimated parameters, we are directly estimating only those coefficients relevant to the control law. For the same reason, because we are not explicitly estimating all the parameters describing the model (2.1), the adaptive control law (8.6)-(8.7) is sometimes called an **implicit** adaptive control scheme.

It is not at all obvious that a control law based on such optimistic reasoning will actually be self-tuning. Let us examine the first order case to see what the possible limit points are. Consider the system

$$y_{k+1} = ay_k + bu_k + cw_k + w_{k+1}, \tag{8.8}$$

where $\{w_k\}$ is a white noise with $var(w_k) = \sigma^2$. Suppose that

$$|c| < 1. \tag{8.9}$$

[Recall from Theorem (5.5.12) that this is not a restriction because the noise $\{cw_k + w_{k+1}\}$ has the same spectrum as the noise $\{\frac{1}{c}v_k + v_{k+1}\}$ when the white noise $\{v_k\}$ has $var(v_k) = \frac{1+c^2}{1+c^{-2}}\sigma^2$.]

Let $\theta^\circ := (a+c, b)^T$. The LSE $\hat{\theta}_k =: (\alpha_k, \beta_k)^T$ is obtained by solving [see Exercise (10.1.6)]:

$$\begin{bmatrix} \frac{1}{k}\sum_{s=0}^{k-1}y_s^2 & \frac{1}{k}\sum_{s=0}^{k-1}y_s u_s \\ \frac{1}{k}\sum_{s=0}^{k-1}y_s u_s & \frac{1}{k}\sum_{s=0}^{k-1}u_s^2 \end{bmatrix}\begin{bmatrix} \alpha_k \\ \beta_k \end{bmatrix} = \begin{bmatrix} \frac{1}{k}\sum_{s=0}^{k-1}y_s y_{s+1} \\ \frac{1}{k}\sum_{s=0}^{k-1}u_s y_{s+1} \end{bmatrix} \tag{8.10}$$

The control law is

$$u_k = -\frac{\alpha_k}{\beta_k} y_k. \tag{8.11}$$

We examine the possible limiting values of $\{\hat{\theta}_k\}$. Suppose

$$\lim_{k \to \infty} \hat{\theta}_k = (\alpha, \beta)^T. \tag{8.12}$$

When (8.12) holds, the control law is asymptotically given by

$$u_k = -\frac{\alpha}{\beta} y_k. \tag{8.13}$$

Substituting (8.13), the left hand side of (8.10) evaluates to

$$\lim_{k \to \infty} \begin{bmatrix} \dfrac{1}{k}\sum_{s=0}^{k-1} y_s^2 & \dfrac{1}{k}\sum_{s=0}^{k-1} y_s u_s \\[2ex] \dfrac{1}{k}\sum_{s=0}^{k-1} y_s u_s & \dfrac{1}{k}\sum_{s=0}^{k-1} u_s^2 \end{bmatrix} \begin{bmatrix} \alpha_k \\[1ex] \beta_k \end{bmatrix}$$

$$= \lim_{k \to \infty} \frac{1}{k}\sum_{s=0}^{k-1} y_s^2 \begin{bmatrix} 1 & -\dfrac{\alpha}{\beta} \\[2ex] -\dfrac{\alpha}{\beta} & \dfrac{\alpha^2}{\beta^2} \end{bmatrix} \begin{bmatrix} \alpha \\[1ex] \beta \end{bmatrix} = \begin{bmatrix} 0 \\[1ex] 0 \end{bmatrix}. \tag{8.14}$$

Taking the limit in (8.10) after substituting from (8.14) gives

$$\lim_{k \to \infty} \begin{bmatrix} \dfrac{1}{k}\sum_{s=0}^{k-1} y_s y_{s+1} \\[3ex] \dfrac{1}{k}\sum_{s=0}^{k-1} u_s y_{s+1} \end{bmatrix} = \begin{bmatrix} 0 \\[1ex] 0 \end{bmatrix}. \tag{8.15}$$

From (8.13),

$$\lim_{k \to \infty} \frac{1}{k}\sum_{s=0}^{k-1} u_s y_{s+1} = -\frac{\alpha}{\beta}\lim_{k \to \infty}\frac{1}{k}\sum_{s=0}^{k-1} y_s y_{s+1},$$

so (8.15) reduces to the single condition

$$\lim_{k \to \infty} \frac{1}{k}\sum_{s=0}^{k-1} y_s y_{s+1} = 0. \tag{8.16}$$

Thus the estimate $\hat{\theta}_k$ converges to $(\alpha, \beta)^T$ only if the lag one sample correlation converges to 0. To evaluate (8.16) it is convenient to denote the closed loop gain

$$\psi := a - \frac{\alpha}{\beta}\, b.$$

The closed loop system that results from applying (8.13) to (8.8) is

$$y_{s+1} = \psi y_s + c w_s + w_{s+1}. \tag{8.17}$$

Multiplying both sides by y_s and taking the average and then the limit gives

$$\lim_{k \to \infty} \frac{1}{k} \sum_{s=0}^{k-1} y_{s+1} y_s$$

$$= \lim_{k \to \infty} \frac{1}{k} \sum_{s=0}^{k-1} (\psi y_s^2 + w_{s+1} y_s + c w_s y_s). \tag{8.18}$$

One can expect that

$$\lim_{k \to \infty} \frac{1}{k} \sum_{s=0}^{k-1} w_{s+1} y_s = 0,$$

since $\{w_k\}$ is white noise and w_{s+1} is uncorrelated with y_s. Moreover, by substituting $y_s = \psi y_{s-1} + w_s + c w_{s-1}$ it is easy to see that

$$\lim_{k \to \infty} \frac{1}{k} \sum_{s=0}^{k-1} w_s y_s = \lim_{k \to \infty} \frac{1}{k} \sum_{s=0}^{k-1} (\psi w_s y_{s-1} + w_s^2 + c w_s w_{s-1})$$

$$= \sigma^2.$$

In the steady state, the variance of y_k equals

$$\gamma^2 := \lim_{k \to \infty} \frac{1}{k} \sum_{s=0}^{k-1} y_s^2. \tag{8.19}$$

Substituting (8.18), (8.19) in (8.16) we see that (α, β) must satisfy

$$\psi \gamma^2 + c \sigma^2 = 0. \tag{8.20}$$

It remains to evaluate γ^2. Squaring both sides of (8.17), taking the average and then the limit gives

$$\lim_{k \to \infty} \frac{1}{k} \sum_{s=0}^{k-1} y_{s+1}^2 = \lim_{k \to \infty} \frac{1}{k} \sum_{s=0}^{k-1} (\psi^2 y_s^2 + 2\psi y_s w_{s+1} + 2c\psi y_s w_s$$

$$+ w_{s+1}^2 + 2c w_{s+1} w_s + c^2 w_s^2).$$

Hence

$$\gamma^2 = \psi^2 \gamma^2 + 2c\psi \sigma^2 + \sigma^2 + c^2 \sigma^2,$$

or

$$\gamma^2 = \frac{1 + 2c\psi + c^2}{1 - \psi^2} \sigma^2. \tag{8.21}$$

Substituting (8.21) in (8.20) gives

$$\psi \frac{1 + 2c\psi + c^2}{1 - \psi^2} + c = 0.$$

Solving this quadratic equation for ψ gives

$$\psi = -c, \quad \text{or } \psi = \frac{-1}{c}.$$

Since $|c| < 1$, the solution $\psi = \frac{-1}{c}$ corresponds to an unstable system in (8.17), and so the only stabilizing solution is $\psi = -c$; since $\psi = a - \frac{\alpha}{\beta} b$,

$$-\frac{\alpha}{\beta} = -\frac{a + c}{b},$$

which is the optimal control law gain for (8.8).

Thus if the adaptive control law (8.10), (8.11) converges to a stabilizing control law, then the limiting control law is optimal and hence self-tuning.

Note that $\hat{\theta}_k = (\alpha_k, \beta_k)^T$ need not converge to $(a + c, b)^T$; only the ratio $\frac{\alpha_k}{\beta_k}$ must converge to $\frac{a + c}{b}$. The parameter estimate may converge to a random multiple of $\theta°$, i.e., $\lim_{k \to \infty} \hat{\theta}_k = \gamma\theta°$, where γ is a scalar random variable.

To analyze the adaptive control law based on the LSE (8.5) is difficult. Hence, as before, we study the corresponding stochastic gradient based algorithm for which the parameter estimate is given by

$$\hat{\theta}_{k+1} = \hat{\theta}_k + \frac{\mu\phi_k}{r_k}(y_{k+1} - \phi_k^T\hat{\theta}_k), \tag{8.22}$$

$$r_k = r_{k-1} + \phi_k^T\phi_k, \tag{8.23}$$

where

$$\phi_k := (y_k, \ldots, y_{k-p}, u_k, \ldots, u_{k-p})^T. \tag{8.24}$$

Again using $\hat{\theta}_k$ in place of $\theta°$ in (8.2), we choose u_k as

$$u_k = -\frac{1}{\hat{\theta}_k(p+2)}[\hat{\theta}_k(1)y_k + \ldots + \hat{\theta}_k(p+1)y_{k-p}$$

$$+ \hat{\theta}_k(p+3)u_{k-1} + \ldots + \hat{\theta}_k(2p+2)u_{k-p}], \tag{8.25}$$

where

$$\hat{\theta}_k =: (\hat{\theta}_k(1), \ldots, \hat{\theta}_k(2p+2))^T. \tag{8.26}$$

As before it is not clear that the adaptive control law (8.22)-(8.26) will work. We first try to analyze it by the ODE method to try to surmise the conditions necessary for the scheme to be self-tuning.

Denote

$$\bar{r}_k := \frac{1}{k} \, r_k.$$

The ODE corresponding to (8.22), (8.23) then is

$$\frac{d\theta(\tau)}{d\tau} = \frac{\mu}{\bar{r}(\tau)} \, E_{\theta(\tau)}[\phi_k y_{k+1}], \tag{8.27}$$

$$\frac{d\bar{r}(\tau)}{d\tau} = E_{\theta(\tau)}[\phi_k^T \phi_k] - \bar{r}(\tau).$$

We need to compute $E_{\theta(\tau)}[y_{k+1}\phi_k]$ Let

$$C(z) := 1 + \sum_{i=0}^{p} c_i z^{i+1},$$

and assume that

> All roots of $C(z)$ are strictly outside the closed unit disc.   (8.28)

Also denote

$$\bar{\theta} := (a_0, \ldots, a_p, b_0, \ldots, b_p)^T.$$

Since $\phi_k^T \theta(\tau) = 0$, the system (8.1) can be rewritten as

$$\begin{aligned}
y_{k+1} &= \phi_k^T \bar{\theta} + C(z)w_{k+1} \\
&= \phi_k^T(\bar{\theta} - \theta^\circ) + \phi_k^T(\theta^\circ - \theta(\tau)) + C(z)w_{k+1} \\
&= -\sum_{i=0}^{p} c_i y_{k-i} + \phi_k^T(\theta^\circ - \theta(\tau)) + C(z)w_{k+1} \\
&= (1 - C(z))y_{k+1} + \phi_k^T(\theta^\circ - \theta(\tau)) + C(z)w_{k+1}.
\end{aligned}$$

Hence

$$C(z)y_{k+1} = \phi_k^T(\theta^\circ - \theta(\tau)) + C(z)w_{k+1}.$$

Multiplying through by $C^{-1}(z)$ gives

$$y_{k+1} = \bar{\phi}_k^T(\theta^\circ - \theta(\tau)) + w_{k+1},$$

where

$$\bar{\phi}_k := C^{-1}(z)\phi_k. \tag{8.29}$$

Therefore

$$E_{\theta(\tau)}[y_{k+1}\phi_k] = E_{\theta(\tau)}[\phi_k \bar{\phi}_k^T][\theta° - \theta(\tau)].$$

Substituting in (8.27), gives

$$\frac{d\theta(\tau)}{d\tau} = \frac{\mu}{\bar{r}(\tau)} E_{\theta(\tau)}[\phi_k \bar{\phi}_k^T][\theta° - \theta(\tau)]. \tag{8.30}$$

and shows that $\theta(\tau) = \theta°$ is an equilibrium point. To examine its stability we consider, as in Section 3, a quadratic Lyapunov function

$$V(\theta) := ||\theta - \theta°||^2.$$

Using (8.30), the total derivative is

$$\frac{dV(\theta(\tau))}{d\tau} = -\frac{\mu}{\bar{r}(\tau)}(\theta(\tau) - \theta°)^T \{E_{\theta(\tau)}[\phi_k \bar{\phi}_k^T]$$

$$+ E_{\theta(\tau)}[\bar{\phi}_k \phi_k^T]\}(\theta(\tau) - \theta°)$$

$$= -\frac{\mu}{\bar{r}(\tau)}(\theta(\tau) - \theta°)^T$$

$$\times [G(\theta(\tau)) + G(\theta(\tau))^T](\theta(\tau) - \theta°), \tag{8.31}$$

where $G(\theta) := E_\theta[\phi_k \bar{\phi}_k^T]$. From (8.31) we see that if the matrix

$$G(\theta) + G(\theta)^T > 0, \tag{8.32}$$

i.e., it is positive definite, then the error $||\theta(\tau) - \theta°||$ decreases with τ.

We now derive some sufficient conditions to guarantee (8.32). From (8.29),

$$G(\theta) + G(\theta)^T = E_\theta[\phi_k(C^{-1}(z)\phi_k^T) + (C^{-1}(z)\phi_k)\phi_k^T].$$

Let x be any real column vector; then

$$x^T[G(\theta) + G(\theta)^T]x = 2E_\theta\{(x^T\phi_k)[C^{-1}(z)(x^T\phi_k)]\}.$$

Define the random sequences $\{\zeta_t\}$ and $\{\bar{\zeta}_t\}$ by

$$\zeta_k := x^T\phi_k \quad \text{and} \quad \bar{\zeta}_k := C^{-1}(z)(x^T\phi_k) = C^{-1}(z)\zeta_k.$$

$\{\bar{\zeta}_k\}$ is the output of a linear system with transfer function $C^{-1}(z)$ and input $\{\zeta_k\}$. Therefore the correlation between ζ_k and $\bar{\zeta}_k$ can be obtained by means of contour integration around the unit circle,

$$E_\theta[\zeta_k \bar{\zeta}_k] = \frac{1}{2\pi} \int_{-\pi}^{\pi} C^{-1}(e^{i\omega})\Phi_\theta(\omega)d\omega,$$

where $\Phi_\theta(\omega)$ is the spectral density of $\{\zeta_k\}$. (It depends on θ because θ determines how $\{\phi_k\}$ is generated.) Since $E_\theta[\zeta_k \bar{\zeta}_k]$ and $\Phi_\theta(\omega)$ are real-valued, we can replace $C^{-1}(e^{i\omega})$ by its real part $\text{Re}[C^{-1}(e^{i\omega})]$. So

$$E_\theta[\xi_k \bar{\xi}_k] = \frac{1}{2\pi} \int_{-\pi}^{\pi} \mathrm{Re}\,[C^{-1}(e^{i\omega})]\Phi_\theta(\omega)\,d\omega. \tag{8.33}$$

Clearly if we assume that

$$\mathrm{Re}\; C^{-1}(e^{i\omega}) > 0 \qquad \text{for } -\pi \le \omega \le \pi,$$

then (8.33) will be positive.

Definition (8.34)
$C(z)$ is **strictly positive real** if $\mathrm{Re}\; C^{-1}(e^{i\omega}) > 0$ for $-\pi \le \omega \le \pi$.

From (8.31)-(8.33) it follows that $||\theta(\tau) - \theta°||$ decreases with τ if $C(z)$ is strictly positive real.

We now prove that the stochastic gradient algorithm (8.22)-(8.26) is self-optimizing if $C(z)$ is strictly positive real. The following property (stated without proof) is useful.

Lemma (8.35)
Let $\{h_k\}$ and $\{g_k\}$ be the output and input (not identically zero) of a linear system with transfer function $C^{-1}(z)$:

$$h_k = C^{-1}(z)\, g_k,$$

Suppose that $C(z)$ is strictly positive real. Then there exists a constant K such that

$$\sum_{k=0}^{N-1} h_k g_k \ge K \qquad \text{for all } N.$$

Theorem (8.36)
Consider the system (8.1) where the noise $\{w_k\}$ satisfies (2.2) and (4.6). Assume that the roots of $C(z) := 1 + \sum_{i=0}^{p} c_i z^{i+1}$ lie outside the closed unit disc and $C(z)$ is strictly positive real, $\mathrm{Re}\;[C^{-1}(e^{i\omega})] > \rho > 0$ for all ω. Suppose the SG based adaptive control law (8.22)-(8.26) is used with $\mu \ne 0$, then the adaptive control law is self-optimizing with respect to the minimum variance criterion, i.e.,

$$\lim_{N \to \infty} \frac{1}{N} \sum_{k=0}^{N-1} y_k^2 = \sigma^2 \; a.s.$$

Proof
We carry out the analysis under the assumption that $0 < \mu < 2$, since it can later be eliminated as in Exercise (4.27). Let $\theta°$ be as in (8.3) and set $\widetilde{\theta}_k := \hat{\theta}_k - \theta°$. Define $V(k) := ||\widetilde{\theta}_k||^2$. Then

$$\widetilde{\theta}_{k+1} = \widetilde{\theta}_k + \frac{\mu\phi_k}{r_k}\, y_{k+1}.$$

Proceeding as in (4.8)-(4.11) we can get

$$E\{V(k+1) \mid F_k\}$$

$$\leq V(k) - \frac{2\mu}{r_k} [\phi_k^T \tilde{\theta}_k - \tfrac{1}{2}(\mu + \gamma)E\{y_{k+1} \mid F_k\}]E\{y_{k+1} \mid F_k\}$$

$$- \frac{\mu\gamma}{r_k} (E\{y_{k+1} \mid F_k\})^2 + \frac{\mu^2}{r_k^2} \phi_k^T \phi_k \sigma^2, \tag{8.37}$$

where $\gamma > 0$ is so chosen that $\mu + \gamma < 2\rho$. Since

$$y_{k+1} - w_{k+1} = \sum_{i=0}^{p} a_i y_{k-i} + b_i u_{k-i} + c_i w_{k-i}$$

and

$$[C(z) - 1](y_{k+1} - w_{k+1}) = \sum_{i=0}^{p} c_i(y_{k-i} - w_{k-i}),$$

we have

$$C(z)E\{y_{k+1} \mid F_k\}$$

$$= C(z)(y_{k+1} - w_{k+1})$$

$$= y_{k+1} - w_{k+1} + [C(z) - 1](y_{k+1} - w_{k+1})$$

$$= \phi_k^T \theta^\circ = -\phi_k^T \tilde{\theta}_k.$$

Hence

$$[C(z) - \tfrac{1}{2}(\mu + \gamma)]E\{y_{k+1} \mid F_k\}$$

$$= -\phi_k^T \tilde{\theta}_k - \tfrac{1}{2}(\mu + \gamma)E\{y_{k+1} \mid F_k\}.$$

Therefore the right-hand side can be viewed as the output of a linear system with a strictly positive real transfer function $[C(z) - \tfrac{1}{2}(\mu + \gamma)]$. From Lemma (8.35) it follows that

$$S_k := 2\mu \sum_{n=1}^{k} [-\phi_n^T \tilde{\theta}_n - \tfrac{1}{2}(\mu + \gamma)$$

$$\times E\{y_{n+1} \mid F_n\}]E\{y_{n+1} \mid F_n\} + K \geq 0,$$

for some appropriate K, and we can rewrite (8.66) as

$$E\{V(k+1) \mid F_k\} \leq V(k) - \frac{S_k - S_{k-1}}{r_k}$$

$$- \frac{\mu\gamma}{r_k} [E\{y_{k+1} \mid F_k\}]^2 + \frac{\mu^2}{r_k^2} \phi_k^T \phi_k \sigma^2.$$

Define

$$M_k := V(k) + \frac{S_{k-1}}{r_{k-1}}.$$

This is a positive near supermartingale since, just as in (4.11),

$$E\{M_{k+1} \mid F_k\} \le M_k + \frac{\mu^2}{r_k^2} \phi_k^T \phi_k \sigma^2 - \frac{\mu\gamma}{r_k} (E\{y_{k+1} \mid F_k\})^2.$$

The rest of the proof is exactly the same as in (4.11)-(4.22).   □

   Arguing along the same lines as in the proof of Theorem (7.32), one obtains the next result.

Theorem   (8.38)
Assume the conditions of Theorem (8.36) and also that the system has no reduced order minimum variance controllers, i.e., there are no $F(z)$ and $G(z)$ both of degree less than or equal to $p - 1$ satisfying (7.3) with a_i replaced by $(a_i + c_i)$. Then for some scalar random variable γ,

$$\lim_{k \to \infty} \hat{\theta}_k = \gamma\theta° \quad a.s.$$

9. The tracking problem

Frequently, the goal of controlling a system such as (8.1) is to **track** a specified reference trajectory $\{y_k^*\}$, rather than to keep the output close to 0. The latter problem, studied so far in this chapter, is called the **regulation** problem.

   Consider the first order system with colored noise

$$y_{k+1} = ay_k + bu_k + cw_k + w_{k+1}, \tag{9.1}$$

and suppose the aim of control is to keep $|y_k - y^*|$ small for all k. Here y^* is a desired constant level.

   If (a, b, c) is known, and if $\{w_k\}$ is observed, then the control law

$$u_k = \frac{y^* - ay_k - cw_k}{b} \tag{9.2}$$

minimizes the average squared tracking error

$$\lim_{N \to \infty} \frac{1}{N} \sum_{k=0}^{N-1} (y_{k+1} - r)^2,$$

and the resulting output would be

$$y_{k+1} = y^* + w_{k+1},$$

so that

$$w_k = y_k - y^*.$$

Substituting this in (9.2) gives

$$u_k = \frac{y^* - (a + c)y_k + cy^*}{b}, \tag{9.3}$$

which is implementable even when $\{w_k\}$ is not observed. Also, substituting (9.2) in (9.1) gives

$$y_{k+1} - y^* = -c(y_k - y^*) + (w_{k+1} + cw_k).$$

Solving this gives

$$y_{k+1} = (-c)^{k+1}(y_1 - y^* - w_0) + (w_{k+1} + y^*).$$

If $|c| < 1$, then

$$\lim_k (y_k - w_k - y^*) = 0,$$

and so (9.3) converges to the optimal control law (9.2). Consequently, we regard (9.3) itself as the optimal control law.

Note that (9.3) can also be expressed as

$$\phi_k^T \theta^\circ = y^*,$$

where

$$\phi_k := (y_k, u_k, -y^*)^T \quad \text{and } \theta^\circ := (a + c, b, c)^T.$$

Call θ° the true parameter. To estimate θ° we proceed as in Section 8, and consider the situation where the optimal control law (9.3) is in place. Then the system (9.1) evolves according to

$$y_{k+1} = \phi_k^T \theta^\circ + w_{k+1},$$

and the corresponding LSE is given by

$$\hat{\theta}_{k+1} = \hat{\theta}_k + R_k^{-1} \phi_k (y_{k+1} - \phi_k^T \hat{\theta}_k),$$

$$R_k := \sum_{i=0}^{k} \phi_i \phi_i^T.$$

We may instead use the SG algorithm

$$\hat{\theta}_{k+1} = \hat{\theta}_k + \frac{\mu}{r_k} \phi_k (y_{k+1} - \phi_k^T \hat{\theta}_k),$$

$$r_{k+1} = r_k + \phi_k^T \phi_k.$$

Both of these are direct adaptive control laws because we are directly estimating the coefficients of the control law.

An important point to note is that now $\theta°$ separately includes the coefficient c, and so we are estimating the spectrum of the colored noise. This need to estimate the spectrum—or, equivalently, the moving average coefficients of the noise—is a key feature distinguishing the tracking problem from the regulation problem.

Now consider the more general ARMAX system:

$$y_{k+1} = \sum_{i=0}^{p}(a_i y_{k-i} + b_i u_{k-i} + c_i w_{k-i}) + w_{k+1}, \quad b_0 \neq 0.$$

The goal is to track a given bounded reference trajectory $\{y_k^*\}$ with minimum variance of the tracking error.

We first derive the optimal control law using the approach of Section (7.8). If $\{w_k\}$ is observed, the minimum variance control law is

$$u_k = \frac{-1}{b_0}[\sum_{i=0}^{p}(a_i y_{k-i} + c_i w_{k-i}) + \sum_{i=1}^{p}b_i u_{k-i} - y_{k+1}^*],$$

and when this control law is applied, the resulting output satisfies

$$y_k = y_k^* + w_k,$$

or, equivalently,

$$w_k = y_k - y_k^*. \tag{9.4}$$

However $\{w_k\}$ is not in fact observed, so we use the right-hand side of (9.4) in its place. The resulting control law is

$$u_k = \frac{-1}{b_0}[\sum_{i=0}^{p}(a_i + c_i)y_{k-i}$$

$$+ \sum_{i=1}^{p}b_i u_{k-i} - \sum_{i=0}^{p}c_i y_k^* - y_{k+1}^*]. \tag{9.5}$$

As in Section (7.8) it can be shown that this control law is either asymptotically optimal, or optimal for the average cost criterion, provided that the polynomial

$$C(z) := 1 + \sum_{i=0}^{p}c_i z^i$$

has all its roots strictly outside the unit circle.

It should be noted that the coefficients occurring in the control law (9.5) are $(a_i + c_i)$, b_i and c_i. Denote

$$\theta° := (a_0 + c_0, \ldots, a_p + c_p, b_0, \ldots, b_p, c_0, \ldots, c_p)^T,$$

and call $\theta°$ the true parameter.

$$\phi_k^T \theta^\circ = y_k^*, \tag{9.6}$$

where

$$\phi_k := (y_k, \ldots, y_{k-p}, u_k, \ldots, u_{k-p}, -y_k^*, \ldots, -y_{k-p}^*)^T$$

(In the regulation problem, the right-hand side of (9.6) is 0.)

To estimate θ° we proceed as in Section 8. Suppose the system is under the control (9.5). Then the resulting system satisfies (9.4). By substituting for w_{k-i} from (9.4), we see that the output is given by

$$y_{k+1} = \sum_{i=0}^{p}(a_i + c_i)y_{k-i} + b_i u_{k-i} - \sum_{i=o}^{p}c_i y_{k-i}^* + w_{k+1},$$

which can be written as

$$y_{k+1} = \phi_k^T \theta^\circ + w_{k+1}.$$

Reasoning as in Section 2, we can propose the SG algorithm

$$\hat{\theta}_{k+1} = \hat{\theta}_k + \frac{\mu\phi_k}{r_k}(y_{k+1} - \phi_k^T\hat{\theta}_k),$$

$$r_{k+1} = r_k + \phi_{k+1}^T\phi_{k+1}.$$

This is again a direct adaptive control law, where $\hat{\theta}_k$ can be regarded as an estimate of θ°.

The behavior of this adaptive control system can be analyzed in the same way as in Section 4 by the use of a stochastic Lyapunov function to obtain the following theorem.

Theorem (9.7)
Suppose $\{y_k^*\}$ is a bounded sequence. Then under the same conditions as in Theorem (8.36), but with $0 < \mu < 2\rho$, the adaptive control law is self-optimizing, i.e.,

$$\lim_{N\to\infty} \frac{1}{N}\sum_{k=0}^{N-1}(y_k - y_k^*)^2 = \sigma^2 \quad a.s.$$

$$\lim_{N\to\infty} \sup \frac{1}{N}\sum_{k=0}^{N-1}u_k^2 < \infty \quad a.s.$$

Thus the direct adaptive control law derived above is self-optimizing. However, since $\phi_k^T\hat{\theta}_k \neq 0$, the orthogonality property utilized so heavily in Section 5 is lost, and so the proofs of Theorems (7.32) and (8.38) cannot be used to show that the adaptive control law is self-tuning. In the next section we modify slightly the adaptive control above in order to guarantee self-tuning.

10. Self-tuning trackers

In this section we modify the algorithm of the preceding section to obtain a self-tuning controller. In the next section we show that this modified adaptive control allows us to adjust the dimension of the parameter estimator according to the degree of richness in the reference trajectory to be followed.

Consider the system

$$y_k = \sum_{i=1}^{p} a_i y_{k-i} + \sum_{i=1}^{q} b_i u_{k-i} + \sum_{i=1}^{s} c_i w_{k-i} + w_k.$$

Here we use different orders p, q, and s because in some of the results to follow, it will be helpful to separate them.

Redefine the true parameter

$$\theta^\circ := (a_1 + c_1, \ldots, a_{p \vee s} + c_{p \vee s}, b_1, \ldots, b_q, 1,$$
$$c_1, \ldots, c_s)^T, \tag{10.1}$$

[where $p \vee s := \max(p, s)$ and we use the convention $a_i := 0$ for $i > p$ and $c_i := 0$ for $i > s$]. Also redefine ϕ_k as

$$\phi_k := (y_k, \ldots, y_{k-p \vee s+1}, u_k \ldots, u_{k-q+1}, -y^*_{k+1}, \ldots,$$
$$-y^*_{k-s+1})^T. \tag{10.2}$$

The key difference between these and the definitions of the previous section is the addition of the components 1 and $-y^*_{k+1}$ in (10.1) and (10.2), respectively. The advantage is that the optimal control law (9.5) now satisfies

$$\phi_k^T \theta^\circ = 0, \tag{10.3}$$

and so we again preserve orthogonality.

To estimate θ° note that when the system is under optimal control, the output $\{y_k\}$ satisfies

$$y_{k+1} = \phi_k^T \theta^\circ + y^*_{k+1} + w_{k+1},$$

or

$$(y_{k+1} - y^*_{k+1}) = \phi_k^T \theta^\circ + w_{k+1},$$

and the argument of Section 2 suggests the SG algorithm

$$\hat{\theta}_{k+1} = \hat{\theta}_k + \frac{\mu \phi_k}{r_k} (y_{k+1} - y^*_{k+1} - \phi_k^T \hat{\theta}_k). \tag{10.4}$$

We use $\hat{\theta}_k$ in place of θ° in (10.3) to get

$$\phi_k^T \hat{\theta}_k = 0, \tag{10.5}$$

which defines u_k implicitly.

We have thus obtained an adaptive control law (10.4) and (10.5), which possesses the orthogonality property (10.5) that we found useful in Sections 5-7.

There is one noteworthy feature of our adaptive control law. $\hat{\theta}_k$ is an estimate of $\theta°$. However, we know that the $(p \vee s + q + 1)$st component is 1, and need not be estimated. Since we are estimating it anyway along with all the other components of $\theta°$ by the recursion (10.4), there is 1 degree of redundancy in what we are doing. Hence the parameter estimator of this section is 1 dimension larger than the parameter estimator of the previous section. In the next section we will see how to use this feature to reduce significantly the dimension of the parameter estimator in some applications.

Using the stochastic Lyapunov function method as before, it can be shown that this adaptive control system is self-optimizing.

Theorem   (10.6)
Suppose $\{y_k^*\}$ is a bounded sequence. Then under the same conditions as in Theorem (8.36) with $\mu > 0$, we have the following:
(1) The algorithm is self-optimizing, i.e.,

$$\lim_{N \to \infty} \frac{1}{N} \sum_{k=0}^{N-1} (y_k - y_k^*)^2 = \sigma^2 \quad a.s. \tag{10.7}$$

(2) In proving this we obtain, as in (4.24)-(4.26),

$$\lim_{N \to \infty} \frac{1}{N} \sum_{k=0}^{N-1} (E\{y_{k+1} - y_{k+1}^* \mid F_k\})^2 = 0 \quad a.s. \tag{10.8}$$

$$\lim_{N \to \infty} \sup \frac{1}{N} \sum_{k=0}^{N-1} u_k^2 < \infty \quad a.s. \tag{10.9}$$

$$\{\|\hat{\theta}_k - \theta°\|\} \text{ converges } a.s. \tag{10.10}$$

Our next goal is to prove the self-tuning property. Because of the orthogonality property (10.5), Lemmas (5.2), (5.3), and (5.5) continue to hold. Hence, as before, in order to prove self-tuning we need only show that there is a subsequence $\{k_n\}$ such that (1) or (2) of Lemma (5.5) holds.

Before embarking on this proof, it is worth noting that in contrast to the regulation problem, the tracking case also requires the estimation of the components $\{c_i\}$ describing the colored noise; see (10.1). Because of the increase in the number of parameters being estimated we need some additional assumptions to prove self-tuning which imply that the reference trajectory is sufficiently rich in an appropriate sense. When, as happens in some situations, the reference trajectory is not sufficiently rich, we will see in the next section that we can reduce the number of components to be estimated.

The proof of the validity of the conditions (1) or (2) of Lemma (5.5) will follow roughly the same lines as the proofs of Sections 6 and 7. We use a slightly more abbreviated notation in order to shorten the proofs. Define the polynomials

$$A(z) := 1 - \sum_{i=1}^{p} a_i z^i, \quad B(z) := \sum_{i=1}^{q} b_i z^{i-1}, \quad \text{and}$$

$$C(z) := 1 + \sum_{i=1}^{s} c_i z^i.$$

and also the time-varying polynomials

$$P(k, z) := \sum_{i=1}^{p \vee s} \alpha_i(k) z^{i-1},$$

$$Q(k, z) := \sum_{i=1}^{q} \beta_i(k) z^{i-1},$$

$$R(k, z) := \sum_{i=0}^{s} \gamma_i(k) z^i.$$

where $\alpha_i(k)$, $\beta_i(k)$ and $\gamma_i(k)$ are the components of $\hat{\theta}_k$,

$$\hat{\theta}_k =: (\alpha_1(k), \ldots, \alpha_{p \vee s}(k), \beta_1(k), \ldots, \beta_q(k),$$
$$\gamma_0(k), \ldots, \gamma_s(k))^T.$$

Here z is the shift operator acting on time dependent quantities. Thus to illustrate the notation,

$$Q(k,z)x(k) := \sum_{i=1}^{q} \beta_i(k) x(k - i + 1),$$

$$Q(k,z)B(z)x(k) := \sum_{i=1}^{q} \beta_i(k) \sum_{j=1}^{q} b_j x(k - i - j + 2),$$

$$B(z)Q(k,z)x(k) := \sum_{j=1}^{q} b_j \sum_{i=1}^{q} \beta_i(k - j + 1) x(k - i - j + 2).$$

Note that $Q(k,z)B(z)x(k) \neq B(z)Q(k,z)x(k)$. However, one useful fact we will repeatedly use is that if $\{ \frac{1}{N} \sum_{k=1}^{N} x^2(k) \}$ is bounded, then

$$\lim_{N} \frac{1}{N} \sum_{k=1}^{N} [Q(k,z)B(z)x(k) - B(z)Q(k,z)x(k)]^2 = 0. \qquad (10.11)$$

[This is verified using the facts that $\lim_k | |\hat{\theta}_k - \hat{\theta}_{k-n}| | = 0$ a.s. and $\{\hat{\theta}_k\}$ is bounded a.s., which follow from Lemmas (5.2) and (5.3).] In the sense of (10.11), therefore, the polynomials $Q(k,z)$ and $B(z)$ nearly commute when operating on such sequences $\{x(k)\}$.

As in Sections 6 and 7, the analysis begins with the consideration of (10.8) which can be written as

$$\lim_N \frac{1}{N} \sum_{k=1}^{N} \{[1 - A(z)]y_{k+1} + zB(z)u_{k+1}$$
$$+ [C(z) - 1]w_{k+1} - y_{k+1}^*\}^2 = 0.$$

Multiplying inside the summation by $Q(k,z)$, we have

$$\lim_N \frac{1}{N} \sum_{k=1}^{N} \{Q(k,z)[1 - A(z)]y_{k+1} + Q(k,z)zB(z)u_{k+1}$$
$$+ Q(k,z)[C(z) - 1]w_{k+1} - Q(k,z)y_{k+1}^*\}^2 = 0 \; a.s.$$

Commuting the polynomials gives

$$\lim_N \frac{1}{N} \sum_{k=1}^{N} \{z^{-1}[1 - A(z)]Q(k,z)y_k) + B(z)Q(k,z)u_k$$
$$+ z^{-1}[C(z) - 1]Q(k,z)w_k - Q(k,z)y_{k+1}^*\}^2 = 0 \; a.s.$$

Noting that the adaptive control law (10.5) can be written as

$$Q(k,z)u_k = -P(k,z)y_k + R(k,z)y_{k+1}^*,$$

we can substitute this expression to get

$$\lim_N \frac{1}{N} \sum_{k=1}^{N} [\{z^{-1}[1 - A(z)]Q(k,z) - B(z)P(k,z)\}y_k$$
$$+ z^{-1}[C(z) - 1]Q(k,z)w_k$$
$$+ \{B(z)R(k,z) - Q(k,z)\}y_{k+1}^*]^2 = 0 \; a.s.$$

Now substituting $y_k = w_k + y_k^* + E\{y_k - y_k^* \mid F_{k-1}\}$, gives

$$\lim_N \frac{1}{N} \sum_{k=1}^{N} [\{z^{-1}[C(z) - A(z)]Q(k,z) - B(z)P(k,z)\}w_k$$
$$+ \{B(z)R(k,z) - zB(z)P(k,z) - A(z)Q(k,z)\}y_{k+1}^*$$
$$+ \{z^{-1}[1 - A(z)]Q(k,z)$$
$$- B(z)P(k,z)\}E\{y_k - y_k^* \mid F_k\}]^2 = 0 \; a.s.$$

By virtue of (10.8) and noting that $\{Q(k,z)\}$ and $\{P(k,z)\}$ are bounded (as a consequence of (10.10)), we can drop the last term above and get

$$\lim_N \frac{1}{N} \sum_{k=1}^{N} [\{z^{-1}[C(z) - A(z)]Q(k,z) - B(z)P(k,z)\}w_{k+1}$$
$$+ \{B(z)R(k,z) - zB(z)P(k,z) - A(z)Q(k,z)\}y^*_{k+1}]^2$$
$$= 0 \ a.s.$$

As in Sections 6 and 7, since $\lim_k | |\hat{\theta}_k - \hat{\theta}_{k-1}| | = 0$ a.s. and $\{y^*_{k+1}\}$ is bounded, we can replace $R(k,z)$, $P(k,z)$ and $Q(k,z)$ by $R(k-l,z)$, $P(k-l,z)$ and $Q(k-l,z)$, respectively, for any integer l. Thus

$$\lim_N \frac{1}{N} \sum_{k=1}^{N} [\{z^{-1}[C(z) - A(z)]Q(k-l,z)$$
$$- B(z)P(k-l,z)\}w_k$$
$$+ \{B(z)R(k-l,z) - zB(z)P(k-l,z)$$
$$- A(z)Q(k-l,z)\}y^*_{k+1}]^2 = 0 \ a.s$$

Choose $l > (p + q + s)$, and apply Exercise (7.6) to get

$$\lim_N \frac{1}{N} \sum_{k=1}^{N} [z^{-1}[C(z) - A(z)]Q(k-l,z)$$
$$- B(z)P(k-l,z)]^2 = 0 \ a.s., \qquad (10.12)$$

and also

$$\lim_N \frac{1}{N} \sum_{k=1}^{N} [\{B(z)R(k-l,z) - zB(z)P(k-l,z)$$
$$- A(z)Q(k-l,z)\}y^*_{k+1}]^2 = 0 \ a.s. \qquad (10.13)$$

Furthermore since $\{y^*_k\}$ is bounded, we can multiply by it in (10.12) to get

$$\lim_N \frac{1}{N} [\{[C(z) - A(z)]Q(k-l,z)$$
$$- zB(z)P(k-l,z)\}y^*_{k+1}]^2 = 0 \ a.s. \qquad (10.14)$$

Subtracting the terms in (10.13) from (10.14) we get

$$\lim_N \frac{1}{N} \sum_{k=1}^{N} \{[C(z)Q(k-l,z)$$
$$- B(z)R(k-l,z)]y^*_{k+1}\}^2 = 0 \ a.s. \qquad (10.15)$$

Changing $k - l$ back to k in (10.12) and (10.15) we have

$$\lim_N \frac{1}{N} \sum_{k=1}^{N} \{z^{-1}[C(z)-A(z)]Q(k,z)-B(z)P(k,z)\}^2 = 0 \ a.s. \quad (10.16)$$

$$\lim_N \frac{1}{N} \sum_{k=1}^{N} \{[C(z)Q(k,z) - B(z)R(k,z)]y_{k+1}^*\}^2$$

$$= 0 \ a.s. \quad (10.17)$$

We now need to consider the amount of excitation that is present in the desired trajectory $\{y_k^*\}$. Note for example that if $y_k^* = 0$ for all k, then (10.17) does not give us any information. So we introduce the following concept that we have already seen in Definition (10.3.19).

Definition (10.18)
The sequence $\{y_k^*\}$ is said to be **sufficiently rich of order** l if there are an integer n and $\epsilon > 0$ such that

$$\sum_{i=k+1}^{k+n} (y_{i-1}^*, \ldots, y_{i-l}^*)^T (y_{i-1}^*, \ldots, y_{i-l}^*) \geq \epsilon I$$

for all large k.

Exercise (10.19)
Show that the constant reference trajectory $y_k^* = y^* \neq 0$ for all k is sufficiently rich of order 1.

We now assume that

$$\{y_k^*\} \text{ is sufficiently rich of order } (q + s) \quad (10.20)$$

to prove the self-tuning property. Define

$$S(k,z) := C(z)Q(k,z) - B(z)R(k,z),$$

and, since $S(k,z)$ is a polynomial of degree $(s + q - 1)$ let $S(k,z) =: \sum_{i=0}^{q+s-1} s_i(k)z^i$. By (10.20) it follows that

$$\frac{1}{n} \sum_{k=jn+1}^{jn+n} [S(jn,z)y_{k+1}^*]^2 \geq \frac{\epsilon}{n} ||S(jn,z)||^2 \quad (10.21)$$

for all large j. But due to (10.17) and $\lim_k ||\hat{\theta}_k - \hat{\theta}_{k-1}|| = 0$ a.s., it follows from (10.21) that

$$\lim_N \frac{1}{N} \sum_{k=1}^{N} ||S(k,z)||^2 = 0 \ a.s.$$

Now we can add the terms of (10.16) to the above to get

$$\lim_N \frac{1}{N} \sum_{k=1}^{N} \{z^{-1}[C(z)-A(z)]Q(k,z) - B(z)P(k,z)\}^2$$

$$+ \{C(z)Q(k,z) - B(z)R(k,z)\}^2 = 0 \ a.s.$$

From this and (10.17) it follows that there is a common subsequence $\{k_n\}$ such that

$$\lim_n \{z^{-1}[C(z) - A(z)]Q(k_n,z) - B(z)P(k_n,z)\} = 0 \ a.s.$$

and

$$\lim_n \{C(z)Q(k_n,z) - B(z)R(k_n,z)\} = 0 \ a.s.$$

Since $\{\hat{\theta}_k\}$ is bounded, we can also assume without loss of generality, that Q, P and R converge along this subsequence, i.e.,

$$\lim_n Q(k_n,z) =: Q(z), \quad \lim_n P(k_n,z) =: P(z), \quad \text{and}$$

$$\lim_n R(k_n,z) =: R(z) \ a.s.$$

Hence

$$z^{-1}[C(z) - A(z)]Q(z) - B(z)P(z) = 0 \ a.s. \tag{10.22}$$

$$C(z)Q(z) - B(z)R(z) = 0 \ a.s. \tag{10.23}$$

At this point we reintroduce the assumption that the system does not have a reduced order minimum variance control law.

Then since $Q(z)$ and $P(z)$ are polynomials of degrees less than or equal to $(q - 1)$ and $(p \vee s - 1)$ respectively, from (10.22) it follows that

$$Q(z) = \gamma B(z) \quad \text{and} \quad P(z) = \gamma z^{-1}[C(z) - A(z)] \tag{10.24}$$

for some random scalar γ. Then (10.23) also shows the $R(z) = \gamma C(z)$. Therefore we have shown (1) of Lemma (5.5) and so part (3) of the Lemma follows.

We summarize this result as a theorem.

Theorem (10.25)

Assume the conditions of Theorem (10.6), and suppose that the reference trajectory is sufficiently rich of order $(q + s)$ and the system has no reduced order minimum variance controllers. Then the adaptive control law is self-tuning wp 1, i.e., there is an almost surely finite non-zero constant γ such that

$$\lim_{k \to \infty} \hat{\theta}_k = \gamma \theta^\circ.$$

[The reader should check that $\gamma \neq 0$ due to Lemma (5.2).]

In particular, since the $(p \vee s + q + 1)$st component of θ° is 1, whereas the corresponding component of $\hat{\theta}_k$ is $\gamma_0(k)$, it follows that $\gamma = \lim_k \gamma_0(k)$. Hence dividing through by $\gamma_0(k)$ gives,

$$\lim_{k} \frac{1}{\gamma_0(k)} (\alpha_1(k) - \gamma_1(k), \ldots, \alpha_p(k) - \gamma_p(k),$$

$$\beta_1(k), \ldots, \beta_q(k), \gamma_1(k), \ldots, \gamma_s(k))$$

$$= (a_1, \ldots, a_p, b_1, \ldots, b_q, c_1, \ldots, c_s) \text{ a.s.}$$

Thus all the parameters $(a_1, \ldots, a_p, b_1, \ldots, b_q, c_1, \ldots, c_s)$ can be consistently estimated. So the adaptive control law yields consistent parameter estimates as well.

11. Linear model following

In the previous section we relied heavily on the assumption that the reference trajectory is sufficiently rich of order $(q + s)$. However this assumption is not satisfied in some situations. For example, in the so-called set-point problem, the reference trajectory is constant,

$$y_k^* = y^* \quad \text{for all k.} \tag{11.1}$$

Such a reference trajectory is sufficiently rich of order 1.

Motivated by this situation, suppose that y_k^* satisfies the homogeneous linear difference equation

$$y_k^* = \sum_{i=1}^{l} h_i y_{k-i}^*. \tag{11.2}$$

The constant reference trajectory satisfies the equation $y_k^* = y_{k-1}^*$ and is a special case, as are all trajectories that are linear combinations of exponentials and sinusoids. Indeed, if the roots of the polynomial $H(z) := 1 - \sum_{i=1}^{l} h_i z^i$ are $\{\lambda_1, \ldots, \lambda_l\}$ and if they are distinct, then the general solution of (11.2) is of the form

$$y_k^* = \sum_{i=1}^{l} d_i \lambda_i^k, \tag{11.3}$$

where the constants d_i are determined by the initial conditions of (11.2).

The mode $\{\lambda_i^k\}$ is said to be excited by the initial conditions of (11.2) if $d_i \neq 0$. We can assume that all modes are excited because otherwise there is a difference equation of lower order that gives the same solution $\{y_k^*\}$. Next, observe that since we are concerned with asymptotic tracking of a bounded reference trajectory, we can disregard modes that are either unbounded or that converge to 0. Combining these two observations leads to the following assumption on $\{y_k^*\}$.

Assumption (11.4)
(1) $\{y_k^*\}$ satisfies no difference equation of lower order than (11.2).
(2) $|\lambda_i| = 1$ for all i, and $\lambda_i \neq \lambda_j$ for $i \neq j$.

A reference trajectory satisfying (11.4) is sufficiently rich of order l.

Now let us see how we can take advantage of reference trajectories satisfying this assumption. Recall from (9.5) that the optimal control law is given by

$$u_k = \frac{-1}{b_1} [\sum_{i=1}^{p \vee s} (a_i + c_i) y_{k+1-i}$$

$$+ \sum_{i=2}^{q} b_i u_{k+1-i} - \sum_{i=1}^{s} c_i y^*_{k-i+1} - y^*_{k+1}].$$

Thus to implement this control law, we really need $(y^*_k + \sum_{i=1}^{s} c_i y^*_{k-i}) = C(z) y^*_k$, where the transfer function $C(z)$ is as before. Suppose the degree l of $H(z)$ is less than the degree s of $C(z)$. Divide $C(z)$ by $H(z)$ and express the result as

$$C(z) = F(z)H(z) + G(z),$$

where $G(z) = \sum_{i=0}^{l-1} g_i z^i$ and $l \le s$. By (11.2), $H(z)y^*_k = 0$, so

$$C(z)y^*_k = G(z)y^*_k.$$

Hence to calculate the optimal control law requires knowledge only of $G(z)y^*_k$; in fact,

$$u_k = \frac{-1}{b_1} [\sum_{i=1}^{p \vee s} (a_i + c_i) y_{k+1-i} + \sum_{i=2}^{q} b_i u_{k+1-i} - \sum_{i=0}^{l-1} g_i y^*_{k-i+1}].$$

Therefore we redefine ϕ_k and θ° as

$$\phi_k := (y_k, \ldots, y_{k-p \vee s+1}, u_k, \ldots, u_{k-q+1},$$
$$- y^*_{k+1}, \ldots, - y^*_{k-l+2})^T,$$
$$\theta^\circ := (a_1 + c_1, \ldots, a_{p \vee s} + c_{p \vee s}, b_1, \ldots, b_q, g_0, \ldots, g_{l-1})^T,$$

and note that the dimensions of θ° and ϕ_k are smaller than in the previous section since $l - 1 \le s - 1$. The optimal control law is thus given implicitly by

$$\phi_k^T \theta^\circ = 0.$$

To estimate θ° we proceed as in Section 8, noting that when the optimal control law is in place, the output y_k is given by

$$y_{k+1} - y^*_{k+1} = \phi_k^T \theta^\circ + w_{k+1}.$$

Hence as in Section 2 we obtain the stochastic gradient algorithm,

$$\hat{\theta}_{k+1} = \hat{\theta}_k + \frac{\mu\phi_k}{r_k} (y_{k+1} - y_{k+1}^* - \phi_k^T\hat{\theta}_k), \tag{11.5}$$

$$r_k := 1 + \sum_{i=1}^{k} \phi_i^T\phi_i. \tag{11.6}$$

With $\hat{\theta}_k$ as an estimate of $\theta°$, the control u_k is then chosen to satisfy

$$\phi_k^T\hat{\theta}_k = 0. \tag{11.7}$$

The adaptive control law is then specified by (11.5)-(11.7).

The important point to note is that $\hat{\theta}_k$ is of smaller dimension than in the previous section. In particular the coefficients $\{c_1, \ldots, c_s\}$ are no longer estimated. Instead, we estimate $\{g_0, \ldots, g_{l-1}\}$ which accounts for as many components of the colored noise, l, as the degree of richness of the reference trajectory. Thus in the set-point problem we estimate only one component g_0 to account for the colored noise and reject it optimally.

By using a stochastic Lyapunov function as in Section 4, making use of the orthogonality property (11.8) as in Section 5, and the richness of the reference trajectory as in Section 10, we can analyze the system as before to obtain the following theorem.

Theorem (11.8)

If the reference trajectory satisfies (11.2) and (11.4) then under the same assumptions as Theorem (8.36),
(1) the adaptive control is self-optimal, i.e.,

$$\lim_N \frac{1}{N} \sum_{k=1}^{N} [y_k - y_k^*]^2 = \sigma^2 \quad a.s.,$$

$$\lim_N \sup \frac{1}{N} \sum_{k=1}^{N} u_k^2 < \infty \quad a.s.;$$

(2) the adaptive control law (11.8) is self-tuning; in particular there is a finite nonzero random scalar γ such that

$$\lim_k \hat{\theta}_k = \gamma\theta° \quad a.s.$$

12. Notes

1. Algorithms of the stochastic gradient type given in Section 2 can be found also in studies on stochastic approximation; see Robbins and Monro (1951) and Kiefer and Wolfowitz (1952).

2. The ordinary differential equation approach to analyzing recursive stochastic algorithms was pioneered by Ljung (1977a) and Ljung and Soderstrom (1983). It is also useful in the analysis of the behavior of

recursive identification algorithms; see Section (10.8) and also Ljung and Soderstrom (1983). The analysis of the self-tuning regulator based on the ODE approach was carried out in Ljung (1977b).

3. Theorem (4.10) can be found in Neveu (1975), Chapter II, Exercise II-4, and also in Robbins and Siegmund (1971). This theorem was used by Solo (1979) in his proof of the convergence of the AML algorithm for the identification of the parameters of an ARMAX process; see Section (10.7).

4. The proofs of self-optimality of the SG algorithm in Sections 4, 8, and 9 are due to Goodwin, Ramadge and Caines (1981). Property (8.35) of positive real transfer functions as well as the property of minimum phase systems [see the equation following Assumption (4.21)] can be found there.

5. The use of the geometric properties of the SG algorithm for minimum variance regulation and the proof of convergence of the parameter estimates to a random multiple of the true parameter are due to Becker, Kumar and Wei (1985). The details of the proof of the colored noise case of Theorem (8.38) may be found there.

6. The pioneering work that stimulated development of self-tuning regulators is Astrom and Wittenmark (1973). A direct adaptive control law, of the type studied in Section 8, was first suggested there, as was the proof that if the adaptive regulator converges, then the limiting regulator must be the minimum variance regulator. That paper also provides the generalization of the analysis of a system closely related to that of (8.8) to the case where the delay can exceed 1.

7. The importance of the positive real condition for the convergence of the self-tuning regulator (and other recursive identification algorithms) in the colored noise case was discovered by Ljung (1977b). The Lyapunov method for the proof of convergence of the ODE was first used there. The positive real condition is also used in the analysis of deterministic adaptive control algorithms; see Landau (1979).

8. For other algorithms that can be analyzed by the stochastic Lyapunov function method, see Goodwin and Sin (1984). The ARMAX system with general delay is examined in Goodwin, Sin and Saluja (1980). A modified version of the least squares based adaptive control algorithm is proved to be self-optimizing in Sin and Goodwin (1982).

9. The results of Sections 10 and 11 on tracking and linear model following are taken from Kumar and Praly (1985). For a general survey of the field of adaptive control as well as many references to the literature see Kumar (1985).

10. When criteria other than of the minimum variance type are used as cost functions, then the self-tuning property may be lost. In

particular, the coincidence of the set of possible limit points of the SG based certainty equivalent adaptive control law with the set of optimal control laws as in (3.13) will not continue to hold; see Lin, Kumar and Seidman (1985).

13. References

[1]   K. Astrom and B. Wittenmark (1973), "On self-tuning regulators," *Automatica,* vol 9, 185-189.

[2]   A. Becker, P.R. Kumar and C.Z. Wei (1985), "Adaptive control with the stochastic approximation algorithm: geometry and convergence," *IEEE Transactions on Automatic Control,* Vol AC-30, 330-338.

[3]   G. Goodwin, P. Ramadge and P. Caines (1981), "Discrete time stochastic adaptive control," *SIAM J. on Control and Optimization,* vol 19, 829-853.

[4]   G. Goodwin and K.S. Sin (1984), *Adaptive filtering, prediction and control,* Prentice Hall, Englewood Cliffs, NJ.

[5]   G. Goodwin, K.S. Sin and K. Saluja (1980), "Stochastic adaptive control and prediction: the general delay-colored noise case," *IEEE Transactions on Automatic Control,* Vol AC-25, 946-949.

[6]   J. Kiefer and J. Wolfowitz (1952), "Stochastic estimation of the maximum of a regression function," *Annals of Mathematical Statistics,* vol 23, 462-466.

[7]   P.R. Kumar (1985), "A survey of some results in stochastic adaptive control," *SIAM J. on Control and Optimization,* Vol 23, 329-380.

[8]   P.R. Kumar and L. Praly (1985), "Self-tuning and convergence of parameter estimates in minimum variance tracking and linear model following," preprint, University of Illinois, Urbana.

[9]   I.D. Landau (1979), *Adaptive control—the model reference approach,* Marcel Dekker, New York.

[10] L. Ljung (1977a). "Analysis of recursive stochastic algorithms," *IEEE Transactions on Automatic Control,* vol AC-22, 551-575.

[11] L. Ljung (1977b), "On positive real functions and the convergence of some recursive schemes," *IEEE Transactions on Automatic Control,* vol AC-22, 539-551.

[12] L. Ljung and T. Soderstrom (1983), *Theory and practice of recursive identification,* MIT Press, Cambridge, MA.

[13] W. Lin, P. R. Kumar and T. Seidman (1985), "Will the self-tuning approach work for general cost criteria?," *Systems and Control Letters,* vol. 6, 77-85.

[14] J. Neveu (1975), *Discrete parameter martingales,* North Holland, Amsterdam.

[15] H. Robbins and S. Monro (1951), "A stochastic approximation method," *Annals of Mathematical Statistics,* vol 22, 400-407.

[16] H. Robbins and D. Siegmund (1971), "A convergence theorem for nonnegative almost supermartingales and some applications," in J.S. Rustagi (ed), *Optimization methods in statistics,* Academic Press, New York, 233-257.

[17] K. Sin and G. Goodwin (1982). "Stochastic adaptive control using a modified least squares algorithm," *Automatica,* vol 18, 315-321.

[18] V. Solo (1979), "The convergence of AML," *IEEE Transactions on Automatic Control,* vol AC-24, 958-962.

REFERENCES

[1] B.D.O. Anderson and J.B. Moore (1979), *Optimal filtering*, Prentice-Hall, Englewood Cliffs, NJ.

[2] K. Astrom (1970), *Introduction to stochastic control theory*, Academic Press, New York.

[3] K.J. Astrom and T. Soderstrom (1974), "Uniqueness of the maximum likelihood estimates of the parameters of an ARMA model," *IEEE Transactions on Automatic Control*, vol AC-19, 769-773.

[4] K. Astrom and B. Wittenmark (1973), "On self-tuning regulators," *Automatica*, vol 9, 185-189.

[5] K.J. Astrom and B. Wittenmark (1984), *Computer controlled systems*, Prentice-Hall, New York.

[6] J. Bather (1973), "Optimal decision procedures for finite Markov chains," *Advances in Applied Probability*, (3 parts), vol. 5, 328-339, 521-540, 541-553.

[7] A. Becker, P.R. Kumar and C.Z. Wei (1985), "Adaptive control with the stochastic approximation algorithm: Geometry and convergence," *IEEE Transactions on Automatic Control*, Vol AC-30, 330-338.

[8] R.E. Bellman and S.E. Dreyfus (1962), *Applied dynamic programming*, Princeton University Press, Princeton, NJ.

[9] D. Bertsekas (1978), *Stochastic optimal control*, Academic Press, New York.

[10] D. Bertsekas and S. Shreve (1978), *Stochastic optimal control: the discrete time case*, Academic Press, New York.

[11] G.J. Bierman (1975), "Measurement updating using the U-D factorization," *Proceedings of the IEEE Conference on Decision and Control*, Houston, 337-346.

[12] G.J. Bierman (1977), *Factorization methods for discrete sequential estimation*, Academic Press, New York.

[13] D. Blackwell (1962), "Discrete dynamic programming," *Annals of Mathematical Statistics*, vol 33, 719-726.

[14] D. Blackwell (1965a), "Discounted dynamic programming," *Annals of Mathematical Statistics,* vol 36, 226-235.

[15] D. Blackwell (1965b), "Positive dynamic programming," *Proceedings of the Fifth Berkeley Symposium on Mathematical Statistics and Probability,* University of California Press, Berkeley, 415-418.

[16] D. Blackwell and M.A. Girshick (1954), *Theory of games and statistical decisions,* John Wiley, New York.

[17] V. Borkar and P. Varaiya (1979), "Adaptive control of Markov chains, I: finite parameter set," *IEEE Transactions on Automatic Control,* vol AC-24, 953-958.

[18] V. Borkar and P. Varaiya (1982), "Identification and adaptive control of Markov chains," *SIAM J. on Control and Optimization,* vol 20, 470-489.

[19] G.E.P. Box and G.M. Jenkins (1970), *Time series analysis, forecasting, and control,* Holden Day, San Francisco.

[20] P. Bremaud (1981), *Point processes and queues, martingale dynamics,* Springer Verlag, New York.

[21] A.E. Bryson, Jr. and Y.-C. Ho (1969), *Applied optimal control,* Ginn, Waltham, MA.

[22] G. Chow (1981), *Econometric analysis by control methods,* John Wiley, New York.

[23] K.L. Chung (1974), *A course in probability theory,* Academic Press, New York.

[24] H. Cramer (1946), *Mathematical methods of statistics,* Princeton University Press, Princeton, NJ.

[25] M.H.A. Davis (1979), "Martingale methods in stochastic control," in M. Kohlmann and W. Vogel (eds), *Stochastic Control Theory and Stochastic Differential Systems,* Lecture Notes in Control and Information Sciences, Vol 16, 85-117, Springer Verlag, Berlin.

[26] M.H.A. Davis and P. Varaiya (1973), "Dynamic programming conditions for partially observable stochastic systems," *SIAM J. Control,* Vol 11, 226-261.

[27] M. H. DeGroot (1970), *Optimal statistical decision,* McGraw Hill, New York.

[28] E. Denardo (1967), "Contraction mappings in the theory underlying dynamic programming," *SIAM Review,* vol 9, 165-177.

[29] C. Derman (1970), *Finite state Markovian decision processes,* Academic Press, New York.

[30] P. Dewilde and H. Dym (1981), "Schur recursions, error formulas, and convergence of rational estimators for stationary stochastic sequences," *IEEE Transactions on Information Theory,* vol IT-27(4), 446-461.

[31] J.L. Doob (1953), *Stochastic processes,* John Wiley, New York.

[32] B. Doshi and S. Shreve (1980), "Strong consistency of a modified maximum likelihood estimator for controlled Markov chains," *Journal of Applied Probability,* vol 17, 726-734.

[33] L. Dubins and L. Savage (1976), *Inequalities for stochastic processes: How to gamble if you must,* Dover, New York.

[34] E. Dynkin and A. Yushkevich (1975), *Controlled Markov processes and their applications,* Springer Verlag, New York.

[35] G. Favier (1982), *Filtrage, modelisation et identification de systemes lineaires stochastiques a temps discret,* Editions du CNRS, Paris.

[36] A. Federgruen, A. Hordijk and H. Tijms (1978), "A note on simultaneous recurrence conditions on a set of denumerable stochastic processes," *Journal of Applied Probability,* vol 15, 356-373.

[37] A.A. Fel'dbaum (1965), *Optimal control systems,* Academic Press, New York.

[38] W. Feller (1957), *An introduction to probability theory and its applications,* Vol 1, John Wiley, New York.

[39] V. Gertz, M. Gevers and E. Hannan (1982), "The determination of optimum structures for the state space representation of multivariate stochastic processes," *IEEE Transactions on Automatic Control,* vol AC-27, 1200-1211.

[40] J.C. Gittins (1979), "Bandit processes and dynamic allocation indices," *Journal of the Royal Statistical Society,* vol 41B, 148-177.

[41] J.C. Gittins and K.D. Glazebrook (1977), "On Bayesian models in stochastic scheduling," *Journal of Applied Probability,* vol 14, 556-565.

[42] J.C. Gittins and D.M. Jones (1972), "A dynamic allocation index for the sequential design of experiments," in J. Gani, K. Sarkadi and I. Vincze (eds), *Colloquia Mathematica Societatis Janos Bolyai,* 9, Progress in Statistics, European Meeting Of Statisticians, North Holland, London, 241-266.

[43] K.D. Glazebrook (1983), "Optimal strategies for families of alternative bandit processes," *IEEE Transactions on Automatic Control,* vol AC-28, 858-861.

[44] K. Glover and J. C. Willems (1974). "Parametrizations of linear dynamical systems: canonical forms and identifiability," *IEEE Transactions on Automatic Control,* vol AC-19, 640-646.

[45] G. Goodwin and R. Payne (1977), *Dynamic system identification: experiment design and data analysis,* Academic Press, New York.

[46] G. Goodwin, P. Ramadge and P. Caines (1981), "Discrete time stochastic adaptive control," *SIAM J. on Control and Optimization,* vol 19, 829-853.

[47] G. Goodwin and K.S. Sin (1984), *Adaptive filtering prediction and control,* Prentice Hall, Englewood Cliffs, NJ.

[48] G. Goodwin, K.S. Sin and K. Saluja (1980), "Stochastic adaptive control and prediction: the general delay - colored noise case," *IEEE Transactions on Automatic Control,* Vol AC-25, 946-949.

[49] M.M. Gupta and E. Sanchez (eds.) (1985a,b), *Approximate reasoning in decision analysis,* and *Fuzzy information and decision processes,* North Holland, Amsterdam.

[50] I. Gustavsson, L. Ljung and T. Soderstrom (1977), "Adaptive processes in closed loop-identifiability and accuracy aspects," *Automatica,* vol 13, 59-79.

[51] B. Hajek (1984), "Optimal control of two interacting service stations," *IEEE Transactions on Automatic Control,* vol. AC-29, 491-498.

[52] E.J. Hannan (1970), *Multiple time series,* John Wiley, New York.

[53] C.M. Harvey (1985), "Preference functions for catastrophe and risk inequity", *Large Scale Systems,* Vol 8(2), April, 131-146.

[54] S. Haykin (1984), *Introduction to adaptive filters,* Macmillan, New York.

[55] A. Hordijk (1974), *Dynamic programming and Markov potential theory,* Mathematical Center Tracts, vol. 51, Amsterdam.

[56] A. Hordijk and L. C. M. Kallenberg (1979), "Linear programming and Markov decision chains," *Management Science,* vol 25, 352-362.

[57] R.A. Howard (1960), *Dynamic programming and Markov processes,* MIT Press, Cambridge, MA.

[58] R.A. Howard (1971), *Dynamic probabilistic systems, Vol I: Markov models,* John Wiley, New York.

[59] F. Itakura and S. Saito (1971), "Digital filtering techniques for speech analysis and synthesis," *Proc. Seventh International Congress on Acoustics,* vol 25-C-1, 261-274, Budapest.

[60] O.L.R. Jacobs and J.W. Patchell (1972), "Caution and probing in stochastic control," *International J. on Control.* Vol. 16(1), 189-199.

[61] T. Kailath (1970), "The innovations approach to detection and estimation theory," *Proceedings of the IEEE,* vol 58, 680-695.

[62] T. Kailath (1974), "A view of three decades of linear filtering theory," *IEEE Transactions on Information Theory,* vol IT-20(2), 146-181.

[63] T. Kailath, M. Morf and G. Sidhu (1974), "Some new algorithms for recursive estimation in constant discrete-time linear systems," *IEEE Transactions on Automatic Control,* Vol AC-19(4), 315-323.

[64] T. Kailath, A. Vieira and M. Morf (1979), "Orthogonal transformation (square-root) implementations of the generalized Chandrasekhar and generalized Levinson algorithms," in A. Bensoussan and J.L. Lions (eds), *International Symposium on Systems Optimization and Analysis,* Lecture Notes in Control and Information Sciences, Vol 14, 81-91, Springer Verlag, Berlin.

[65] G. Kallianpur and R.L. Karandikar (1983), "Some recent developments in nonlinear filtering theory," *Acta Applicandae Mathematicae,* Vol 1(4), 399-434.

[66] F.P. Kelly (1981), "Multi-armed bandits with discount factor near one: the Bernoulli case," *Annals of Statistics,* vol 9, 987-1001.

[67] J.G. Kemeny and J.L. Snell (1960), *Finite Markov chains,* Van Nostrand, Princeton, NJ.

[68] J.G. Kemeny, J.L. Snell and A.W. Knapp (1976), *Denumerable Markov chains,* Springer Verlag, New York.

[69] M. G. Kendall and A. Stuart (1964), *The advanced theory of statistics,* Griffin, London.

[70] D. Kendrick (1981), *Stochastic control for economic models,* McGraw-Hill, New York.

[71] J. Kiefer and J. Wolfowitz (1952), "Stochastic estimation of the maximum of a regression function," *Annals of Mathematical Statistics,* vol 23, 462-466.

[72] L. Kleinrock (1975), *Queueing Systems, Vol I: Theory, Vol II: Computer applications,* Wiley-Interscience, New York.

[73] P.R. Kumar (1982), "Adaptive control with a compact parameter set," *SIAM J. on Control and Optimization,* vol. 20, 9-13.

[74] P.R. Kumar (1983a), "Simultaneous identification and adaptive control of unknown systems over finite parameter sets," *IEEE Transactions on Automatic Control,* vol AC-28, 68-76.

[75] P.R. Kumar (1983b), "Optimal adaptive control of linear quadratic Gaussian systems," *SIAM J. on Control and Optimization*, vol 21, 163-178.

[76] P.R. Kumar (1985), "A survey of some results in stochastic adaptive control," *SIAM J. on Control and Optimization*, vol 23(3), 329-380.

[77] P.R. Kumar and A. Becker (1982), "A new family of optimal adaptive controllers for Markov chains," *IEEE Transactions on Automatic Control*, vol AC-27, 137-146.

[78] P.R. Kumar and W. Lin (1982), "Optimal adaptive controllers for Markov chains," *IEEE Transactions on Automatic Control*, vol AC-27, 765-774.

[79] P.R. Kumar and L. Praly (1985), "Self-tuning and convergence of parameter estimates in minimum variance tracking and linear model following," preprint, University of Illinois, Urbana.

[80] S. Y. Kung, H. J. Whitehouse and T. Kailath, (eds.) (1985), *VLSI and modern signal processing*, Prentice Hall, Englewood Cliffs, NJ.

[81] H. Kushner (1971), *Introduction to stochastic control*, Holt, New York.

[82] H. Kushner (1977), *Probability methods for approximations in stochastic control and for elliptic equations*, Academic Press, New York.

[83] H. Kushner (1984), *Approximation and weak convergence methods for random processes*, MIT Press, Cambridge.

[84] H. Kushner and D. Clark (1978), *Stochastic approximation methods for constrained and unconstrained systems*, Applied Math. Science Series 26, Springer, Berlin.

[85] H. Kushner and A. Schwartz (1984), "An invariant measure approach to the convergence of stochastic approximations with state dependent noise," *SIAM J. on Control and Optimization*, vol 22, 13-27.

[86] T.L. Lai and H. Robbins (1984), "Asymptotically efficient adaptive allocation rules," Advances in Applied Mathematics, vol 5, 1-19.

[87] T.L. Lai and C.Z. Wei (1982), "Least squares estimates in stochastic regression models with applications to identification and control of dynamic systems," *Annals of Statistics*, vol 10(1), 154-166.

[88] I.D. Landau (1979), *Adaptive control - the model reference approach*, Marcel Dekker, New York.

[89] L. Ljung (1976a). "Consistency of the least-squares identification method," *IEEE Transactions on Automatic Control*, vol AC-21,

779-781.

[90] L. Ljung (1977a), "Analysis of recursive stochastic algorithms," *IEEE Transactions on Automatic Control,* vol AC-22, 551-575.

[91] L. Ljung (1977b), "On positive real functions and the convergence of some recursive schemes," *IEEE Transactions on Automatic Control,* vol AC-22, 539-551.

[92] L. Ljung and P.E. Caines (1979), "Asymptotic normality of prediction error estimation for approximate system models," *Stochastics,* vol 3, 29-46.

[93] L. Ljung and T. Soderstrom (1983), *Theory and practice of recursive identification,* MIT Press, Cambridge.

[94] N. Levinson (1946), "The Wiener RMS (root mean square) error criterion in filter design and prediction," *Journal of Mathematics and Physics,* vol XXV, 261-278.

[95] W. Lin and P. R. Kumar (1984), "Optimal control of a queueing system with two heterogeneous servers," *IEEE Transactions on Automatic Control,* vol AC-29, 696-703.

[96] W. Lin, P. R. Kumar and T. Seidman (1985), "Will the self-tuning approach work for general cost criteria?" *Systems and Control Letters,* to appear.

[97] A. Lindquist (1974), "A new algorithm for optimal filtering of discrete-time stationary processes," *SIAM J. Control,* Vol 12(4), 736-746.

[98] S. Lippman (1973), "Semi-Markov processes with unbounded rewards," *Management Science,* vol 19, 717-731.

[99] M. Loeve (1960), *Probability theory,* 2nd Ed., Van Nostrand, Princeton, NJ.

[100] R.D. Luce and H. Raiffa (1957), *Games and decisions,* John Wiley, New York.

[101] P. Mandl (1974), "Estimation and control in Markov chains," *Advances in Applied Probability,* vol 6, 40-60.

[102] R.K. Mehra (1976), "Synthesis of optimal inputs for multiinput multioutput systems with process noise," in R.K. Mehra and D.G. Lainiotis (eds), *System identification: advances and case studies,* Academic Press, New York, 211-250.

[103] D.G. Messerschmitt and M. Honig (1984), *Adaptive filters: structures, algorithms and applications,* Kluwer, Hingham, MA.

[104] M. Metivier and P Priouret (1984), "Applications of a Kushner and Clark Lemma to general classes of stochastic algorithms,"

IEEE Transactions on Information Theory, vol IT-30, 140-151.

[105] B. Miller and A. Veinott (1969), "Discrete dynamic programming with a small interest rate," *Annals of Mathematical Statistics,* vol 40, 366-370.

[106] M. Morf (1974), "Fast algorithms for multivariable systems," *Ph. D. Thesis,* Department of Electrical Engineering, Stanford University.

[107] M. Morf and T. Kailath (1975), "Square root algorithms for least squares estimation," *IEEE Transactions on Automatic Control,* Vol AC-20(4), 487-497.

[108] J. Neveu (1975), *Discrete parameter martingales,* North Holland, Amsterdam.

[109] D. Ornstein (1969), "On the existence of stationary optimal strategies," *Proceedings of the American Mathematical Society,* vol 20, 563-569.

[110] A. Papoulis (1985), "Levinson's algorithm, Wold's decomposition, and spectral estimation," *SIAM Review,* vol 27(3), 405-441.

[111] V. Peterka (1972), "On steady state minimum variance control strategy," *Kybernetica,* vol. 8, pp. 219-232.

[112] C. R. Rao (1973), *Linear statistical inference and its applications,* Wiley, New York.

[113] D. Revuz (1975), *Markov chains,* American Elsevier, New York.

[114] H. Robbins (1952), "Some aspects of the sequential design of experiments," *Bulletin of the American Mathematical Society,* vol 58, 527-537.

[115] H. Robbins and S. Monro (1951), "A stochastic approximation method," *Annals of Mathematical Statistics,* vol 22, 400-407.

[116] H. Robbins and D. Siegmund (1971), "A convergence theorem for nonnegative almost supermartingales and some applications," in J.S. Rustagi (ed), *Optimization methods in statistics,* Academic Press, New York, 233-257.

[117] V. K. Rohatgi (1976), *An introduction to probability theory and mathematical statistics,* John Wiley, New York.

[118] Z. Rosberg, P.P. Varaiya and J.C. Walrand, "Optimal control of service in tandem queues," *IEEE Transactions on Automatic Control,* Vol. AC-27(3), 600-609.

[119] S. Ross (1970), *Applied probability models with optimization applications,* Holden-Day, San Francisco.

[120] S. Ross (1983), *Introduction to stochastic dynamic programming,* Academic Press, New York.

[121] M. Rothschild (1974), "A two-armed bandit theory of market pricing," *Journal of Economic Theory,* vol 9, 185-202.

[122] L.J. Savage (1954), *The foundations of statistics,* John Wiley, New York.

[123] M. Schal (1973), "Dynamic programming under continuity and compactness assumptions," *Advances in Applied Probability,* vol 5, 24-25.

[124] M. Schal (1975), "Conditions for optimality in dynamic programming and for the limit of N-stage optimal policies to be optimal," *Z. Wahrscheinlichkietstheorie vew. Gebiete,* vol 32, 179-196.

[125] U. Shaked and P. R. Kumar (1986), "Minimum variance control of multivariable ARMAX systems," *SIAM J. Control and Optimization,* to appear.

[126] H. Simon (1975), "Theories of decision-making in economic and behavioral science," reprinted in E. Mansfield (ed), *Microeconomics: Selected Readings,* 2nd Ed., Norton, New York.

[127] K. Sin and G. Goodwin (1982). "Stochastic adaptive control using a modified least squares algorithm," *Automatica,* vol 18, 315-321.

[128] D. Snyder (1975), *Random point processes,* Wiley-Interscience, New York. vol 2, 91-102.

[129] V. Solo (1979), "The convergence of AML," *IEEE Transactions on Automatic Control,* vol AC-24, 958-963.

[130] V. Solo (1982), *Topics in advanced time series analysis,* Springer Verlag Lecture Notes, to appear.

[131] S. Stidham, Jr. (1985), "Optimal control of admission to a queueing system," *IEEE Transactions on Automatic Control,* Vol. AC-30(8), 705-713.

[132] W. F. Stout (1974), *Almost sure convergence,* Academic Press, New York.

[133] R. Strauch (1966), "Negative dynamic programming," *Annals of Mathematical Statistics,* vol 3, 871-890.

[134] C. Striebel (1965), "Sufficient statistics in the optimum control of stochastic systems," *J. Mathematical Analysis and Applications,* Vol 12(3), 576-592.

[135] C.L. Thornton and G.J. Bierman (1977), "Gram-Schmidt algorithms for covariance propogation," *International Journal of Control,* Vol 25(2), 243-260.

[136] E. Tse, Y. Bar-Shalom and L. Meier (1973), "Wide sense adaptive dual control of stochastic nonlinear systems," *IEEE Transactions on Automatic Control,* Vol. AC-18(2), 98-108.

[137] N.H. van Dijk (1984), *Controlled Markov processes: time discretization,* CWI Tract 11, Center for Mathematics and Computer Science, Amsterdam.

[138] H. Van Trees (1968), *Detection, estimation and modulation theory,* John Wiley, New York.

[139] P. Varaiya, J. Walrand and C. Buyukkoc (1985), "Extensions of the multi-armed bandit problem: the discounted case," *IEEE Transactions on Automatic Control,* vol AC-30, 426-439.

[140] A. Veinott (1969), "Discrete dynamic programming with sensitive optimality criteria," *The Annals of Mathematical Statistics,* vol 40, 1635-1660.

[141] J. von Neumann and O. Morgenstern (1947), *Theory of games and economic behavior,* 2nd Ed., Princeton University Press, Princeton, NJ.

[142] A. Wald (1950), *Statistical decision functions,* John Wiley, New York.

[143] N. Wiener (1949), *Extrapolation, interpolation and smoothing of stationary time series,* MIT Press, Cambridge, MA.

[144] P. Whittle (1980), "Multi-armed bandits and the Gittins index," *Journal of the Royal Statistical Society,* vol 42B, 143-149.

[145] P. Whittle (1982), *Optimization over time,* John Wiley, New York.

[146] J. Wijngaard (1971), "Stationary Markovian decision problems and perturbation theory of quasi-compact linear operators," *Mathematics of Operations Research,*

[147] E. Wong (1983), *Introduction to random processes,* Springer, New York.

[148] Z. You-Hong (1982), "Stochastic adaptive control and prediction based on a modified least squares—the general delay-colored noise case," *IEEE Transactions on Automatic Control,* vol AC-27, 1257-1260.

AUTHOR INDEX

SUBJECT INDEX